AF773972

Regional Development and Land Use Change

Regional Development and Land Use Change

Edited by:
Quan Cui

SOCIETY Publishing

www.societypublishing.com

Regional Development and Land Use Change

Quan Cui

Society Publishing

2010 Winston Park Drive,

2nd Floor

Oakville, ON L6H 5R7

Canada

www.societypublishing.com

Tel: 001-289-291-7705
001-905-616-2116
Fax: 001-289-291-7601
Email: orders@arclereducation.com

© 2018 Society Publishing

ISBN: 978-1-77361-115-0 (Hardcover)

This book contains information obtained from highly regarded resources. Reprinted material sources are indicated. Copyright for individual articles remains with the authors as indicated and published under Creative Commons License. A Wide variety of references are listed. Reasonable efforts have been made to publish reliable data and views articulated in the chapters are those of the individual contributors, and not necessarily those of the editors or publishers. Editors or publishers are not responsible for the accuracy of the information in the published chapters or consequences of their use. The publisher assumes no responsibility for any damage or grievance to the persons or property arising out of the use of any materials, instructions, methods or thoughts in the book. The editors and the publisher have at- tempted to trace the copyright holders of all material reproduced in this publication and apologize to copyright holders if permission has not been obtained. If any copyright holder has not been acknowledged, please write to us so we may rectify.

Notice: Registered trademark of products or corporate names are used only for explanation and identification without intent of infringement.

Society Publishing publishes wide variety of books and eBooks. For more information about Society Publishing and its products, visit our website at www.societypublishing.com.

DECLARATION

Some content or chapters in this book are open access copyright free published research work, which is published under Creative Commons License and are indicated with the citation. We are thankful to the publishers and authors of the content and chapters as without them this book wouldn't have been possible.

ABOUT THE EDITOR

QUAN CUI

Quan Cui obtained her PhD from Beijing Normal University in 2015. She worked as environmental impact assessment engineer for years. Aiming at linking science to policy, her study interest spans from ecosystem modelling to water resources management. She published peer-reviewed articles about using machine learning models in inflow forecasting, CO_2 management and watershed ecosystem health assessment.

TABLE OF CONTENTS

LIST OF CONTRIBUTORS

Hansong Chen
Xingzhi College, Zhejiang Normal University, Jinhua, Zhejiang Province, China

Shaoyuan Zhang
Rural Energy Office, Jinhua Agricultural Bureau, Jinhua, Zhejiang Province, China

Qian Yang
School of Resource and Environmental Sciences, Wuhan University, Wuhan, Hubei Province, China

Shaocheng Zheng
Xingzhi College, Zhejiang Normal University, Jinhua, Zhejiang Province, China

Xiaozhong Li
Xingzhi College, Zhejiang Normal University, Jinhua, Zhejiang Province, China

Quan Cui
School of Environment, Beijing Normal University, Beijing, China

Alexander Popp
Potsdam Institute for Climate Impact Research (PIK), 14412 Potsdam, Germany

Katherine Calvin
Pacific Northwest National Laboratory, Joint Global Change Research Institute at the University of Maryland–College Park, 5825 University Research Court, Suite 3500, College Park, MD 20740, USA

Shinichiro Fujimori
National Institute for Environmental Studies (NIES), Japan

Petr Havlik
International Institute for Applied Systems Analysis (IIASA), Austria

Florian Humpenöder
Potsdam Institute for Climate Impact Research (PIK), 14412 Potsdam, Germany

Elke Stehfest
PBL Netherlands Environmental Assessment Agency, Postbus 30314, 2500 GH The Hague, The Netherlands

Benjamin Leon Bodirsky
Potsdam Institute for Climate Impact Research (PIK), 14412 Potsdam, Germany
Commonwealth Scientific and Industrial Research Organisation, St. Lucia, Australia

Jan Philipp Dietrich
Potsdam Institute for Climate Impact Research (PIK), 14412 Potsdam, Germany

Jonathan C. Doelmann
PBL Netherlands Environmental Assessment Agency, Postbus 30314, 2500 GH The Hague, The Netherlands

Mykola Gusti
International Institute for Applied Systems Analysis (IIASA), Austria
Lviv Polytechnic National University, 12 Bandera street, 79013 Lviv, Ukraine

Tomoko Hasegawa
National Institute for Environmental Studies (NIES), Japan

Page Kyle
Pacific Northwest National Laboratory, Joint Global Change Research Institute at the University of Maryland–College Park, 5825 University Research Court, Suite 3500, College Park, MD 20740, USA

Michael Obersteiner
International Institute for Applied Systems Analysis (IIASA), Austria

Andrzej Tabeau
Wageningen Economic Research part of Wageningen University & Research, The Netherlands

Kiyoshi Takahashi
National Institute for Environmental Studies (NIES), Japan

Hugo Valin
International Institute for Applied Systems Analysis (IIASA), Austria

Stephanie Waldhoff
Pacific Northwest National Laboratory, Joint Global Change Research Institute at the University of Maryland–College Park, 5825 University Research Court, Suite 3500, College Park, MD 20740, USA

Isabelle Weindl
Potsdam Institute for Climate Impact Research (PIK), 14412 Potsdam, Germany
Leibniz Institute for Agricultural Engineering and Bioeconomy (ATB), Max-Eyth-Allee 100, 14469 Potsdam, Germany

Marshall Wise
Pacific Northwest National Laboratory, Joint Global Change Research Institute at the University of Maryland–College Park, 5825 University Research Court, Suite 3500, College Park, MD 20740, USA

Elmar Kriegler
Potsdam Institute for Climate Impact Research (PIK), 14412 Potsdam, Germany

Hermann Lotze-Campen
Potsdam Institute for Climate Impact Research (PIK), 14412 Potsdam, Germany
Humboldt-University of Berlin, 10099 Berlin, Germany

Oliver Fricko
International Institute for Applied Systems Analysis (IIASA), Austria

Keywan Riahi
International Institute for Applied Systems Analysis (IIASA), Austria
Graz University of Technology, Austria

Detlef P. van Vuuren
Graz University of Technology, Austria

Ilpo Tammi
Council of Tampere Region, 33101 Tampere, Finland

Kaisa Mustajärvi
Ramboll Finland Oy, Tampere, Finland

Jussi Rasinmäki
Simosol Oy, Riihimäki, Finland

Gianbattista Bussi
School of Geography and the Environment, University of Oxford, South Parks Road, OX1 3QY Oxford, UK

Simon J. Dadson
School of Geography and the Environment, University of Oxford, South Parks Road, OX1 3QY Oxford, UK

Christel Prudhomme
Centre for Ecology and Hydrology (CEH), Benson Lane, Crowmarsh Gifford, OX10 8BB Wallingford, UK
Department of Geography, Loughborough University, Loughborough LE11 3TU, UK

Paul G. Whitehead
School of Geography and the Environment, University of Oxford, South Parks Road, OX1 3QY Oxford, UK

I.P. Holman
School of Water, Energy and Environment, Cranfield University, Cranfield, Bedford MK43 0AL, UK

C Brown
School of Geosciences, University of Edinburgh, Edinburgh EH8 9XP, UK

V Janes
School of Water, Energy and Environment, Cranfield University, Cranfield, Bedford MK43 0AL, UK

D Sandars
School of Water, Energy and Environment, Cranfield University, Cranfield, Bedford MK43 0AL, UK

Lourdes López-Merino
Institute of Environment, Health and Societies, Brunel University London, Uxbridge UB8 3PH, UK

Oscar Serrano
Centre for Marine Ecosystems Research, Edith Cowan University, 270 Joondalup Drive, Joondalup, WA 6027, Australia
UWA Oceans Institute, University of Western Australia, 35 Stirling Highway, Crawley 6009, Australia

María Fernanda Adame
Australian Rivers Institute, Griffith University, 170 Kessels Road, Nathan, QLD 4111, Australia

Miguel Ángel **Mateo**
Centre for Marine Ecosystems Research, Edith Cowan University, 270 Joondalup Drive, Joondalup, WA 6027, Australia
Centro de Estudios Avanzados de Blanes, Consejo Superior de Investigaciones Científicas, Acceso a la Cala S. Francesc 14, 17300 Blanes, Spain

Antonio Martínez Cortizas
Departamento de Edafología y Química Agrícola, Facultad de Biología, Universidad de Santiago de Compostela, Campus Sur, 15782 Santiago de Compostela, Spain

C. Gutzler
Directorate, Leibniz Centre for Agricultural Landscape Research (ZALF), Germany

K. Helming
Directorate, Leibniz Centre for Agricultural Landscape Research (ZALF), Germany

D. Balla
Institute of Landscape Hydrology, Leibniz Centre for Agricultural Landscape Research (ZALF), Germany

R. Dannowski
Institute of Landscape Hydrology, Leibniz Centre for Agricultural Landscape Research (ZALF), Germany

D. Deumlich
Institute of Soil Landscape Research, Leibniz Centre for Agricultural Landscape Research (ZALF), Germany

M. Glemnitz
Institute of Land Use Systems, Leibniz Centre for Agricultural Landscape Research (ZALF), Germany

A. Knierim
Department of Rural Sociology, University of Hohenheim, Germany
Institute of Socio-Economics, Leibniz Centre for Agricultural Landscape Research (ZALF), Germany

W. Mirschel
Institute of Landscape Systems Analysis, Leibniz Centre for Agricultural Landscape Research (ZALF), Germany

C. Nendel
Institute of Landscape Systems Analysis, Leibniz Centre for Agricultural Landscape Research (ZALF), Germany

C. Paul
Institute of Socio-Economics, Leibniz Centre for Agricultural Landscape Research (ZALF), Germany

S. Sieber
Institute of Socio-Economics, Leibniz Centre for Agricultural Landscape Research (ZALF), Germany

U. Stachow
Institute of Land Use Systems, Leibniz Centre for Agricultural Landscape Research (ZALF), Germany

A. Starick
Institute of Socio-Economics, Leibniz Centre for Agricultural Landscape Research (ZALF), Germany

R. Wieland
Institute of Landscape Systems Analysis, Leibniz Centre for Agricultural Landscape Research (ZALF), Germany

A. Wurbs
Institute of Land Use Systems, Leibniz Centre for Agricultural Landscape Research (ZALF), Germany

P. Zander
Institute of Socio-Economics, Leibniz Centre for Agricultural Landscape Research (ZALF), Germany

AlessandroDe Pinto
International Food Policy Research Institute, USA

ManLi
International Food Policy Research Institute, USA

AkikoHaruna
International Food Policy Research Institute, USA

Glenn GrahamHyman
International Center for Tropical Agriculture, Colombia

Mario Andrés LondoñoMartinez
Universidad de los Andes, Colombia

BernardoCreamer
International Center for Tropical Agriculture, Colombia

Ho-YoungKwon
International Food Policy Research Institute, USA

Jhon Brayan ValenciaGarcia
International Center for Tropical Agriculture, Colombia

JeimarTapasco
International Center for Tropical Agriculture, Colombia

Jesus DavidMartinez
International Center for Tropical Agriculture, Colombia

Kerstin Engstrom
Department of Physical Geography and Ecosystem Science, Lund University, Solvegatan 12, 22362 Lund, Sweden

Mark D.A. Rounsevell
School of GeoSciences, University of Edinburgh, Geography Building, Drummond Street, EH89XP Edinburgh, UK

Dave Murray-Rust
School of GeoSciences, University of Edinburgh, Geography Building, Drummond Street, EH89XP Edinburgh, UK

Catherine Hardacre
School of GeoSciences, University of Edinburgh, Crew Building, The King's Buildings, West Mains Road, EH93JN Edinburgh, UK

Peter Alexander
School of GeoSciences, University of Edinburgh, Geography Building, Drummond Street, EH89XP Edinburgh, UK

Xufeng Cui
School of GeoSciences, University of Edinburgh, Geography Building, Drummond Street, EH89XP Edinburgh, UK

Paul I. Palmer
School of GeoSciences, University of Edinburgh, Crew Building, The King's Buildings, West Mains Road, EH93JN Edinburgh, UK

Almut Arneth
Karlsruhe Institute of Technology, Institute of Meteorology and Climate Research, Atmospheric Environmental Research (IMK-IFU), Kreuzeckbahnstr. 19, 82467 Garmisch-Partenkirchen, Germany

PREFACE

Land use/cover change (LUCC), especially conversion and degradation of natural land, is the key factor causing terrestrial carbon stocks declines in the world. As we noticed the main driving force of LUCC is human activities and socio-economic development, many types of research have been conducted. This book contains reviews on regional land-use change and socio-economic relations, special issues like deforestation and urbanization, environmental impacts of land-use change, LUCC on energy flow, specific elements stocks related with LUCC, carbon storage and LUCC, et al. Combing studies from different angles, this book give its readers an up-to-date concept about what regional development and land-use change can do to our society and nature.

Editor

1 RESEARCH PROGRESS ON THE ECO-ENVIRONMENTAL IMPACT OF REGIONAL LAND-USE CHANGE

Hansong Chen[a], Shaoyuan Zhang[b], Qian Yang[c], Shaocheng Zheng[a], Xiaozhong Li[a], and Quan Cui[d]

[a]Xingzhi College, Zhejiang Normal University, Jinhua, Zhejiang Province, China

[b]Rural Energy Office, Jinhua Agricultural Bureau, Jinhua, Zhejiang Province, China

[c]School of Resource and Environmental Sciences, Wuhan University, Wuhan, Hubei Province, China

[d]School of Environment, Beijing Normal University, Beijing, China

ABSTRACT

Land-use change as a major cause of environmental change has become the hotspot in global change research. Here we introduced the current researches of the impact of regional land-use change on the ecological and environmental factors such as atmosphere, water, soil and biological system, and summarized the research progress of the effect of land-use change on material energy flow and ecological service value. We also analyzed the feedback of the eco-environmental effect on the land-use.

Scientific, rational and sustainable land-use policy should be based on the comprehensive understanding of the mutual relationship between land-use and eco-environmental factors. The existing problems in the land-use change field, such as limited research scale, lack of dynamic simulation, poor comparability among the results, imperfect quantitative indicators, can be overcome by employing inter-disciplinary and multi-dimensional researches using complete modelling and the unified standard.

Keywords: Regional land-use change; Eco-environmental impact; Ecological factor

Land-use change, which refers to the land utilization mode and degree alternations caused by human actions, expresses the interaction between human activity and natural eco-environment (Tuner li, 1995). Since 1995, when the International Geosphere-Biosphere Program (IGBP) and International Human Dimensions Programme on Global Environmental Change (IHDP) jointly proposed the land-use / cover change (LUCC) scientific research program, land-use change has become one of the important components in global environmental change (Lambin, 1999; Ojima, 2005; Tuner li, 1995). Land-use change researches have extended from the early land-use and land cover type classification, driving force analysis, to land change processes at different time scales and spatial scales, assessments of resources, environment, ecological effects, et al. The eco-environmental impact of land-use change is increasingly concerned by international organizations and countries, and becomes the forefront topic in global change research.

Recent research has shown that land-use change is closely related to land surface matter cycling and life processes (Meyfroidt, 2013; Mooney, 2013). As a result of human economic-social development and urbanization, natural land-use types (forests, grasslands, wetlands, etc.) are rapidly changing into human-dominated land-use types such as agricultural land, urban construction land, etc. (Liu, 2014; Rockstrom, 2009). Land-use change not only brings great changes in the surface structure but also significantly affects the regional atmospheric environment, hydrology and water resources, soil environment, biodiversity and biogeochemical cycles (Hernández, 2016; Jin, 2015; Sterling, 2012; Tian, 2012). The combined effects of these eco-environmental factors will further affect the structure and function of the whole ecosystem, which means will cause the fragmentation of the landscape, the disturbance of community components and their succession process, and the changes in the structure, material energy flow and function of the ecosystem (Gross, 2007; Li, 2003; Song, 2015; Wan, 2005; Winck, 2017). Land-use change can be influenced by eco-environmental constraints. The resulting eco-environmental effects will feedback to the driving force of land-use, promote the change of the original land-use types and structure responsively (Figure 1).

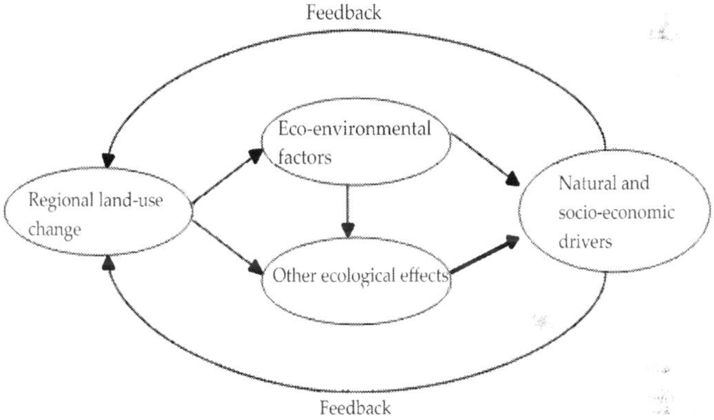

Figure 1: Overall relationship between regional land-use change and eco-environment factors

The research on the eco-environmental impact of land-use change can not only reveal the inherent mechanism of the interaction between man and land system, but also can predict the future development trend of land-use pattern. This is of great theoretical significance and application value to the development of policies for land-use and eco-environmental protection. In this paper, the impact of land-use change on the main eco-environment factors and on the structure/function of ecosystems is summarized, the feedback of eco-environment on the driving force of land-use is analyzed.

IMPACT OF LAND-USE CHANGE ON ECO-ENVIRONMENTAL FACTORS

Impact on atmosphere

Land-use change can alter the atmospheric chemical properties and processes. Most of the land-use changes increase the carbon content in the atmosphere, mainly due to large-scale deforestation, agricultural development, and urbanization. The natural ecosystem to artificial ecosystem conversion will reduce plant biomass, accelerate the decomposition of soil organic matter and the release of large amounts of CO_2, CH_4 and other gases into the atmosphere (Murty, 2002; Virtuosic, 1997). According to Olofsson and Hickler (2008), the carbon emission during 4000 B.C.-1990 by the sourced from permanent farmland is about 194 GtC, of which 74% is contributed by the conversion from forest to permanent farmland. Second, farmland has been changed into construction land. This process is usually accompanied by the removal of surface vegetation, replacement the original land with impervious ground which could weaken the ability of carbon absorption, resulting in higher carbon content in the atmosphere. Land-use also enhances nitrogen oxide emissions. It was shown that land-use change might account for 80% of overall N_2O sources (Cicerone, 1988), greatly affected the components of tropospheric photochemical smog. Winkler (1988) and Cusack (2013) found that the land-use in the industrialized area did lead to a significant increase in photochemical smog. In addition, land-use patterns have a significant impact on atmospheric particulate concentrations, for example, severe regional air pollution often correlates to the higher degree of urbanization (Sun, 2016).

Land-use change does not only affect the quality of ambient air, but it also alters the air temperature, humidity, wind speed and rainfall by changing the underlying surface properties such as reflectivity, roughness, vegetation leaf area, and vegetation coverage, leading to regional climate change. Land-use pattern also affects net radiation, the ratio of sensible heat and latent heat, and the distribution of rainfall in soil water, evapotranspiration and runoff (Foley, 2005; Hua, 2013). The urban heat island effect is the most typical example of regional climate performance affected by land-use change. The increase in the urban construction area and building density, together with the reduction of the green area have led to an increase of heat stored on the surface and a decrease in evaporative cooling (Stathopoulou, 2007).

In the context of deterioration of regional atmospheric quality and global warming, it is particularly important to quantitatively assess the relative contribution of various factors that lead to changes in the atmospheric environment. The study of the land-use change on the atmospheric environment, especially the synthetical influence of climate on geophysical effects and chemical action has drawn more and more attention. The research method is mainly a comprehensive simulation of coupled model. The role of land-use change in climate change and CO_2 concentrations over the past 150 years has been analyzed using historical land-use reconstruction data, as described by Brovkin et al.(2004). Findell et al. (2007) used the GFDL model to study the climatic effects of land-use change caused by human activities. But limited by their spatial resolution, these models cannot be applied to the study of the atmospheric environment at regional scale. Thus high-resolution regional models (RegCM3, RAMS, RIEMS, etc.) have been extensively studied in recent years. However, the lack of sufficient consideration of material circulation and dynamic vegetation process has limited the application of these models. The development of fully coupled high-resolution model will be in great need in the future.

Impact on water

The impact of land-use change on water environment is mainly because they change the hydrological processes and water quality. On the regional scale, the land-use change which affects the water environment mainly includes vegetation change (such as deforestation and afforestation, grassland reclamation, etc.), agricultural development activities (such as farmland reclamation, crop cultivation and management, etc.), urbanization process (Bronstert, 2002; Gebrehiwot, 2010; Julian, 2015; Zhou, 2017).

When vegetation changes the hydrological process, the results vary greatly according to the regional scale size of study. On the small watershed scale, the deforestation increases the flow rate, meantime the afforestation on hollowness ground reduces the total runoff (Bosch, 1982; Gebrehiwot, 2010; Sahin, 1996). However, in the large watershed scale, due to the multiplicity of the land use types and temporal-spatial distribution of rainfall, conclusions will be different (Gebrehiwot, 2015; Malmer, 2010; Marcos, 2003). In the process of agricultural development and urbanization, the reduction of trees and vegetation reduces the surface evaporation and closure. The agricultural land and the impervious areas such as industrial, residential land and roads have reduced surface penetration and groundwater levels, increased surface runoff and the amount of river deposits, correspondingly exacerbating the threat of potential floods (Doerra, 2006; Moussa, 2002). At the same time, surface runoff will carry large amounts of sediment into rivers or lakes, causing soil erosion (Keesstra, 2016). According to the study, 1.6 billion tons of sediment flows into the Yellow River each year on the Loess Plateau in China, due to serious soil erosion caused by unreasonable land-use, especially since the historical period (Peng, 2013). In addition, a large number of studies have shown that irrational land-use has also caused serious water pollution problems (Table 1).

Table 1: Impact of land-use change on water quality

Types of land-use change	**Main impact**	**typical region**	**reference**
Forest to farm-land	Nonpoint source water pollution by use of fertilizer	Mississippi River Basin USA	Turner (2003)
Forest to urban construction land	Construction land will lead to deterioration of water quality, river non-point source pollution intensified	Mondego river basin (Portugal)	Teixeira (2014)
Farmland to grassland	Significant negative correlation between water nutrient concentration, suspended sediment and grass cover	Southern New Zealand	Galbraith (2007)
Grassland to farmland	Increased loss of nitrogen fertilizer, and deteriorated water quality	The Black Creek watershed USA	Sarkar (2014)

Pasture to bare land	Deterioration of water quality due to grass-land degradation	The Zayandehroud River (Iran)	(Bateni, 2013)
Farmland and forest to urban construction land and waste-land	Change of water quality indicators such as nitrogen (N), phosphorus (P), sediment concentration	Choctawhatchee basin in southern Alabama USA	Bhattarai (2008)
Forest to mine	The ecological environment of the river basin was polluted by heavy metals	Chambishi and Mwambashi Streams Zambia	Ntengwe (2006)
Forest to grass-land	Nitrogen content in runoff was reduced, but the concentrations of phosphorus and sediment were increased.	Central Queensland Australia	Elledge (2017)

Research on the impact of land-use change on the water environment is usually conducted by using the comparative watershed method, but the test period is longer in small watersheds. Those method is difficult to operate on larger scale watersheds, and its results are difficult to apply to other watersheds (Li, 2007). With the development of Geographic Information System (GIS) and Remote Sensing (RS) technology, hydrological models, such as SWAT, YOPLATS, DEM, DDRM, have been widely used to quantitatively simulate and predict the impact of land-use change on all aspects of water cycle (Crokea, 2004; Giri, 2012; Loosvelt, 2014). However, these models require a lot parameters and a complete databases; otherwise it will be difficult to assess the uncertainty of simulation results. So the models should be simplified in operation, and effectively coupled with other methods to improve the accuracy of the simulation, which will be a major trend in this field.

Impact on soil

Changing land-use pattern into irrational land may change the physical and chemical properties (such as soil erosion, soil hardening, etc.) of the soil, results in soil pollution and soil fertility decline.

Land-use change directly alters plant cover, while vegetation and litter can change the physical, chemical and biological properties of topsoil, such as soil pH, conductivity, soil texture, soluble organic matter and soil enzyme activity (De Vries, 2007; Jiang, 2015; Remon, 2005; Xia, 2009). Land-use change can cause soil pollution, for example, Motelay-Massei et al. (2004) investigated the spatial distribution of toxic organic pollutants such as polychlorinated biphenyls (PCBs) and polycyclic aromatic hydrocarbons (PAHs) in the Seine River Basin, France. They found that the content of the urban soil was significantly higher than the agricultural soil. Land-use change will also cause the migration and transformation of soil nutrients, thus affecting soil fertility. Jiang et al. (2015) studied the effects of conversion between various land types on the C, N, P, K and other nutrients in the Heihe River Basin, Northwest China. They found that the degradation of swamp land was the most significant contribution to soil fertility decline.

Models such as RUSLE, SEDD, PLOAD, SWAT model, et al. have been developed to investigate the effect of land-use change on the soil environment (Kemanian, 2011; Lin, 2016; Yang, 2012). These models using RS and GIS technologies acquire large amounts of basic data, combining spatial coordinates with soil quality attribute values to describe the spatial pattern of soil quality attributes, finally predict the spatial characteristic of soil quality and reveal the regional spatial structure of parameters.

Impact on biodiversity

Biodiversity affects the stability and sustainability of ecosystems. Land-use change leads to habitat fragmentation and loss, which leads to biodiversity change (de Vries, 2013; Rossum, 2002). At the genetic level, habitat fragmentation results in a reduction of propagules of given species, accompanied with the loss of some genetic variants (Cunnigham, 2000; Diaz, 1999; dos Santos, 2007). On the other hand, habitat fragmentation can also enhance genetic differentiation by decreasing the chance of gene exchange between species and increasing probability of inbreeding and rare allele loss (Cascante, 2002; Dixo, 2009; Tomimatsu, 2002). At the community level, the species have formed a relatively stable relationship between each other and with their environment during the process of long-term evolution, and the land-use change changes these relationships. Habitat fragmentation makes some species more adaptable to the local environment, increasing their population density, and eliminates some other species, thereby affecting species dominance and abundance (Fahrig, 2002; Wu, 2013). Telleria et al. (2003) found that fragmentation of forests in Europe caused by land-use change led to differences in bird diversity and ecosystem composition. Ross et al. (2010) indicated that habitat fragmentation is more likely to cause alien species to invade, thereby threatening the indigenous species.

Land-use change can also cause changes in human production and living activities. People have selected variants of species, affecting the diversity of genetics and species.

Species growth and reproduction have close relationship with climate, hydrology and soil and other environmental factors. Thus changes in these environmental factors caused by land-use change must have a consequence in species diversity. Currently studies in this area remain relatively insufficient.

Due to the lack of well-defined indicators of quantitative biodiversity, there is a lack of reasonable evaluation system and method to determine the impact of land-use change on biodiversity. How to quantify the impact of land-use change on biodiversity and then accurately predict the impact of land-use change on biodiversity will be an important direction in this research field.

LAND-USE CHANGE AND ECOSYSTEM

Material flow and energy flow

Ecosystems maintain themselves by cycling energy and nutrients obtained from external sources. The process is known as inseparable material flow and energy flow. Studies on the regional land-use change on ecosystems is mainly focused on carbon cycle. Most of carbon exchange process are carbon-emitting process. Its impact on the dramatic increase in atmospheric CO_2 is second to combustion fossil fuels (Bolin, 2000). Land-use change is considered to be an important driver of the terrestrial ecosystem carbon cycle. The soil carbon pool is the largest carbon pool in terrestrial ecosystems, which is 3.3 times the atmospheric carbon pool and 4.5 times the bio-carbon pool (Lal, 2004; Murty, 2002). Small changes in soil carbon pool will greatly ease or accelerate its exchange flux with other carbon pools, thereby altering the global carbon cycle (Yang, 2009).

Some studies have shown that land-use change, especially between agriculture, forestry, and animal husbandry types, will inevitably lead to changes in soil carbon storage. For example, the conversion of agricultural land to grassland and woodland will increase soil carbon storage, and the soil carbon storage capacity decreases with time (the confidence interval is lower than 50%) (Guo, 2002; Morris, 2007; Ostle, 2009). This reflects that the recovery rate of soil carbon stocks in the early stages of land-use change is greater than in the late (Coleman, 1997). Soil carbon pool reservation decreases after reclamation of grassland (Dawson, 2007; Guo, 2002; Murty, 2002), but the conversion of grassland to forest land has a greater uncertainty, it may either increase or decrease carbon stock (Davis, 2007; Soussana, 2004). The conversion of woodland to grassland is likely to result in a slight decline in soil carbon pool, but may also lead to an increase in carbon pool (Post, 2000; Soussana, 2004), Forest land to agricultural land conversion significantly reduces carbon pool (Gamboa, 2011; Guo, 2002). The carbon pool declines after converting wetland into agricultural land, while wetland restoration can increase carbon pool (Bolin, 2000; Freibauer, 2004).

There are three main categories of research models for the ecosystem material and energy flow during land-use change. The first are statistical models, such as bookkeeping models. Despite that these model are counting models, they express many basic ecological processes into various disturbance curves, and have many features of the ecological process model (Houghton, 2003). The second are the ecosystem models, including CASA, BIOME, EPIC, LPJ-DEVM and TEM, etc. These models can simulate the flux variation of water, carbon, nitrogen and other substances in the ecosystem by using a series of spatial distribution parameters of different regions such as climate, vegetation, water capacity, soil properties, altitude and so on (Felzer, 2015; Gao, 2005; Izaurralde, 2006; Ni, 2002). The third are coupled models, such as IMAGE, LPJML, etc. These models fully take into account the influence mechanism of land-use change on ecosystems, thus can improve the accuracy of the simulation results (Lapola, 2009; Schneider, 2007).

Ecosystem service

Ecosystem services are obtained directly or indirectly through the structure, processes and functions of the ecosystem. They form and sustain the environmental conditions for human survival and development, and are critical to human well-being (Millennium Ecosystem Assessment, 2005). Human activities exert influence on ecosystems through different land-use strategies, altering ecosystem structures and processes, and influencing the ability of regional ecosystems to provide products and services to society.

Land-use area is an important influencing factor of ecosystem services. Decreased land area with a high value of ecosystem services will result in a significant reduction in the value of regional ecosystem services (Costanza, 1997). Estoque et al.(2012) have shown that the value of ecosystem services in Baguio City in the Philippines decreased by more than $ 3.2 million from 1988 to 2009. The loss is 60% of the total value of the ecosystem services in the city, of which about 98% is due to a reduction in forest area.

Land-use types can also change the temporal and spatial distribution of biological habitats and resources, thereby affect ecosystem services (Carpenter, 2009; Le Maitre, 2007; Polask, 2011). The close relationship between land-use types and ecosystem services is mainly reflected in the differences in individual ecosystem services of different land-use styles. For example, farm ecosystems focus on food production services; forest ecosystems focus on biodiversity maintenance, soil conservation, climate regulation and other service functions; water and wetland ecosystems focus on hydrological regulation, waste disposal and other service functions. Changes in land-use types have led to changes in the value of individual ecosystem services, thus affecting the total value of ecosystem services.

In addition, land-use spatial pattern also affect regional ecosystem services. The spatial pattern of land-use affect the movement of energy, matter and living organisms in the landscape space, inevitably affecting or restricting the movement of species, the migration of water and nutrients, the formation of non-point source pollution, and the dynamics of the population and biodiversity (Fiquepron, 2013; Peng, 2006; Song, 2017). The fragmentation of the spatial pattern could adversely affect ecosystem services as it may lead to a decline in habitat quality that affects wildlife survival and soil microbial movement, resulting in destruction of habitat corridors in forests and croplands, reduced forest and farm productivity, and impeded carbon fixation (Su, 2012).

The evaluation methods of ecosystem services can be divided into three categories: energy analysis, material evaluation and value evaluation. The value evaluation method has good operability and applicability because it represented as monetary value, which could compare the same ecosystem service with different ecosystems, and could also combine the individual service functions of an ecosystem (Costanza, 1997). Thus, models based on this method, such as SoIVES, InVest, ARIES are widely used. However, due to differences in ecosystem classification and parameter selection, these models may produce significantly different results.

FEEDBACK OF ECO-ENVIRONMENTAL EFFECTS

Land-use change causes the regional eco-environmental changes which become the driving factor of land-use change. The driving factors of land-use process is complex, which can be natural, institutional, technical or economic (Quan, 2006) at different spatial and temporal scales.

In a specific region, land-use change caused by climate, hydrology, soil, biological and other natural factors, can directly restrain the land-use behavior. This feedback is relatively stable and has a cumulative effect. It is the endogenous driving factor of regional land-use change. In addition, this feedback can also indirectly influence the regional land-use though the change of human eco-environment consciousness, and return the benefit to ecological value.

At present, the academic research on this issue is scattered in different fields. Due to the complexity and diversity of the problem, few scholars have put the above problems in a unified framework to study. This to a certain extent has led to the lack of comprehensive understanding of the relationship between land-use change and ecological environment. To improve current researches in this field, the feedback effect of eco-environmental effects should be combined with the impact of land-use on the ecological environment.

SUMMARY

With the intensification of human activities, LUCC has become the dominant factor affecting the global environmental change. The impact of land-use change on the ecological environment has become a frontier topic. Considering the complexity of process and mechanism of land-use change on the ecological environment, many issues still needs to be studied.

(1) Interdisciplinary studies

The research on the impact of land-use change on the ecological environment involves many disciplines, and the underlying mechanisms are very complicate. The existing research focuses on the study of the influence of individual environmental factors or certain indexes. The theory and method are limited by the researcher's academic background, and the results often lack certainty and comprehensive evaluation. Therefore, interdisciplinary studies that can fully reveal the ecological and environmental effects of land-use change should be stressed in the future.

(2) Multi-scale investigation

Environmental effects are a complex dynamic process. The occurrence, distribution and evolution of land-use change and its coupling with the ecological environment are interdependent with the scale. The study of single scale cannot provide a complete understanding and solution for ecological and environmental problems. As most studies emphasize on small-scale areas, they cannot be extrapolated to large and medium-scale areas. In the future, it is not only necessary to study the small-scale studies such as community and city, but also necessary to strengthen the study areas sensitive to environmental change, ecological fragile zones, intense farming and economic lands at regional and global scales, so that we can comprehensively analyze the impact of land-use change on ecological environment.

(3) Model analysis

Model analysis has always been one of the core in LUCC research. The study of eco-environmental impact of land-use change will put forward higher requirements for modeling work. At present, RS and GIS play only the role of data source and mass data management in modeling and simulation, and their connection with environmental process model is not enough, particularly dynamic simulation research is weak. By effectively coupling the various models with different methods to dynamically simulate the possible ecological environment effects of land-use change and assess the uncertainty of the results, we should be able to provide scientific decision support for sustainable management of land at the global, national and regional levels.

(4) Standard system

Modeling and simulation cannot be separated from the relevant theories. At present, there are many models for the related research, but the difference between the underlying theories leads to the lack of comparability between the results or even big difference. Therefore, it is necessary to constantly enrich and improve the theories of land change and establish a unified standard system, so that the results of different studies are more compatible and comparable.

(5) Feedback involvement

Regional land-use changes not only affect the ecological environment; it also receives feedback from it. Their interaction is complicate and dynamic. However, it will help develop scientific, rational and sustainable land-use policy if the feedback of the eco-environmental can be well combined with the study of land-use change and ecological environment.

REFERENCES

1. Bateni, F., Fakheran, S., Soffianian, A. 2013. Assessment of land cover changes & water quality changes in the Zayandehroud River Basin between 1997–2008. *Environmental Monitoring and Assessment*, 185(12): 10511-10519.
2. Bhattarai, G., Srivastava, P., Marzen, L. 2008. Assessment of Economic and Water Quality Impacts of Land-useChange Using a Simple Bioeconomic Model. *Environmental Management*, 42(1): 122-131.
3. Bolin, B., Sukumar, R. 2000. *Global perspective//Watson R T, Noble*

I R, Ravindranath N H, Dokken D J. Land-use, Land-use Change, and Forestry: Cambridge: Cambridge University Press.

4. Bosch, J. M., Hewlett, J.D.,. 1982. A review of catchment experiments to determine the effect of vegetation changes on water yield and evapotranspiration. *Journal of Hydrology*, 55(1-4): 3-23.
5. Bronstert, A., Daniel, N., Gerd, B.,. 2002. Effects of climate and land-use change on storm runoff generation: present knowledge and modelling capabilities. *Hydrological Processes*, 16(2): 509-529.
6. Brovkin, V., Sitch, S.B.W., Claussen, M., Bauer, E., Cramer, W.,. 2004. Role of land cover changes for atmospheric CO2 increase and climate change during the last 150 years. *Global Change Biology*, 10(8): 1253-1266.
7. Carpenter, S. R., Mooney, H.A., Agard, J., Capistrano, D., Defries, R.S., Diaz, S., Dietz, T., Duraiappah, A.K., Oteng-Yeboah, A., Pereira, H.M., Perrings, C., Reid, W.V., Sarukhan, J., Scholes, R.J., Whyte, A. 2009. Science for managing ecosystem services: Beyond the Millennium Ecosystem Assessment. *Proceedings of the National Academy of Sciences of the United States of America*, 106(5): 1305-1312.
8. Cascante, A., Quesada, M., Lobo, J.J., Fuchs, E.A.,. 2002. Effects of dry tropical forest fragmentation on the reproductive success and genetic structure of the tree samanea saman. *Conservation Biology*, 16(1): 137-147.
9. Cicerone, R. J., Oremland, R. S.,. 1988. Biogeochemical aspects of atmospheric methane. *Global Biogeochemical Cycles*, 2(4): 299-327.
10. Coleman, K., Jenkinson, D.S., Crocker, G.J., Grace, P.R., Klir, J., Korschens, M., Poulton, P.R., Richter, D.D.,. 1997. Simulating trends in soil organic carbon in long-term experiments using RothC-26. *Geoderma*, 81(1-2): 29-44.
11. Costanza, R., Darge, R., De Groot, R., Farber, S., Grasso, M., Hannon, B., Limburg, K., Naeem, S., Oneill, R., Paruelo, J., Raskin, R., Sutton, P., Belt, M.V.D. 1997. The value of the word's ecosystem services and natural capital. *Nature*, 387: 253-260.
12. Crokea, B. F. W., Merrittc, W., SJakemana, A.J.,. 2004. A dynamic model for predicting hydrologic response to land cover changes in

gauged and ungauged catchments. *Journal of Hydrology*, 291(1-2): 115-131.

13. Cunnigham, S. A. 2000. Effects of habitat fragmentation on the reproductive ecology of four plant species in mallee woodland. *Conservation Biology*, 14(3): 758-768.
14. Cusack, M., Pérez, N., Pey, J., Alastuey, A.,. 2013. Source apportionment of fine PM and sub-micron particle number concentrations at a regional background site in the western Mediterranean: a 2.5 yr study. *Atmospheric Chemistry&Physics*, 13(10): 5173-5187.
15. Davis, M., Nordmeyer, A., Henley, D., Watt, M.,. 2007. Ecosystem carbon accretion 10 years after afforestation of depleted subhumid grassland planted with three densities of Pinus nigra. *Global Change Biology*, 13(7): 1414-1422.
16. Dawson, J. J. C., Smith, P.,. 2007. Carbon losses from soil and its consequences for land-use management. *Science of the Total Environment*, 382(2-3): 165-190.
17. de Vries, F. T., Thébault, E., Liiri, M., Birkhofer, K., Tsiafouli, M.A., Bjørnlund, L., Bracht Jørgensen, H., Brady, M.V., Christensen, S., de Ruiter, P.C., d'Hertefeldt, T., Frouz, J., Hedlund, K., Hemerik, L., Hol, W.H.G., Hotes, S., Mortimer, S.R., Setälä, H., Sgardelis, S.P., Uteseny, K., van der Putten, W.H., Wolters, V., Bardgett, R.D. 2013. Soil food web properties explain ecosystem services across European Land-use systems. *Proceedings of the National Academy of Sciences*, 110(35): 14296-14301.
18. De Vries, W., Lofts, S., Tipping, E., Meili, M., Groenenberg, J.E., Schütze, G.,. 2007. Impact of soil properties on critical concentrations of cadmium, lead, copper, zinc, and mercury in soil and soil solution in view of ecotoxicological effects. *Reviews of Environmental Contamination and Toxicology*, 191(4): 47-89.
19. Diaz, M., Santos, T., Telleria, J.,. 1999. Effects of forest fragmentation on the winter body condition and population parameters of an habitat generalist,the wood mouse apodemus sylvaticus: a test of hypotheses. *Acta Oecologica*, 20(1): 39-49.
20. Dixo, M., Metzger, J.P., Morgante, J.S. 2009. Habitat fragmentation

reduces genetic diversity and connectivity among toad populations in the Brazilian Atlantic Coastal Forest. *Biological Conservation*, 142(8): 1560-1569.

21. Doerra, S. H., Shakesbya, R.A., Blakeb, W.H., Chaferc, C.J.,. 2006. Effects of differing wildfire severities on soil wettability and implications for hydrological response. *Journal of Hydrology*, 319(1-4): 295-311.
22. dos Santos, K., Kinoshita, L.S., Santos, F.A.M. 2007. Tree speciescomposition and similarity in semi deciduous forest fragments of southeastern Brazil. *Biological Conservation*, 135(2): 268-277.
23. Elledge, A., Thornton, C. 2017. Effect of changing Land-use from virgin brigalow (Acacia harpophylla) woodland to a crop or pasture system on sediment, nitrogen and phosphorus in runoff over 25 years in subtropical Australia. *Agriculture, Ecosystems and Environment*, 239(1): 119-131.
24. Estoque, R. C., Murayama, Y.,. 2012. Examining the potential impact of Land-use/cover changes on the ecosystem services of Baguio city, the Philippines: A scenario-based analysis. *Applied Geography*, 35(1): 316-326.
25. Fahrig, L. 2002. Effect of habitat fragmention on the extinction threshold: a synthesis. *Ecological Applications*, 12(2): 346-353.
26. Felzer, B. S., Cronin, T.W., Melillo, J.M., Kicklighter, D.W., Schlosser, C.A.,. 2015. Importance of carbon-nitrogen interactions and ozone on ecosystem hydrology during the 21st century. *Journal of Geophysical Research: Biogeosciences*, 114(G3): 141-153.
27. Findell, K. L., Shevliakova, E., Milly, P.C.D., Stouffer, R.J.,. 2007. Modeled Impact of Anthropogenic Land Cover Change on Climate. *Journal of Climate*, 20(14): 3621-3634.
28. Fiquepron, J., Garcia, S., Stenger, A. 2013. Land-use impact on water quality: valuing forest services in terms of the water supply sector. *J. Environ. Manag*, 126(14): 113-121.
29. Foley, J. A., Defries, R., Asner, G.P., Barford, C., Bonan, G., Carpenter, S.R., Chapin, F.S., Coe, M.T., Daily, G.C., Gibbs, H.K., Helkowski, J.H., Holloway, T., Howard, E.A., Kucharik, C.J., Monfreda, C., Patz, J.A., Prentice, I.C., Ramankutty, N., Snyder,

P.K.,. 2005. Global consequences of Land-use. *Science*, 309(5734): 570-574.

30. Freibauer, A., Rounsevell, M.D.A., Smith, P., Verhagen, J.,. 2004. Carbon sequestration in the agricultural soils of Europe. *Geoderma*, 122(1): 1-23.
31. Galbraith, L. M., Burns, C.W. 2007. Linking land-use Water Body Type and Water Quality in Southern New Zealand. *Landscape Ecology*, 22(2): 231-241.
32. Gamboa, A. M., Galicia, L. 2011. Differential influence of Land-use/cover change on topsoil carbon and microbial activity in low-latitude temperate forests. *Agriculture Ecosystems & Environment*, 142(3): 280-290.
33. Gao, Z. Q., Liu, J.Y., Cao, M.K., Li, K.R., Tao, B. 2005. Impacts of land-use and climate changes on ecosystem productivity and carbon cycle in the cropping-grazing transitional zone in China. *Science in China Series D: Earth Sciences*, 48(9): 1479-1491.
34. Gebrehiwot, S. G. 2015. Forests, water and food security in the northwestern highlands of Ethiopia: Knowledge synthesis. *Environmental Science & Policy*, 48(4): 128-136.
35. Gebrehiwot, S. G., Ayele, T., Bishop, K. 2010. Forest cover and stream flow in a headwater of the Abbay: complementing observational data analysis with community perception. *AMBIO*, 39(4): 284-294.
36. Giri, S., Nejadhashemi, A.P., Woznicki, S.A. 2012. Evaluation of targeting methods for implementation of best management practices in the Saginaw River Watershed. *J. Environ. Manag*, 103(7): 24-40.
37. Gross, N., Suding, K.N., Lavorel, S.,. 2007. Leaf dry matter content and lateral spread predict response to Land-use change for six sub alpine grassland species. *Journal of Vegetation Science*, 18(2): 289-300.
38. Guo, L. B., Gifford, R.M.,. 2002. Soil carbon stocks and Land-use change: a meta analysis. *Global Change Biology*, 8(4): 345-360.
39. Hernández, A., Arellano, E.C., Morales-Moraga, D., Miranda, M.D. 2016. Understanding the effect of three decades of Land-use change on soil quality and biomass productivity in a Mediterranean

landscape in Chile. *Catena*, 140(5): 195-204.

40. Houghton, R. A. 2003. Revised estimates of the annual net flux of carbon to the atmosphere from changes in Land-use and land management 1850-2000. *Tellus B*, 55(2): 378-390.

41. Hua, W. J., Chen H.S. 2013. Impacts of Regional-Scale Land-use/ Land Cover Change on Diurnal Temp erature Range. *Advances in Climate Change Research*, 4(3): 166-172.

42. Izaurralde, R. C., Williams, J.R., McGill, W.B., Rosenberg, N.J., Jakas, M.C.Q. . 2006. Simulating soil C dynamics with EPIC: model description and testing against long-term data. *Ecological Modelling*, 192(3-4): 362-384.

43. Jiang, P., Cheng, L., Li, M., Zhao, R., Duan, Y. 2015. Impacts of LUCC on soil properties in the riparian zones of desert oasis with remote sensing data: a case study of the middle Heihe River basin, China. *Science of the Total Environment*, 506-507(1): 259-271.

44. Jin, G., Wang, P., Zhao, T., Bai, Y.P., Zhao, C.H., Chen, D.D. 2015. Reviews on Land-use change induced effects on regional hydrological ecosystem services for integrated water resources management. *Physics & Chemistry of the Earth Parts A/b/c*, 89-90(1): 33-39.

45. Julian, J. P. 2015. Long-term impacts of land cover changes on stream channel loss. *Science of the Total Environment*, 537(2015): 399-410.

46. Keesstra, S., Pereira, P., Novara, A., Brevik, E.C., Azorin-Molina, C. Parras-Alcantara, L., Jordan, A., Cerda, A. 2016. Effects of soil management techniques on soil water erosion in apricot orchards. *Science of the Total Environment*, 551-552(1): 357-366.

47. Kemanian, A. R., Julich, S., Manoranjan, V.S., Arnold, J.R. 2011. Integrating soil carboncycling with that of nitrogen and phosphorus in the watershed model SWAT:theory and model testing. *Ecol. Model.*, 222(12): 1913-1921.

48. Lal, R. 2004. Soil carbon sequestration impacts on global climate change and food security. *science*, 304(5677): 1623-1627.

49. Lambin, E. F., Baulies, X., Bockatael, N.. 1999. Land-use and land-cover change implementation strategy . IGBP Report No.48 &

IHDP Report No.10.

50. Lapola, D. M., Priess, J.A., Bondeau, A.,. 2009. Modeling the land requirements and potential productivity of sugarcane and jatropha in Brazil and India using the LPJmL dynamic global vegetation model. *Biomass & Bioenergy*, 33(8): 1087-1095.
51. Le Maitre, D. C., Milton, S.J., Jarmain, C., Colvin, C.A., Saayman, I., Vlok, J.H.,. 2007. Linking ecosystem services and water resources: landscape-scale hydrology of the Little Karoo. *Frontiers in Ecology and the Environment*, 5(5): 261-270.
52. Li, L. J., Jiang, D.J., Li, J.Y.,. 2007. Advances in Hydrological Response to Land-use/Land Cover Change. *Jouranal of Natural Resources*, 22(2): 211-224.
53. Li, X. Y., Fang, J.L., Piao, S.L.,. 2003. Land-use change and its implication to the ecological consequences in Lower Yangtze Region. *Acta Geographica Sinica*, 58(5): 659-667.
54. Lin, C., Wu, Z., Ma, R., Su, Z. 2016. Detection of sensitive soil properties related to non-point phosphorus pollution by integrated models of SEDD and PLOAD. *Ecological Indicators*, 60(1): 483-494.
55. Liu, Y., Chen, J., Cheng, W., Sun, C., Zhao, S., Pu Y. 2014. Spatiotemporal dynamics of the urban sprawl in a typical urban agglomeration: a case study on Southern Jiangsu, China (1983–2007). *Frontiers of Earth Science*, 8(4): 490-504.
56. Loosvelt, L., Baets, B.D., Pauwels, V.R.N., Verhoest, N.E.C. 2014. Assessing hydrologic prediction uncertainty resulting from soft land cover classification. *Journal of Hydrology*, 517(2): 411-424.
57. Malmer, A., Murdiyarso, D., Bruijnzeel, L.A., Ilstedt, U. 2010. Carbon sequestration in tropical forests and water: a critical look at the basis for common used generalization. *Glob. Change Biol*, 16(599-604).
58. Marcos, H. C., Aurélie, B., Jeffrey, A.C.,. 2003. Effects of large-scale changes in land cover on the discharge of the Tocantins River, Southeastern Amazonia. *Journal of Hydrology*, 283(1-4): 206-217.
59. Meyfroidt, P., Lambin, E. F., Erb, K. H.,Hertel T.W. 2013. Globalization of Land-use: Distant drivers of land change and geographic displacement of Land-use. *Current Opinion in*

Environment Sustainability, 5(5): 438-444.

60. Millennium Ecosystem Assessment. 2005. *Ecosystem and human well-beings: A framework for assessment*. Washington D C: Island press.
61. Mooney, H. A., Duraiappah, A., Larigauderiea, A. 2013. Evolution of natural and social science interactions in global change research programs. *PNAS*, 110(Suppl 1): 3665-3672.
62. Morris, S. J., Bohm, S., Haile-Mariam, S., Paul, E.A.,. 2007. Evaluation of carbon accrual in afforested agricultural soils. *Global Change Biology*, 13(6): 1145-1156.
63. Motelay-Massei, A., Ollivon, D., Garban, B., Teil, M.J., Blanchard, M.,. 2004. Distribution and spatial trends of PAHs and PCBs in soils in the Seine River basin, France. *Chemosphere*, 55(4): 555.
64. Moussa, R., Voltz, M., Andrieux, P.,. 2002. Effects of the spatial organization of agricultural management on the hydrological behaviour of a farmed catchment during flood events. *Hydrological Processes*, 16(2): 393-412.
65. Murty, D., Kirschbaum, M.U.F., Mcmurtrie, R.E., Mcgilvray, H.,. 2002. Does conversion of forest to agricultural land change soil carbon and nitrogen? A review of the literature. *Global Change Biology*, 8(2): 105-123.
66. Ni, J. 2002. Biome models: main principles and applications. *Acta Phytoecologica Sinaica*, 26(4): 481-488.
67. Ntengwe, F. W., Maseka, K.K. 2006. The Impact of Effluents Containing Zinc and Nickel Metals on Stream and River Water Bodies: The Case of Chambishi and Mwambashi Streams in Zambia. *Physics and Chemistry of the Earth*, 31(15-16): 814-820.
68. Ojima, D. S., Moran, E. F., McConnell, W.,. 2005. Global land project: science plan and implementation strategy.
69. Olofsson, J., Hickler, T.,. 2008. Effects of human land-use on the global carbon cycle during the last 6000 years. *Vegetation History and Archaeobotany*, 17(5): 605-615.
70. Ostle, N. J., Levy, P.E., Evans, C.D., Smith, P.,. 2009. UK Land-use and soil carbon sequestration. *Land-use Policy*, 26(S1): S274-S283.
71. Peng, J., Wang, Y.L.,Wu, J.S., Yue, J., Zhang, Y.A., Li,W.F. 2006.

Ecological effects associated with land-use change in China's southwest agricultural landscape. *Int. J. Sustain. Dev. World*, 13(4): 315-325.

72. Peng, K. S. 2013. The loess plateau area water and soil loss characteristics, management stage and thinking research. *Journal of Capital Normal University*, 34(5): 82-90.
73. Polask, Y. S., Nelson, E., Pennington, D., Johnson, K.A. 2011. The impact of land-use change on ecosystem services, biodiversity and returns to landowners: A case study in the state of Minnesota. *48*, 2(219-242).
74. Post, W. M., Kwon, K.C.,. 2000. Soil carbon sequestration and land-use change: processes and potential. *Global Change Biology*, 6(3): 317-327.
75. Quan, B., Chen, J.F., Qiu, H.L.,. 2006. Spatial-Temporal Pattern and Driving Forces of Land-use Changes in Xiamen. *Pedosphere*, 16(4): 477-488.
76. Remon, E., Bouchardon, J.L., Cornier, B., Guy, B., Leclerc, J.C., Faure, O.,. 2005. Soil characteristics, heavy metal availability and vegetation recovery at a former metallurgical landfill: implications in risk assessment and site restoration. *Environmental Pollution*, 137(2): 316-323.
77. Rockstrom, J., Steffen, W., Noone, K., Persson, A., Chapin, F.S.3rd, Lambin, E.F., Lenton, T.M., Scheffer, M., Folke, C., Schellnhuber, H.J., Nykvist, B., de Wit, C.A., Hughes, T., van der Leeuw, S., Rodhe, H., Sörlin, S., Snyder, P.K., Costanza, R., Svedin, U., Falkenmark, M., Karlberg, L., Corell, R.W., Fabry, V.J., Hansen, J., Walker, B., Liverman, D.,Richardson, K., Crutzen, P., Foley, J.A. 2009. A safe operating space for humanity. *Nature*, 461(7263): 472-475.
78. Ross, K. A., Fox, B.J., Fox, M.D.,. 2010. Changes to plant species richness in forest fragments: fragment age, disturbance and fir history may be as important as area. *Journal of Biogeography*, 29(29): 749-765.
79. Rossum, F. V., Echchgadda, G., Szabadi, I.,. 2002. Commonness and long-term survival in fragmented habitats: primula elatior as a study case. *Conservation Biology*, 16(5): 1286-1295.
80. Sahin, V., Hall, M.J., . 1996. The effects of afforestation and

deforestation on water yields. *Journal of Hydrology*, 178(1-4): 293-309.

81. Sarkar, S., Miller, S.A. 2014. Water quality impacts of converting intensively-managed agricultural lands to switchgrass. *Biomass & Bioenergy*, 68(5): 32-43.
82. Schneider, U.A. 2007. Soil organic carbon changes in dynamic Land-use decision models. *Agriculture Ecosystems and Environment*, 119(3-4): 359-367.
83. Song, W., Deng, X.Z. 2017. Land-use/land-cover change and ecosystem service provision in China. *Science of the Total Environment* 576(1): 705-719.
84. Song, W., Deng, X.Z., Yuan, Y.W., Wang, Z., Li, Z.H. 2015. Impacts of land-use change on valued ecosystem service on the in rapidly urbanized North China Plain. *Ecological Modelling*, 318(1): 245-253.
85. Soussana, J. F., Loiseau, P., Vuichard, N., Ceschia, E., Balesdent, J., Chevallier, T., Arrouays, D.,. 2004. Carbon cycling and sequestration opportunities in temperate grasslands. *Soil Use and Management*, 20(2): 219-230.
86. Stathopoulou, M., Cartalis, C.,. 2007. Daytime urban heat islands from Landsat ETM+ and Corine land cover data: An application to major cities in Greece. *Solar Energy*, 81(3): 358-368.
87. Sterling, S. M., Ducharne, A., Polcher, J.,. 2012. The impact of global land-cover change on the terrestrial water cycle. *Nature Climate Change*, 3(4): 385-390.
88. Su, S. L., Xiao, R., Jiang, Z.L., Zhang, Y.,. 2012. Characterizing landscape pattern and ecosystem service value changes for urbanization impacts at an eco-regional scale. *Applied Geography*, 34(1): 295-305.
89. Sun, L., Wei, J., Duan, D.H., Guo, Y.M., Yang, D.X., Jia, C., Mi, X.T. . 2016. Impact of Land-use and Land-Cover Change on urban air quality in representative cities of China. *Journal of Atmospheric and Solar-Terrestrial Physics*, 142(5): 43-54.
90. Teixeira, Z., Teixeira, H., Marques, J.C. 2014. Systematic processes of Land-use/land cover change to identify relevant driving forces: Implications on water quality. *Science of the Total Environment*,

470-471(C2): 1320-1335.

91. Telleria, J. L., Baquero, R., Santos, T.,. 2003. Effects of forest fragmentation on European birds: implications of regional differences in species richness. *Journal of Biogeography*, 30(4): 621-628.
92. Tian, H. Q., Chen, G.S., Zhang, C., Liu, M.L., Sun, G., Chappelka, A., Ren, W., Xu, X.F., Lu, C.Q., Pan, S.F., Chen, H., Hui, D.F., Mcnulty, S., Lochaby, G., Vance, E.,. 2012. Century-scale response of ecosystem carbon storage to multifactorial global change in the Southern United States. *Ecosystems*, 15(4): 674-794.
93. Tomimatsu, H., Ohara, M.,. 2002. Effects of forest fragmentation on seed production of the understory herb trillium camschatcense. *Conservation Biology*, 16(5): 1277-1285.
94. Tuner li, B. L., Skole, D., Sanderson, S., Fischer, G., Fresco, L.O., Leemans R.. 1995. Land-use and land-cover change science/ research plan. IGBP Report No.35 & IHDP Report No.7.
95. Turner, R. E., Rabalais, N.N. 2003. Linking Landscape and Water Quality in the Mississippi River Basin for 200 Years. *Bioscience*, 53(6): 563.
96. Virtuosic, P. M., Mooney, H.A., Lubchenco, J., Melillo, J.M.,. 1997. Human domination of earth's ecosystems. *science*, 277(277): 494-499.
97. Wan, R. R., Yang, G.S.,. 2005. Changes of Land-use and landscape pattern in Taihu Lake Basin. *Chinese Journal of Applied Ecology*, 16(3): 475-480.
98. Winck, B. R., Sá, E., Rigotti, V.M., Chauvat, M. 2017. Relationship between land-use types and functional diversity of epigeic Collembola in Southern Brazil. *Applied Soil Ecology*, 109(2017): 49-59.
99. Winkler, P., Kaminski, U.,. 1988. Increasing Sub-micron Particle Mass Concentration at Humburg-I. Observations. *Atmospheric Environment*, 22(12): 2871-2878.
100. Wu, J., Liu, Z.M., Qian, J.Q. 2013. Non-linear effect of habitatfragmentation on plant diversi ty: Evidence from a sand dune field in a desertified grassland in northeastern China. *Ecological Engineering*, 54(5): 90-96.

101. Xia, W. B., Li, X., Gao, H., Huang, B.R., Zhang, H.Z., Liu, Y.G., Zeng, G.M., Fan, T.. 2009. Influence factors analysis of removing heavy metals from multiple metal -contaminated soils with different extractants. *Journal of Central South University of Technology*, 16(1): 108-111.

102. Yang, M., Li, X., Hu, Y.M., He, Y.Y. 2012. Assessing effects of landscape pattern on sedi-ment yield using sediment delivery distributed model and a landscape indicator. *Ecol. Indic*, 22(17): 38-52.

103. Yang, Y. S., Xie, J.S., Sheng, H., Chen, G.S., Li, X., Yang, Z.J.,. 2009. The impact of Land-use /cover change on storage and quality of soil organic carbon in midsubtropical mountainous area of southern China. *Journal of Geographical Sciences*, 19(1): 49-57.

104. Zhou, Y., Xu, J.F., Yin, W., Ai, L., Fang, N.F. 2017. Hydrological and environmental controls of the stream nitrate concentration and flux in a small agricultural watershed. *Journal of Hydrology*, 545(1): 355-366.

2 LAND-USE FUTURES IN THE SHARED SOCIO-ECONOMIC PATHWAYS

Alexander Popp[a], Katherine Calvin[b], Shinichiro Fujimori[c], Petr Havlik[d], Florian Humpenöder[a], Elke Stehfest[e], Benjamin Leon Bodirsky[a,h], Jan Philipp Dietrich[a], Jonathan C. Doelmann[e], Mykola Gusti[d,i], Tomoko Hasegawa[c], Page Kyle[b], Michael Obersteiner[d], Andrzej Tabeau[g], Kiyoshi Takahashi[c], Hugo Valin[d], Stephanie Waldhoff[b], Isabelle Weindl[a,j], Marshall Wise[b], Elmar Kriegler[a], Hermann Lotze-Campen[a,k], Oliver Fricko[d], Keywan Riahi[d,f], Detlef P. van Vuuren[f]

[a]Potsdam Institute for Climate Impact Research (PIK), 14412 Potsdam, Germany

[b] Pacific Northwest National Laboratory, Joint Global Change Research Institute at the University of Maryland–College Park, 5825 University Research Court, Suite 3500, College Park, MD 20740, USA

[c] National Institute for Environmental Studies (NIES), Japan

[d] International Institute for Applied Systems Analysis (IIASA), Austria

[e]PBL Netherlands Environmental Assessment Agency, Postbus 30314, 2500 GH The Hague, The Netherlands

[f]Graz University of Technology, Austria

[g]Wageningen Economic Research part of Wageningen University & Research, The Netherlands

[h] Commonwealth Scientific and Industrial Research Organisation, St. Lucia, Australia

[i] Lviv Polytechnic National University, 12 Bandera street, 79013 Lviv, Ukraine

[j] Leibniz Institute for Agricultural Engineering and Bioeconomy (ATB), Max-Eyth-Allee 100, 14469 Potsdam, Germany

[k]Humboldt-University of Berlin, 10099 Berlin, Germany

[l]Copernicus Institute for Sustainable Development, Utrecht University, Heidelberglaan 2, 3584 CS Utrecht, The Netherlands

ABSTRACT

In the future, the land system will be facing new intersecting challenges. While food demand, especially for resource-intensive livestock based commodities, is expected to increase, the terrestrial system has large potentials for climate change mitigation through improved agricultural management, providing biomass for bioenergy, and conserving or even enhancing carbon stocks of ecosystems. However, uncertainties in future socio-economic land use drivers may result in very different land-use dynamics and consequences for land-based ecosystem services. This is the first study with a systematic interpretation of the Shared Socio-Economic Pathways (SSPs) in terms of possible land-use changes and their consequences for the agricultural system, food provision and prices as well as greenhouse gas emissions. Therefore, five alternative Integrated Assessment Models with distinctive land-use modules have been used for the translation of the SSP narratives into quantitative projections. The model results reflect the general storylines of the SSPs and indicate a broad range of potential land-use futures with global agricultural land of 4900 mio ha in 2005 decreasing by 743 mio ha until 2100 at the lower (SSP1) and increasing by 1080 mio ha (SSP3) at the upper end. Greenhouse gas emissions from land use and land use change, as a direct outcome of these diverse land-use dynamics, and agricultural production systems differ strongly across SSPs (e.g. cumulative land use change emissions between 2005 and 2100 range from −54 to 402 Gt CO_2). The inclusion of land-based mitigation efforts, particularly those in the most ambitious mitigation scenarios, further broadens the range of potential land futures and can strongly affect greenhouse gas dynamics and food prices. In general, it can be concluded that low demand for agricultural commodities, rapid growth in agricultural productivity and globalized trade, all most pronounced in a SSP1 world, have the potential to enhance the extent of natural ecosystems, lead to lowest greenhouse gas emissions from the land system and decrease food prices over time. The SSP-based land use pathways presented in this paper aim at supporting future climate research and provide the basis for further regional integrated assessments, biodiversity research and climate impact analysis.

Keywords: Scenarios, Land use, Emissions, Mitigation, Food prices, Integrated assessment, and SSP

INTRODUCTION

Agricultural land-use activities have significantly reshaped our planet, as approximately 40% of the terrestrial surface is currently under agricultural use, either as cropland or pasture (Kareiva et al., 2007; FAO 2014). Historically, the food and material needs of a growing population have been met through expansions in cultivated agricultural area and an increase in agricultural productivity through increases in modern inputs such as new crop varieties, machinery, irrigation water, fertilizer and pesticides (Foley et al., 2005; Rudel et al., 2009; Tilman et al., 2011; Burney et al., 2010; Steinfeld and Gerber, 2010). While the benefits of agriculture to society have been immense, the increase in agricultural production has also been a source of environmental degradation including the loss of biodiversity and wildlife habitats (Carpenter et al., 2009), nutrient run-off (Bodirsky et al., 2014), and emissions of greenhouse gases (GHGs) (Smith et al., 2013).

In the future, the world will be facing new interconnected challenges. Global income levels are expected to increase, as will global population at least through 2050. As a result, an increase in food consumption and a greater demand for livestock-based products is likely (Bodirsky et al., 2015), increasing environmental pressure from agriculture even further. Currently, land use and land-use change are responsible for approximately a quarter of global GHG emissions, largely from tropical deforestation, methane emissions from livestock and rice cultivation, and nitrous oxide emissions from fertilized soils and manure management (Tubiello et al., 2015). This also means that land use management plays a key role in mitigation strategies. In addition to potential emission control mechanisms, such as avoided deforestation and improved agricultural management, the land system could also contribute to climate change mitigation by enhanced carbon dioxide removal (CDR) through afforestation and bioenergy crop production combined with carbon capture and sequestration (BECCS) (Obersteiner et al., 2001; Smith et al., 2013; Humpenöder et al., 2014). Such additional pressures could pose huge challenges for the sustainability of future land systems.

There is uncertainty as to how the demand for agricultural goods will evolve in the future (Valin et al., 2014) and how land use dynamics will respond to an anticipated increase in the demand for ecosystem services (Schmitz et al., 2014). These demands depend strongly on future trends in population growth, dietary preferences, trade, demand for non-food products such as bioenergy, future developments in agricultural yields, and relevant policies. Over time, these uncertainties may result in very different land-use patterns, associated emissions and food prices (von Lampe et al., 2014). Scenario analysis of alternative plausible futures is often used as a tool to explore and evaluate the extensive uncertainties associated with possible future developments (van Vuuren et al., 2012). Most studies of the future environment, including the climate change scenarios of the Intergovernmental Panel on Climate Change (IPCC) summarised in the Special Report on Emissions Scenarios (Nakicenovic and Swart, 2000) and the Millennium Ecosystem Assessment (Carpenter and Pingali, 2005), have used storylines in combination with models to provide scenarios of plausible alternative futures. The objective of these storylines and scenarios is to assess the variation in possible futures and to provide insights into the magnitude and uncertainty of future changes.

Recently, a new set of scenarios has been proposed that are organized around two important dimensions: the extent of climate change and possible future socio-economic conditions. The amount of climate change in the future was explored through the development of different representative concentration pathways (RCPs, (Van Vuuren et al., 2011)). The possible future socio-economic conditions are described in the Shared Socio-economic Pathways (SSPs (O'Neill et al., this Special Issue)), that can be combined with the RCPs in a scenario matrix architecture (Van Vuuren et al., 2016). The SSPs provide five different stories of future socio-economic development, including possible trends in agriculture and land use (O'Neill et al., this Special Issue). In each of the SSPs, climate policies can be introduced to reduce emissions and to enhance carbon uptake to reach radiative forcing level targets consistent with the RCP pathways (Kriegler et al., 2014).

Clearly, a new set of scenarios relevant for climate research will need to address development in the land-use system in detail. In this paper, we describe how a set of Integrated Assessment Models (IAMs) have elaborated the initial storylines of the SSPs and how the results of these models compare to each other. To this end, we will first present relevant aspects of the SSP framework for the land system. Then, we describe possible future pathways of land use, including the resulting GHG emissions and food prices, under different SSPs and climate policy assumptions based on the implementation of the narratives and quantitative elements of the SSPs into the IAMs.

METHODS—SHORT DESCRIPTION OF MODELS & MARKERS

SSP storylines for the land use sector

The SSPs provide a framework for developing new socio-economic scenarios for use in global climate change studies but also for assessments of the broader sustainable development context (Riahi, this Special Issue), (Ebi et al., 2013; Vuuren et al., 2013; O'Neill et al., 2013). The SSPs depict five different global futures (SSP 1–5) with substantially different socio-economic conditions that aim to reflect different socio-economic challenges to mitigation and adaptation. On the most fundamental level, each SSP is described by a narrative (O'Neill et al., this Special Issue) including their challenges to adaptation and mitigation. SSP1 describes a future pathway with low challenges for adaptation and mitigation, whereas in SSP3 both challenges are high. In addition, two "asymmetric cases" are designed, comprising a future in which high challenges to mitigation is combined with low challenges to adaptation (SSP5), and a case where the opposite is true (SSP4). A fifth narrative (SSP2) describes medium challenges of both kinds and is intended to represent a future in which development trends are not extreme in any of the dimensions, but rather follow middle-of-the-road pathways.

These SSP baseline storylines, and their respective implementations in five IAMs, describe socio-economic developments without the assumption of climate policies and excluding climate change and CO_2 fertilization effects. The exclusion of climate policy and climate change is consistent with the idea that these baseline pathways should be used in subsequent studies of mitigation, adaptation and climate impacts. In a next step, SSP-specific socio-economic baseline conditions can be combined with climate policy to achieve RCP-specific climate forcing levels and determine the ability and efforts to mitigate climate change (Vuuren et al., 2013) and the associated contributions of the land use sector. In each of the SSPs, climate policies can be introduced to reduce emissions or to enhance carbon uptake to reach radiative forcing level targets consistent with the RCPs (Kriegler et al., 2012). The climate policies vary across SSPs in terms of international cooperation, timing and sectoral participation in an effort to be consistent with the general and sector-specific SSP storylines. To this end, shared climate Policy Assumptions (SPAs, (Kriegler et al., 2014)), capturing key climate policy attributes such as targets, instruments and obstacles have been developed to guide the implementation of climate policies in the IAMs. In this section we present summaries of the five narratives focusing on the land sector (see also Table 1 for overview):

Table 1: Overview on land use description for the 5 SSP scenarios

	SSP 1	**SSP 2**	**SSP 3**	**SSP 4**	**SSP 5**
Land-use change regulation	Strong regulation to avoid environmental tradeoffs	Medium regulation; slow decline in the rate of deforestation	Limited regulation; continued deforestation	Highly regulated in MICs and HICs; lack of regulation in LICs lead to high deforestation rates	Medium regulation; slow decline in the rate of deforestation

Land productivity growth	High improvements in agricultural productivity; rapid diffusion of best practices	Medium pace of technological change	Low technology development	Productivity high for large scale industrial farming, low for small-scale farming	Highly managed, resource-intensive; rapid increase in productivity
Environmental Impact of food consumption	Low growth in food consumption, low-meat diets	Material-intensive consumption, medium meat consumption	Resource-intensive consumption	Elites: high consumption lifestyles; Rest: low consumption	Material-intensive consumption, meat-rich diets
International Trade	Moderate	Moderate	Strongly constrained	Moderate	High, with regional specialization in production
Globalization	Connected markets, regional production	Semi-open globalized economy	De-globalizing, regional security	Globally connected elites	Strongly globalized
Land-based mitigation policies	No delay in international cooperation for climate change mitigation. Full participation of the land use sector	Delayed international cooperation for climate change mitigation. Partial participation of the land use sector	Heavily delayed international cooperation for climate change mitigation. Limited participation of the land use sector	No delay in international cooperation for climate change mitigation. Partial participation of the land use sector	Delayed international cooperation for climate change mitigation. Full participation of the land use sector

SSP1: sustainability—taking the green road

The world shifts gradually, but pervasively, toward a more sustainable path, emphasizing more inclusive development that respects perceived environmental boundaries. Land use is strongly regulated, e.g. tropical deforestation rates are strongly reduced. Crop yields are rapidly increasing in low- and medium-income regions, leading to a faster catching-up with high income countries. Healthy diets with low animal-calorie shares and low waste prevail. In an open, globalized economy, food is traded internationally. In SSP1, international cooperation for climate change mitigation starts early (after 2020). All land use emissions are priced at the level of carbon prices in the energy sector.

SSP2: middle of the road

The world follows a path in which social, economic, and technological trends do not shift markedly from historical patterns. Land use change is incompletely regulated, i.e. tropical deforestation continues, although at slowly declining rates over time. Rates of crop yield increase decline slowly over time, but low-income regions catch up to a certain extent. Caloric consumption and animal calorie shares converge slowly towards high levels. International trade remains to large extent regionalized. In SSP2, international cooperation for climate change mitigation is delayed due to a transition phase to a uniform carbon price until 2040. In this transition phase, emissions from agricultural production are priced at the level of energy sector emissions, while avoided deforestation and afforestation are not incentivized before 2030.

SSP3: regional rivalry—a rocky road

A resurgent nationalism, concerns about competitiveness and security, and regional conflicts push countries to increasingly focus on domestic or, at most, regional issues, including food and energy security. Land use change is hardly regulated. Rates of crop yield increase decline strongly over time, especially due to very limited transfer of new agricultural technologies to developing countries. Unhealthy diets with high animal shares and high food waste prevail. A regionalized world leads to reduced trade flows for agricultural goods. In SSP3, forest mitigation activities and abatement of agricultural GHG emissions are limited due to major implementation barriers such as low institutional capacites in developing countries. In addition, they are delayed as a consequence of low international cooperation. In 2020, high income countries start the transition to a unifom carbon price until 2040, whereas low income countries start in 2030 and converge until 2050.

SSP4: Inequality—A road divided

Highly unequal investments in human capital, combined with increasing disparities in economic opportunity and political power, lead to increasing inequalities and stratification both across and within countries. Land use change is strongly regulated in high income countries, but tropical deforestation still occurs in poor countries. High income countries achieve high crop yield increases, while low income countries remain relatively unproductive in agriculture. Caloric consumption and animal calorie shares converge towards medium levels. Food trade is globalized, but access to markets is limited in poor countries, increasing vulnerability for non-connected population groups. In SSP4, international cooperation for climate change mitigation starts early (after 2020). But emissions from agricultural and land use are incompletely priced, with limited incentives for avoided deforestation and afforestation before 2030.

SSP5: fossil-fueled development—taking the highway

Driven by the economic success of industrialized and emerging economies, this world places increasing faith in competitive markets, innovation and participatory societies to produce rapid technological progress and development of human capital as the path to sustainable development. Land use change is incompletely regulated, i.e. tropical deforestation continues, although at slowly declining rates over time. Crop yields are rapidly increasing. Unhealthy diets with high animal shares and high waste prevail. Barriers to international trade are strongly reduced, and strong globalization leads to high levels of international trade. In SSP5, all land use emissions are priced at the level of carbon prices in the energy sector. But in contrast to SSP1, international cooperation for climate change mitigation is delayed due to a transition phase to a uniform carbon price until 2040.

Scenario quantification

For this SSP assessment, five integrated assessment models (IAMs) have been applied to derive a quantitative description of the storylines, namely AIM (Fujimori et al., 2014, 2016), GCAM (Wise et al., 2014), IMAGE (Stehfest et al., 2014), MESSAGE-GLOBIOM (Kindermann et al., 2006; Havlík et al., 2014; Riahi et al., 2012) and REMIND/MAgPIE (Popp et al., 2011, 2014). All of these models have in common that they contain land-use modules, which differ, however, in their representation and parametrization of biogeochemical, biophysical and socio-economic processes. More detailed descriptions of the individual models can be found in the SOM (see Section 1: Overview on integrated modeling frameworks), as well as in the respective papers on the SSP implementations by the different teams within this Special Issue (Calvin et al., this Special Issue; Fricko et al., this Special Issue; Fujimori et al., this Special Issue; Kriegler et al., this Special Issue; van Vuuren et al., this Special Issue). These models are driven by the same projections of economic growth (Dellink et al., this special issue; Dietrich et al., 2014) and population (KC and Lutz, this special issue), developed for the SSPs. For other characteristics of the storylines (Table 1), the modeling teams made their own assumptions on how to best represent the described trends of these drivers (see table on IAM implementations in the SOM).

As future development of parameters like crop yields and livestock intensification are known to strongly influence future land use, this limited harmonization inevitably will lead to a rather wide range in the outcomes across IAMs for one specific SSP. However, this strategy of limited quantitative harmonization was intentional, in order to explore uncertainty in scenario implementation and model behavior. In addition, the different structures of the models would not allow for a precise harmonization in this respect. For each of the five SSPs, the implementation by one specific IAM has been selected as the so-called marker scenario. The selection of the markers was guided by two main considerations: the internal consistency of the full set of SSP markers, and the ability of the different models to represent distinct characteristics of the storyline. Identifying the markers involved an iterative process with multiple rounds of internal and external reviews of energy system, emissions and land use model representations (Riahi et al., this special issue). Furthermore, additional realizations of the SSPs have also been computed by non-marker IAMs (see Table 2 for an overview) since they provide insights into possible alternative projections of the same qualitative storyline, including a first-order estimate of uncertainties attending to model structure and interpretation/implementation of the storylines. In this paper, we concentrate on the detailed presentation and discussion of the SSP marker scenarios for the baseline case without climate change mitigation, as well as the RCP4.5 and the RCP2.6 mitigation cases. Additionally, we indicate the range across all quantifications (marker and non-markers). Finally, we note throughout the paper where the selection of marker models may influence the qualitative conclusions drawn.

Table 2: Overview of SSPs and marker teams

Identifier	Descriptor	Marker model (institution)	Also computed by
SSP1	Sustainability	IMAGE-MAGNET (PBL, LEI)	All
SSP2	Middle-of-the-Road	MESSAGE-GLOBIOM (IIASA)	All
SSP3	Regional Rivalry	AIM/CGE (NIES)	IMAGE, MESSAGE-GLOBIOM, GCAM (only baseline)
SSP4	Inequality	GCAM (PNNL)	AIM/CGE
SSP5	Fossil-fueled Development	REMIND-MAgPIE (PIK)	AIM/CGE, GCAM

RESULTS

In this section, we describe SSP-specific dynamics of agricultural demand, production and trade, land use, agricultural intensification, GHG emissions and food prices at the global and regional level. Due to differences in base year data sources, and large uncertainties in historical data, land use outputs differ more between the IAMs than for example energy outputs (Bauer et al., this special issue). Therefore, in this paper, we focus on changes over time compared to 2005, as opposed to absolute values. All data can also be accessed over the interactive public SSP web-database hosted by IIASA (https://secure.iiasa.ac.at/web-apps/ene/SspDb). Data is shown and discussed in this paper for five aggregate regions: (1) OECD90 countries and new EU member states and candidates (OE0CD), (2) reforming economies of the Former Soviet

Union (excluding EU member states; REF), (3) countries of the Middle East and Africa (MAF), (4) countries of Latin America and the Caribbean (LAM) and (5) Asian countries with the exception of the Middle East, Japan and Former Soviet Union states (ASIA). We mainly describe results along the baseline and two selected mitigation cases (RCP4.5 & RCP2.6). Please note that SSP3 was infeasible for RCP2.6, and therefore does not show up in the respective results. While most of the discussion focuses on quantitative results from the marker scenarios, we indicate where results differ across models throughout this section.

Demand, production & trade

Global population has risen from 3.3 billion in 1965 to 6.5 billion in 2005 (Worldbank, 2015). This growth has been accompanied by increased per capita food and feed demand (2310 kcal/cap/day in 1965 to 2763 kcal/cap/day in 2005 (FAOSTAT 2015)), resulting in total crop production of 3750 million t DM in 2005 and 250 million t DM in 2005 of livestock products. Fig. 1 shows this historical development (FAO 2014), as well as the change of global demand for food and feed crops and livestock products in the different SSPs from 2005 to 2100. Future demand depends on population dynamics and per capita demand, which in turn depends on income, preferences, and food price sensitivities.

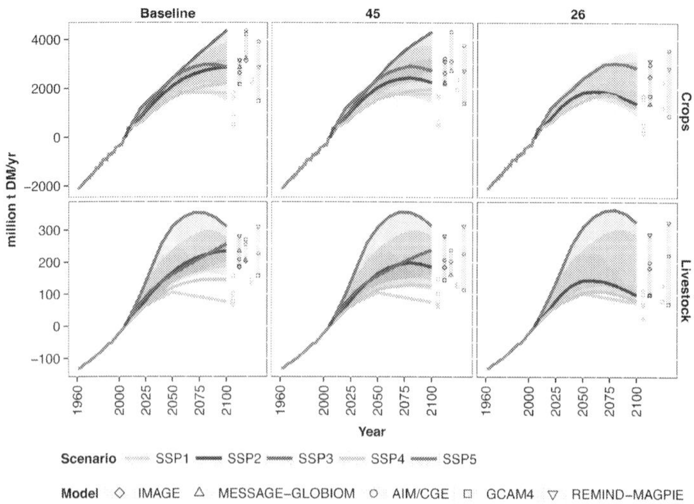

Figure 1: Change in global demand for crops (upper row) and livestock products (lower row) of the five SSP marker scenarios for the baseline (left column),

RCP4.5 (middle column) and RCP2.6 (right column) cases. Colored lines indicate the marker model results for each SSP. Colored bars indicate the range of data in 2100 across all marker and non-marker projections for each SSP (models are depicted by icon). Grey line shows historical trends based on FAO data (FAO 2014).

In the *SSP2* baseline scenario, population dynamics, per capita caloric consumption and animal calorie shares increase moderately. As a consequence, global demand for crop (plus 2860 mio t DM in 2100) and livestock products (plus 235 mio t DM in 2100) increases moderately in SSP2 with the highest shares and increases in demand over time in ASIA. Production remains fairly regionalized and trade with agricultural goods grows slowly. In *SSP1*, due to low population increases and healthy diets with low animal-calorie shares and low food waste, food demand for crops (plus 1820 mio t DM in 2100) and livestock (plus 76 mio t DM in 2100) products increases slightly until the mid of the century and then decreases. Additionally, trade of agricultural products increases only moderately, although markets are globally connected. Here, an emphasis on domestic production reduces the incentive for specialization in agricultural production and hence limits the increase in trade volumes. In *SSP3*, very high population increases, but low economic growth drive increases in global demand for crops to higher levels than in SSP2 (plus 4384 mio t DM in 2100, mainly in MAF), and to similar levels as in SSP2 for livestock products (plus 256 mio t DM in 2100, mainly in MAF and ASIA). In such a de-globalized world, a small share of agricultural goods is traded, and when it is, highly populated regions like MAF are importers. Compared to SSP2, *SSP4* shows relatively low increases in demand of both crops (plus 2201 mio t DM in 2100) and livestock products (plus 147 mio t DM in 2100) (mainly in MAF), despite having global population growth very similar to SSP2. This difference in demand is because the increase in population in *SSP4* is mainly in low-income regions such as MAF with limited access to markets, and even more importantly because the GDP growth in these already poor regions is even slower under SSP4 than under SSP2. The high demand for crops in MAF and ASIA is met with both local production and imports from OECD and LAM. *SSP5* reaches similar levels of crop demand as SSP2 (plus 2870 mio t DM in 2100), but much higher demand for livestock products occurs in SSP5, especially in the middle of the century (plus 354 mio t DM in 2070) due to unhealthy diets with high animal shares and high shares of food waste. Part of the crop demand increases in SSP5 are associated with intensified livestock production systems and higher feed crop use. In the strongly globalized world of SSP5, agricultural products are not necessarily produced domestically. Instead, ASIA becomes the most important exporter for crops in the latter half century and MAF the most important exporter for livestock products due to high increases in agricultural productivity in both, livestock and crops systems.

Global results in the non-marker baseline and mitigation scenarios are largely consistent with these findings for SSP1, SSP2, and SSP3, as indicated by the limited overlap in uncertainty bars for these three scenarios. All models find lower demand in 2100 for SSP1 than SSP2, and most find higher demand in SSP3 than in SSP2 (MESSAGE-GLOBIOM has slightly smaller demand for livestock in SSP3). Models diverge with respect to crop demand in SSP4, with GCAM showing roughly equal crop demand in SSP2 and SSP4, while AIM has lower crop demand in SSP4 than in SSP2. Additionally, differences across models emerge in SSP5. Both GCAM and REMIND show lower crop demand than SSP2, while AIM shows higher. For livestock, both REMIND and AIM show higher demand in SSP5, while GCAM shows lower.

As shown in Fig. 2, demand for dedicated 2nd generation bioenergy crops plays a critical role not so much in the baseline scenario but in nearly all mitigation scenarios because it provides an option to reduce emissions in the electricity and transport sectors and allows for active carbon dioxide removal from the atmosphere if combined with carbon capture and sequestration (BECCS). Particularly in SSP5, a large amount of carbon dioxide removal is needed for limiting climate forcing to levels of 4.5 W/m^2 and 2.6 W/m^2 by the end of the century, as a result of high exploitation of abundant fossil fuel resources (Bauer et al., this special issue), the adoption of resource and energy intensive lifestyles around the world and associated high levels of GHG emissions in the baseline case. Due to strong reliance on technical solutions in SSP5, land-based CDR in SSP5 rely on BECCS leading to large scale 2nd generation bioenergy crop demand of 9364 t DM in 2100 in RCP4.5 and 23142 t DM in 2100 in RCP2.6. This dependence on bioenergy in SSP5 is consistent across marker and non-marker models (non-marker models show 20775 and 21132 million t DM in 2100 in RCP2.6).

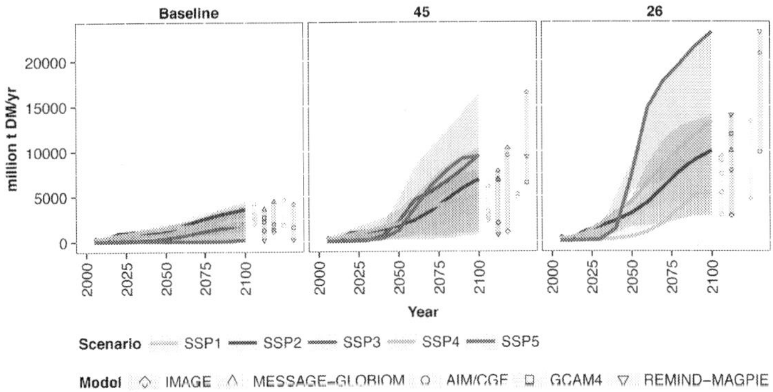

Figure 2: Global demand for dedicated 2nd generation bioenergy crops of the five SSP marker scenarios for the baseline (left column), RCP4.5 (middle column) and RCP2.6 (right column) cases. Colored lines indicate the marker model results for each SSP. Colored bars indicate the range of data in 2100 across all marker and non-marker projections for each SSP (models are depicted by icon).

The response of food demand to climate policy differs across marker and non-marker models, particularly in the RCP2.6 mitigation case. This results mainly from explicit model assumptions on high (e.g. MESSAGE-GLOBIOM) or low (e.g. REMIND-MAgPIE, IMAGE) sensitivities of food demand to increased food prices from mitigation pressure on land. Regional mitigation pressures on the land system, such as 2nd generation bioenergy production in LAM, MAF and ASIA displace agricultural production particularly in the most globalized SSP5 scenario. For example, in the SSP5 RCP2.6, trade of food and feed crops is further expanded compared to the Ref scenario, and livestock production is shifted from MAF mainly to ASIA and LAM.

Trends in land use and agricultural intensification

According to the Food and Agriculture Organization of the United Nations (FAO 2014), 4.9 billion hectares (approximately 40% of the land surface) was under agricultural use in 2005, either as cropland (1.5 billion hectares) or pasture (3.4 billion hectares). Historically, in the era before industrial fertilizers, increases in agricultural production were mainly achieved by expanding cropland and pasture land into forests and natural ecosystems. Today, in some regions such as Sub-Sahara Africa and Latin America, agricultural expansion has continued, at the cost of 110 million hectares of natural forests between 1990 and 2005 (Fig. 4).

However, over the last few decades, agricultural intensification was the major source of increases in global crop production, with world average cereal yield having increased by factor of 2.5 from 1.3 t/ha in 1960 to 3.3 t/ha in 2005.

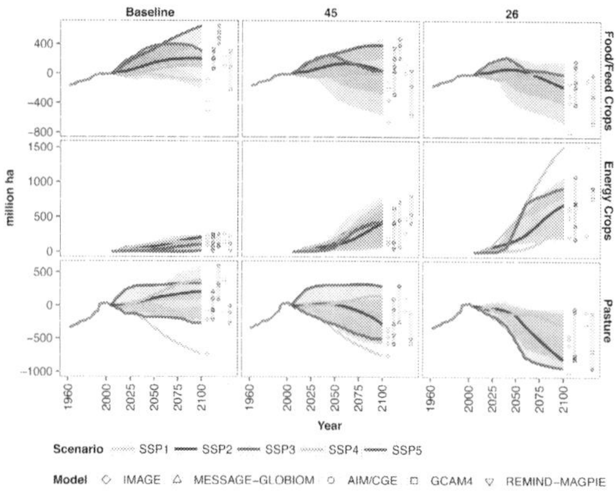

Figure 3: Change in global land for food and feed crops (upper row), energy crops (middle row) and pasture (lower row) of the five SSP marker scenarios for the baseline (left column), RCP4.5 (middle column) and RCP2.6 (right column) cases. Colored lines indicate the marker model results for each SSP. Colored bars indicate the range of data in 2100 across all marker and non-marker projections for each SSP (models are depicted by icon). Grey line shows historical trends based on FAO data (FAO 2014).

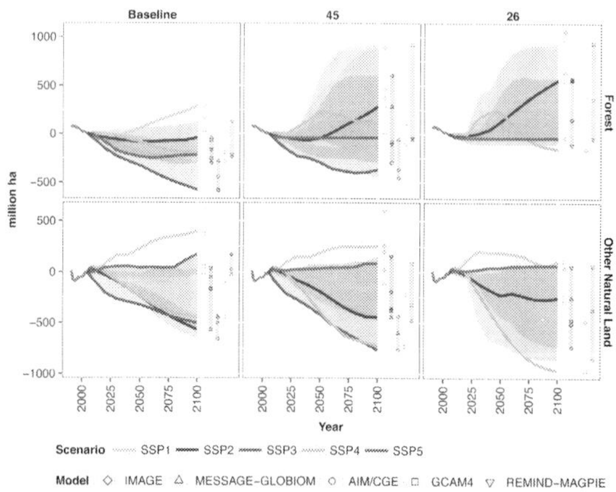

Figure 4: Change in global land for forest (upper row) and other natural land (lower row) of the five SSP marker scenarios for the baseline (left column), RCP4.5 (middle column) and RCP2.6 (right column) cases. Colored lines indi-

cate the marker model results for each SSP. Colored bars indicate the range of data in 2100 across all marker and non-marker projections for each SSP (models are depicted by icon). Grey line shows historical trends based on FAO data (FAO 2014).

The baseline SSPs cover a very wide-range of land-use futures, illustrated by the trends shown in Fig. 3 (agricultural land), Fig. 4 (forests and other natural land), and Fig. 5(agricultural yield increases). The models have similar global land cover at the initial time, but notably differ in their regional classification of land use types, as well as in their regional allocation of total land.

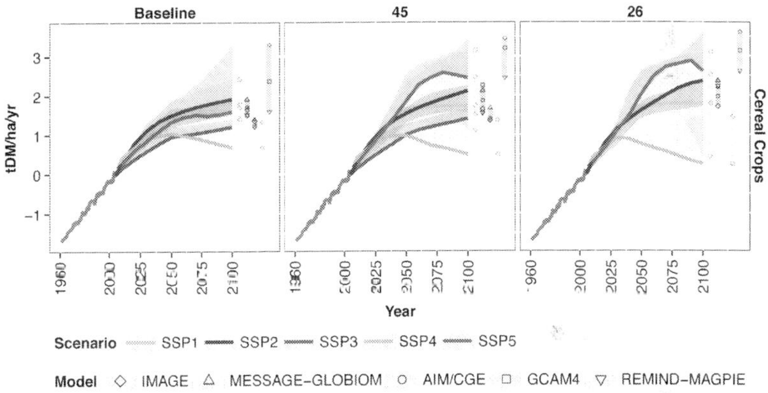

Figure 5: Change in global cereal crop yields of the five SSP marker scenarios for the baseline (left column), RCP4.5 (middle column) and RCP2.6 (right column) cases. Colored lines indicate the marker model results for each SSP. Colored bars indicate the range of data in 2100 across all marker and non-marker projections for each SSP (models are depicted by icon). Grey line shows historical trends based on FAO data (FAO 2014).

The use of cropland for food and feed production increases moderately in *SSP2* (plus 231 mio ha between 2005 and 2100), due to relatively high demand for food and feed crops, combined with high yield increases (by a factor of 1.6 between 2005 and 2100). Pasture area increases strongly in SSP2 (plus 204 mio ha until 2100). Agricultural expansion mainly happens in MAF and LAM as a result of medium demand for livestock products satisfied mostly through rather extensive livestock production systems. These increases in agricultural land happen at the expense of forest areas (LAM) and other natural land (MAF). In *SSP1*, with one of the lowest demand for agricultural goods and high intensification of agricultural production, agricultural land decreases and is significantly lower than in the other SSPs. As a consequence of such agricultural abandonment and regrowth of natural vegetation, other natural land and forests expand strongly in all regions. The highest increases in pasture and cropland for food and feed production (mainly in MAF and LAM at the cost of forests and other natural land) are observed in *SSP3*, mostly driven by an increasing global population combined with low agricultural intensification. *SSP4* shows minor increases of cropland and strong increases of pastureland (mainly in MAF) at the expense of forests. These increases in agricultural land are caused by a similar reasons to those noted for SSP3, as low income regions such as MAF in SSP4 show high population growth while remaining relatively unproductive in agriculture. As a result of production increases in unproductive regions, global average crop yields decline the second half of the century, despite continued yield improvements in all regions. *SSP5* shows an increasing use of cropland until 2050 (mainly in ASIA, LAM & MAF) which then decreases towards medium levels in 2100 and a decline in pasture throughout the century. Contraction of agricultural land in SSP5 occurs as population decreases and consumption stabilizes at high levels per capita and at the same time production of livestock products is met through a shift from extensive to more intensive animal husbandry.

These findings are fairly consistent across baseline marker and non-marker scenarios, with all models showing lower cropland area in SSP1 and higher cropland area in SSP3 than in SSP2. Forest area, in contrast, is largest in SSP1 and smallest in SSP3 in all models (with the exception of IMAGE SSP3). Land cover in SSP4 and SSP5 are more similar to SSP2 in most cases; however, models diverge in some instances. For example, GCAM has significantly larger pasture area in SSP4 than in SSP2, while AIM shows roughly equal areas in the two scenarios.

In most of the mitigation cases, dynamics of agricultural land for food and feed production are affected by land demanding mitigation options such as bioenergy, avoided deforestation or afforestation. The clear exception is in SSP3 where major implementation barriers for land-based mitigation are assumed to occur and where no climate stabilization levels of RCP 2.6 can be attained. Avoided deforestation restricts agricultural expansion in most of the SSPs, with the exception of SSP3 and the low-income regions of SSP4 (MAF and ASIA) due to weak land use change regulation. The land system can also contribute to climate change mitigation by increasing carbon stocks. This trend is observed particularly in the very ambitious mitigation target of RCP2.6, which relies on land-based carbon dioxide removal options such as afforestation or BECCS. Here, afforestation for carbon sequestration increases global forest areas in 2100, compared to Ref, by 262 mio ha in SSP1 and by 601 mio ha in SSP2 (mainly in MAF and LAM). Bio-energy plays a critical role in nearly all mitigation but also in the baseline scenarios. The SSPs allocate between 121 million ha (SSP1) and 473 million ha (SSP4) in the RCP4.5 and between 245 million ha (SSP1) and 1517 million ha (SSP4) in the RCP 2.6 mitigation scenario to ligno-cellulosic bioenergy crop production in 2100. Both land-based CDR strategies (afforestation and BECCS) happen at the expense of other natural land (SSP4), unprotected forests (SSP3), land for food and feed crops (SSP2, SSP4 and SSP5) and pastureland (SSP2, SSP4 and SSP5). Generally, as a result of land needed for large scale bioenergy production and afforestation programs in the mitigation scenarios, the use of land for food and feed production and pasture is reduced, following considerable agricultural intensification (SSP5) and dietary changes (SSP2) compared to the baseline scenarios.

Projections of GHG emissions

Land-use change, mainly the conversion of tropical forests to agricultural land, is a significant source of carbon emissions accounting for approximately 12% of all anthropogenic carbon emissions from 1990 to 2010. The agricultural sector is also the largest contributor to global anthropogenic non-CO_2 GHGs. In total non-CO_2 GHG emissions from agriculture (CH_4 from enteric fermentation, rice production and animal waste management systems; N_2O from synthetic fertilizer application and animal waste management systems) are estimated to account for about 10–12% of global anthropogenic emissions in 2010 (Smith et al., 2013). Between 1970 and 2005, global agricultural CH_4emissions grew from 128.5 Mt to 144.0 Mt CH_4/yr and global agricultural N_2O emissions from 3.3 to 5.7 Mt N_2O/yr (FAO 2014).

GHG dynamics as an outcome of land use dynamics and agricultural production systems span a broad range of potential futures with SSP3 being highest and SSP1 lowest for all GHG categories considered. In the baseline scenario of *SSP2*, global CO_2 emissions from land use change amount to 219 Gt CO_2 cumulatively between 2005 and 2100 (Fig. 6). Annual CO_2 emissions decrease steadily until the end of the century and are negative from 2080 onwards. Emissions occur mainly in MAF and LAM as a result of cropland and pasture expansion and the associated loss of forests and other natural land. Carbon uptake happens from mid-century onwards, mainly due to regrowth of vegetation in ASIA. Annual CH_4 emissions from agricultural production increase by 41 Mt CH_4 between 2005 and 2050 and then remain fairly constant due to lower increases in demand, especially for livestock products, and more intensified livestock production systems associated with lower emission factors (Fig. 7). Annual N_2O emissions increase by 3.5 Mt N_2O through 2100. The lowest projected CO_2, CH_4 and N_2O emissions across all SSPs occurs in *SSP1*. Global cumulative CO_2 emissions are even negative in 2090 already (-54 Gt CO_2 until 2100) due to abandonment of agricultural land and associated carbon uptake through vegetation regrowth in MAF, LAM and OECD. Low CH_4 and N_2O emissions in SSP1 are associated with generally low demand for agricultural goods, especially livestock products, and livestock production systems with high conversion efficiencies. In contrast to SSP1, the highest cumulative CO_2 emissions from land use change occur in ***SSP3*** (402 Gt CO_2 through 2100), as a result of very high agricultural expansion (mainly in ASIA and MAF). CH_4 emissions increase steadily until 2100 (by 238 Mt CH_4 compared to 2005) with highest contributions from ASIA and MAF driven by a combination of population growth and associated demand increases and low intensification of agricultural production. Also N_2O emissions (mainly from MAF and ASIA) show highest increases across all scenarios (by 9.0 Mt N_2O compared to 2005). In *SSP4*, global cumulative CO_2 emissions (mainly due to forest losses in MAF) reach 183 Gt CO_2 in 2100. Annual CH_4 emissions reach 202 Mt CH_4/yr and N_2O emissions 8.2 Mt CH_4/yr in 2100. Finally, *SSP5* shows increasing cumulative CO_2 emissions until 2050 (190 Gt CO_2) that remain constant afterwards as land-use changes in ASIA, LAM and MAF and related carbon emissions come to a halt. Annual CH_4 emissions in SSP5 show highest increases until 2060 (177 Mt CH_4) mainly due to livestock production for export in MAF and decrease strongly afterwards as global livestock demand decreases and agricultural management becomes more intensified. Similarly, N_2O emissions increase slightly until 2040 (3.5 Mt N_2O) and then decrease due to improved agricultural management systems with high nitrogen use efficiencies.

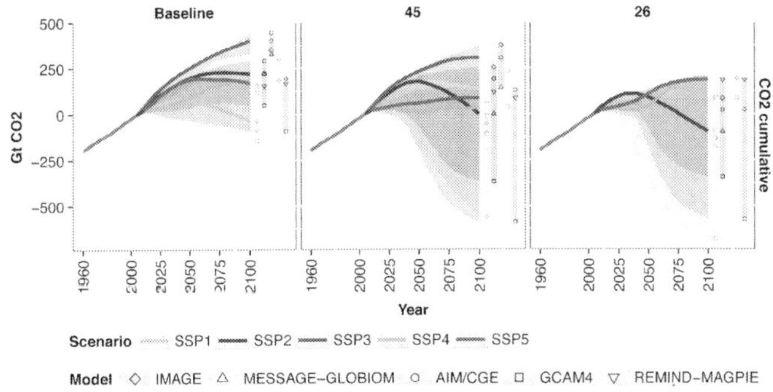

Figure 6: Change in global cumulative land-use change emissions since 2005 of the five SSP marker scenarios for the baseline (left column), RCP4.5 (middle column) and RCP2.6 (right column) cases. Colored lines indicate the marker model results for each SSP. Colored bars indicate the range of data in 2100 across all marker and non-marker projections for each SSP (models are depicted by icon). Grey line shows historical trends based on RCP data (van Vuuren et al., 2012).

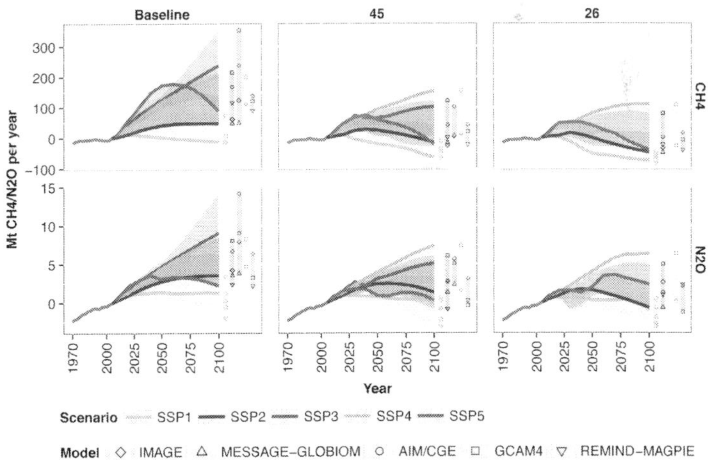

Figure 7: Change in global agricultural CH_4 (upper row) and N_2O emissions (lower row) of the five SSP marker scenarios for the baseline (left column), RCP4.5 (middle column) and RCP2.6 (right column) cases. Colored lines indicate the marker model results for each SSP. Colored bars indicate the range of data in 2100 across all marker and non-marker projections for each SSP (models are depicted by icon). Grey line shows historical trends based on EDGAR data (Edgar 2015).

Baseline cumulative CO_2 emissions in 2100 are highest in SSP3 and lowest in SSP1 across all models (marker and non-marker). Emissions in SSP4 lie somewhere between SSP2 and SSP3 levels in both models, and emissions in SSP5 lie somewhere between SSP1 and SSP2 for most models (emissions in REMIND-MAGPIE are slightly above SSP2).

In the mitigation case RCP4.5, avoided deforestation strongly reduces CO_2 emissions in SSP5, SSP4 and SSP2 compared to the baseline scenarios. However, as a result of weak land use change regulation, CO_2 emissions from land use change still occur in SSP3 (307 Gt CO_2 cumulatively until 2100). In the RCP2.6 mitigation case, emissions are again higher in SSP4 and SSP5, due to displacement effects into pasture land caused by high bioenergy production combined with forest protection only (Popp et al., 2014), and (for SSP4) due to additional land demand for bioenergy crop production in low income regions like MAF and ASIA without forest protection. Afforestation increases terrestrial C sequestration especially in SSP2 in LAM, MAF and ASIA and in the high income regions (OECD, LAM and REF) of SSP4 where land use change is successfully regulated. CH_4 emissions in the mitigation cases are remarkably lower compared to the baseline cases in all SSPs due to improved agricultural management (such as improved water management in rice production, improved manure management by e.g. covering of storages or adoption of biogas plants, better herd management and better quality of livestock through breeding and improved feeding practices). Dietary shifts away from emissions-intensive livestock products (SSP2) also lead to decreased CH_4 emissions. N_2O emissions are significantly lower particularly in the RCP4.5 scenario due to improvement of N-efficiency and improved manure management. However, high levels of bioenergy production result in increased N_2O emissions in SSP5 due to N fertilization of dedicated grassy bioenergy crops such as Miscanthus.

Food price dynamics

SSP-specific changes in population, income, international trade, agricultural expansion and technological change as discussed above are the major drivers for long-term changes in world food prices. For the baseline scenario, *SSP2*, *SSP4* and *SSP5* show either flat or slightly falling world market prices for crops and livestock products by 2100, compared to 2005 (Fig. 8). While average world market price effects are rather modest in these SSPs, there is more variation especially in SSP3 at the regional level. As access to international markets is limited in poor regions in SSP4, the highest price effects are projected for ASIA and MAF. By contrast, *SSP1* shows global price decreases of about 60% by 2100 consistent with lowest demand for agricultural commodities, more rapid growth in agricultural productivity and globalized trade. The opposite is true for *SSP3*, where world market prices increase by about 50% as a result of large population increases, very low productivity increases, and restricted trade of agricultural commodities. Here, the highest price effects are projected for MAF and ASIA. The effect of different SSPs on global food prices is robust across marker and non-marker models, with most models showing higher food prices in SSP3 and lower prices in SSP1 in 2100 (AIM shows little difference in food prices between SSP1 and SSP2). Food prices in SSP4 and SSP5 are similar to SSP2, with small declines in SSP5 in GCAM & REMIND-MAGPIE and small increases in the AIM SSP4.

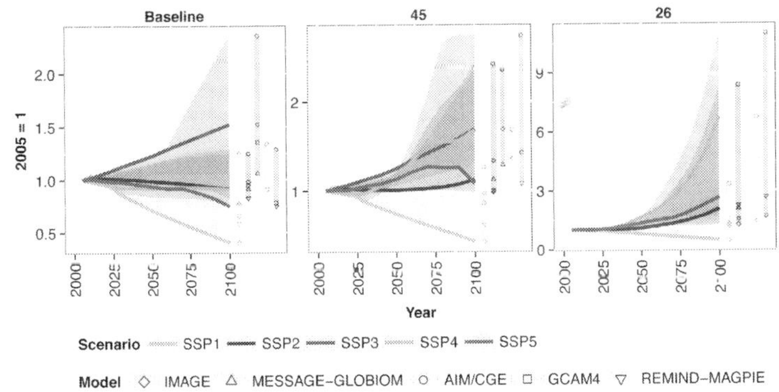

Figure 8: Change in world market prices [2005 = 1] aggregated across all crop and livestock commodities of the five SSP marker scenarios for the baseline (left column), RCP4.5 (middle column) and RCP2.6 (right column) cases (Note that baseline, RCP4.5 and RCP2.6 have indivudal scales). Colored lines indicate the marker model results for each SSP. Colored bars indicate the range of data in 2100 across all marker and non-marker projections for each SSP (models are depicted by icon).

In the mitigation scenarios, particularly in RCP2.6, land based mitigation measures cause world market prices to increase relative to 2005 in the SSP2 (+110%), SSP5 (+170%) and SSP4 (+570%) scenarios as a result of the carbon tax, changes in agricultural management, increased bioenergy production, and land used for afforestation. As bioenergy production, forest mitigation activities, and the abatement of agricultural GHG emissions are limited in the SSP3 due to major implementation barriers, the food price is barely affected in RCP 4.5. In SSP1 mitigation does hardly influence food prices due to a general 'food first' policy, which can restrict agricultural expansion to avoid deforestation, but futher only allows bio-energy on areas not needed for food and feed production. In general, considerable agricultural intensification (such as in SSP5), responses in agricultural trade (such as in SSP4 and SSP5), and changes in total production and consumption (such as in SSP2) have the capability to diminish food price reactions. The uncertainty across models for food prices is significant, with GCAM projecting much larger increases in the mitigation cases than other models. These price effects in GCAM are due to the strong dependence on afforestation and bioenergy as mitigation options, leading to significant land competition. Due to this uncertainty, the selection of marker models strongly influences the ranking of this variable, unlike previous results. For food prices in the mitigation cases, all models show food prices that are lower in the SSP1 and higher in SSP3 than the SSP2 in 2100. Food prices in SSP4 are less than (GCAM) or equal to (AIM) prices in SSP2. Food prices in the SSP5 are higher than the SSP2 in all models. While the qualitative ordering is robust across models, the magnitude of change differs significantly across models, with GCAM showing higher increases due to mitigation than any other model.

CONCLUSION

Future development of the agricultural system depends strongly on population dynamics and economic growth. But also other basic socio-economic conditions such as technological change in the crop and livestock sector (Havlík et al., 2013; Weindl et al., 2015), investment in agricultural technology (Robinson et al., 2014), dietary patterns and food demand (Smith et al., 2013; Popp et al., 2010; Stehfest et al., 2009), trade of agricultural goods (Schmitz et al., 2012; Verburg et al., 2009) and interactions with other sectors (e.g., the energy system via bioenergy demand) (Popp et al., 2013) strongly influence land-use dynamics, GHG emissions, potential of land-based mitigation and impacts. It is therefore of key importance to improve our understanding of the agricultural system and land use under different sets of future socio-economic settings, such as the newly developed Shared Socio-economic pathways. IAMs with distinctive land-use modules, like those applied in this exercise, are well-suited tools for doing so as these models represent interconnections between the land sector and a whole suite of human systems. In the previous sections, we have shown the land-use sector as simulated by the SSP marker models reflecting the general storylines of the SSPs and their positioning with respect to socio-economic challenges to mitigation.

All in all, the SSP marker scenarios provide a broad range of potential land-use futures. In the SSP baseline cases without climate change mitigation, global cropland (including bioenergy crops) decreases between 2005 and 2100 by 3 mio ha in SSP1 on the low and by up to 753 mio ha in SSP3 on the high end. Pastureland shows even more divergence with global area increases of up to 380 mio in SSP4 at the high and decreases of 742 mio ha in SSP1 at the lower end. The inclusion of mitigation efforts, particularly those in the most ambitious mitigation scenarios of RCP 2.6, broaden the range of potential future agricultural area, as total cropland expands by up to 1413 mio ha until 2100 in SSP4 due to expansion of bioenergy cropland, and pastureland decreases until 2100 by up to 940 mio ha in SSP5. These outcomes are consistent with the range of estimates reported in the recent literature. The AgMIP model comparison of global economic land use models projects global cropland changes between − 75 and + 450 mio ha between 2005 and 2050 under constant climate (Schmitz et al., 2014). The FAO projects an increase of 69 million ha for the same time span (Alexandratos and Bruinsma, 2012). Smith et al. (Smith et al., 2010) provided a review of studies on land-use projections of the past two decades, indicating a range between 90 and 470 million ha in 2050. For 2030, Lambin and Meyfroidt (Lambin and Meyfroidt, 2011) report projected cropland changes between 125 and 265 million ha. However, all of these assessments do not focus so much on the variability of future socio-economic drivers, they rather examine business-as-usual scenarios, or a limited variation in drivers, and do not include land-based mitigation, thus excluding several sources for diverging land-use futures. In the broader scenario literature (Nakicenovic and Swart, 2000; Carpenter and Pingali, 2005; Van Vuuren et al., 2011) projections for cropland range from a decrease of 140 mio ha to an increase of 1130 mio ha in 2100 compared to 2005, whereas pasture land ranges between increases of 217 mio ha and decreases of 1550 mio ha.

The SSP land dynamics have consequences for sustainable development. First, the highest losses of forest and other natural land is observed in SSP3, mainly in MAF and LAM. These losses could have consequences for biodiversity. In contrast, SSP1 shows abandonment of agricultural land and associated regrowth of natural vegetation in the baseline and mitigation cases. Similarly, long-term changes in food prices are affected very differently, with SSP1 indicating global price decreases whereas the opposite is true for SSP3 with highest price effects projected for countries of the Middle East and Africa as well as Asia. So in general, we see that the SSP1 has the most positive effects for sustainable development, with its sustainable food consumption (i.e. low food waste, diets with low shares of animal products), rapid growth in agricultural productivity and globalized trade.

But the wide range of possible land-use futures, both in literature and in this study, shows that there are major uncertainties in the global agricultural system, whose dynamics are still poorly understood.

Differences in land use projections result from a combination of different model architectures and philosophies, inherent uncertainties on modeled processes such as for example irrigation of cropland, and differences about how to parameterize these processes along various storylines such as the SSPs. While GDP and population trends were explicitly prescribed per SSP, spanning an uncertainty range across these dimensions, other important drivers were prescribed in qualitative terms (Table 2), and quantification was left to the model teams to cover the uncertainty of these drivers under a defined storyline (see table on IAM implementations in the SOM). In the interplay with different model structures and representation of processes across the IAMs, the final effect of such model drivers parameterization on agricultural production, trade, land use, GHG emissions and food prices introduces an additional level of uncertainty, and as a result, the range of outcomes across the multiple IAM realizations per SSP and RCP is rather wide. Indivdual behavior of marker and non-marker models, however, is similar for SSP1, SSP2, and SSP3: Demand for agricultural products is lowest in SSP1 and highest in SSP3. Cropland area is lowest in SSP1 and highest in SSP3, leading to higher forest area in SSP1 and lower forest area in SSP3. Cumulative

CO_2 emissions from land use change are lowest in SSP1 and highest in SSP3. Food prices are largest in SSP3 and smallest in SSP1. Models are less consistent with their representations of SSP4 and SSP5. Agricultural demand in SSP4 is less than (AIM) or equal to (GCAM) that of SSP2. Pastureland is significantly higher in SSP4 than in SSP2 in GCAM, but comparable in AIM. Agricultural demand in SSP5, however, is either lower or higher than SSP2 depending on the model. These differences do result in diverging cropland areas, but the differences between SSP5 and SSP2 are small in all models. Despite these differences, cumulative CO_2 emissions in the baseline scenarios represent the SSP storylines well, with SSP5 emissions falling between that of SSP1 and SSP2 and SSP4 emissions falling between SSP2 and SSP3. While much of the discussion in this article focused on comparing among marker scenarios, the dependence on these marker scenarios does not affect the qualitative conclusions in most cases. That is, using a single model across SSPs would not change the ranking of scenarios in most cases. One notable exception is in the interpretation of food prices under a mitigation case. Here, model differences dominate scenario differences. Interestingly, the ranking within models is consistent, but the absolute values are strikingly different and mixing and matching models can be misleading.

One major uncertainty is the assumed increase in crop yields, which amounts at least to a doubling of current levels in most SSPs. While this would be a continuation of past trends, and is also in line with FAO's projections until 2050 (Alexandratos and Bruinsma, 2012), it is increasingly questioned whether past increases can be continued in the future for already high yielding crops and regions, as some important sources of yield improvements like increasing the harvested index, might have reached their limits. On the other hand, it is widely accepted that closing the yield gaps to currently attainable yields in low-income regions like Sub-Sahara Africa pose a huge potential for higher crop production (Mueller et al., 2012). With respect to the mitigation scenarios, there are several important uncertainties. For example, the large contribution of the land-use sector to climate change mitigation via delivering bio-energy and removing CO_2 from the atmosphere via BECCS relies on major innovations (significant yield increases, implementation of BECCS). Likewise, emission reduction of non-CO_2 gases in the agricultural sector rely on high adoption rates worldwide. Furthermore, some prominent land based mitigation options such as soil carbon management with an economic potential estimated at 3.5 $GtCO_2eq$ per year by 2030 (Smith et al., 2013) are not included in this assessment. Finally, the IAMs only account for the emissions and carbon balance of land-based mitigation, while recent studies demonstrate the importance of land cover as well as land management changes (e.g. (Luyssaert et al., 2014)) and as a consequence that the inclusion of biophysical consequences of land-based mitigation within an IAM (Jones et al., 2015), could have significant consequences for climatic conditions at local and global levels. Another important caveat of our multi-model uncertainty analysis is that the uncertainty ranges for the individual SSPs are based on different sample sizes as not all modeling teams developed a scenario for each of the SSPs due to specific model characteristics (see Table 2 for an overview). For example SSP4, covered only by 2 IAMs, shows lowest uncertainty for most of the indicators compared to the other SSPs, especially in the cases without climate change mitigation. This fact complicates the assessment of scenario versus structural (model) uncertainty. Hence, in a next step, a much higher participation of additional IAM modeling teams in the quantification of the SSPs is needed wherefore the modeling protocol for this study has been made available to the broader IAM and land use modeling (e.g. AgMIP) community.

At present, SSP land use information is provided at the level of 5 world regions (https://secure.iiasa.ac.at/web-apps/ene/SspDb). However, information at much finer spatial resolution is needed, especially for climate model projections, impacts, vulnerability and adaption assessments (Preston et al., 2011), for the assessment of gross land use changes with great importance for example for biodiversity assessments which are not visible at the regional scale and also for the assessment of sustainable development. This work is planned in subsequent phases of the scenario development. Additionally, these SSP outcomes will also be harmonized with the most recent data on historic land use, allowing for a smooth transition from the historical periods to the scenario period while conserving the original underlying IAM scenario signal, in a similar way as done for the RCPs (Hurtt et al., 2011), but with an updated methodology, more emphasis on land management, and a higher resolution of 0.25°.

The land-use futures of the SSPs, as elaborated by the five IAMs, shown in this study serve multiple purposes, and are expected to be used by a broad range of different communities. First, climate modeling is increasingly interested to study the multiple effects of land use and land-use change, including mitigation, on both biogeochemical and biophysical processes (Pitman et al., 2009), as land use change not only affects CO_2 concentrations (biogeochemical) but also albedo changes and evapotranspiration (biophysical) (Bonan, 2008; Jackson et al., 2008; Anderson et al., 2010). Previous efforts through the RCPs, resulting in a large number of experiments within the Coupled Model Intercomparison Project (CMIP5), were hampered by a lack of consistency across land-use trajectories, as the RCPs had been developed with a pure focus on radiative forcing targets in 2100 (Van Vuuren et al., 2011). The land-use information developed through the IAM implementation of the SSPs provide additional opportunities for assessment. For example, Earth System Models can better study the effects of deforestation and afforestation on climate by contrasting the land-cover dynamics in an SSP3 baseline scenario (deforestation) with those of an SSP1 RCP2.6 scenario (afforestation), an effort planned in the Land Use Model Intercomparison Project portion of CMIP6 (Lawrence et al., 2016). Second, in general the SSP framework is very well suited for climate change impacts, adaptation, and vulnerability (IAV) research, and in fact, it was designed to serve the needs of this community (Wilbanks and Ebi, 2013). The matrix architecture of combining SSPs with RCPs into new scenarios and comparing those with the respective baseline scenario without climate change impacts allows for a consistent assessment of climate change impacts under different socio-economic conditions, as well as vulnerability and adaptive capacity. In this study, climate change and CO_2 fertilization impacts are not considered on purpose, to allow a pure assessment of climate change mitigation pathways, without climate change and CO_2fertilization impacts and adaptation already accounted for in the IAM scenario. In addition to assessing the single climate change and CO_2 fertilization impacts, the SSP framework can also be applied to investigate the combined effects of climate change and land-based mitigation in a next step. Third, the IAM SSP scenarios which focus on large-scale trends and therefore ignore dynamics within countries or regions, could be combined with regional integrated assessments. For example, the marker scenarios presented here could serve for example the AgMIP community by providing the global Representative Agricultural Pathways (RAPs) framing the regional scenario developments (Valdivia et al., 2015). Such regional and country-level integrated assessments and scenario developments are of great importance for developing concrete policy options towards sustainable development. In this, the link between global processes and regional drivers is an important consideration, and global results might serve as explicit boundary conditions for smaller scale assessments. However, this link remains a huge challenge due to the multiple feedbacks between the two scales. For example, the SSP scenarios delivered by the IAMs contain long-term projections of world market prices for agricultural commodities. Aggregate price indices at the global level reflect global drivers, such as production and consumption in major world regions as well as international trade (Nelson et al., 2014). But such aggregate results may neglect significant distributional consequences at the national and sub-national level, and therefore miss important impacts of future climate change as well as ambitious mitigation polices (Hussein et al., 2013).

Last but not least, the elaboration of the SSPs in the agricultural and land-use components of the IAMs has spurred activities of model comparison, harmonization and improvement, within and outside the IAM community. Most notably, it is carried further in AgMIP's global agricultural economics team (Wiebe et al., 2015), and is being picked up by other global and regional assessments and research projects. But also for biodiversity and ecosystem services scenario assessments such as IPBES extended SSPs could be used as a staring point (Kok et al., 2016). In the end, these scientific communities working in a parallel and iterative process, as envisioned by the SSP design, will hopefully lead to a more complete understanding of land-use change dynamics at the global, regional and local scale, and its complex interactions with climate, impacts, adaptation and sustainable development.

ACKNOWLEDGEMENTS

PBL and PIK gratefully acknowledge funding from the European Union's Seventh Framework Program under grant agreement no. 603542 (LUC4C). IIASAs analysis contributing to this study was partly conducted in partnership with the CGIAR Research Program on Climate Change, Agriculture and Food Security (CCAFS) and supported by the European Union-funded project An integration of mitigation and adaptation options for sustainable livestock production under climate change (ANIMALCHANGE) (Grant 266018). NIES is grateful for the research support of the "Global Environmental Research Fund" (2-1402) provided by the Ministry of the Environment, Japan. PNNL gratefully acknowledges research support provided by the Integrated Assessment Research Program in the Office of Science of the U.S. Department of Energy.

REFERENCES

1. Alexandratos and Bruinsma, 2012 N. Alexandratos, J. Bruinsma World Agriculture Towards 2030/2050: The 2012 Revision Food and Agriculture Organization of the United Nations, Rome (2012)
2. Anderson et al., 2010 R.G. Anderson, J.G. Canadell, J.T. Randerson, R.B. Jackson, B.A. Hungate, D.D.Baldocchi, G.A. Ban-Weiss, G.B. Bonan, K. Caldeira, L. Cao, N.S. Diffenbaugh, K.R. Gurney, L.M.Kueppers, B.E. Law, S. Luyssaert, T.L. O'Halloran Biophysical considerations in forestry for climate protection Front. Ecol. Environ., 9 (2010), pp. 174-182, 10.1890/090179
3. Bauer et al., 2016 N. Bauer, Calvin Katherine, Emmerling Johannes, Fricko Oliver, Fujimori Shinichiro, Hilaire Jérôme, Eom Jiyong, Krey Volker, Kriegler Elmar, Mouratiadou Ioanna, de Boer Harmen Sytze, van den Berg Maarten, Carrara Samuel, Drouet Laurent, Edmonds Jae, Gernat David, Harmsen Mathijs, Havlik Petr, Johnson Nils, Kyle Page, Luderer Gunnar, Marangoni Giacomo, Masui Toshihiko, Strefler Jessica, Strubegger Manfred, Wise Marshall, Riahi Keywan, D.P.v. Vuuren Shared socio-economic pathways of the energy sector—quantifying the narratives Global Environ. Change (2016), 10.1016/j.gloenvcha.2016.05.008 (this special issue)
4. Bodirsky et al., 2014 B.L. Bodirsky, A. Popp, H. Lotze-Campen, J.P. Dietrich, S. Rolinski, I. Weindl, C. Schmitz, C. Müller, M. Bonsch, F. Humpenöder, A. Biewald, M. StevanovicReactive nitrogen requirements to feed the world in 2050 and potential to mitigate nitrogen pollution Nat. Commun., 5 (2014), 10.1038/ncomms4858
5. Bodirsky et al., 2015 B.L. Bodirsky, S. Rolinski, A. Biewald, I. Weindl, A. Popp, H. Lotze-CampenGlobal food demand scenarios for the 21st century PLoS One, 10 (2015), p. e0139201, 10.1371/

journal.pone.0139201

6. Bonan, 2008 G.B. BonanForests and climate change: forcings feedbacks, and the climate benefits of forests Science, 320 (2008), pp. 1444-1449, 10.1126/science.1155121
7. Burney et al., 2010 J.A. Burney, S.J. Davis, D.B. LobellGreenhouse gas mitigation by agricultural intensification Proc. Natl. Acad. Sci., 107 (2010), pp. 12052-12057, 10.1073/pnas.0914216107
8. Calvin et al., 2016 Calvin, K., Bond-Lamberty, B., Clarke, L., Edmonds, J., Eom, J., Hartin, C., Kim, S., Kyle, P., Link, R., Moss, R., Mcjeon, H., Patel, P., Smith, S., Waldhoff, S., Wise, M., 2016. SSP4: A World of Inequality. Global Environ. Change, this Special Issue.
9. Carpenter and Pingali, 2005 S. Carpenter, P. PingaliMillennium Ecosystem Assessment—Scenarios Assessment Island Press, Washington, DC (2005)
10. Carpenter et al., 2009 S.R. Carpenter, H.A. Mooney, J. Agard, D. Capistrano, R.S. DeFries, S. Díaz, T. Dietz, A.K. Duraiappah, A. Oteng-Yeboah, H.M. Pereira, C. Perrings, W.V. Reid, J. Sarukhan, R.J. Scholes, A.WhyteScience for managing ecosystem services: beyond the millennium ecosystem assessment Proc. Natl. Acad. Sci., 106 (2009), pp. 1305-1312, 10.1073/pnas.0808772106
11. Dellink et al., 2016 Dellink, R., Chateau, J., Lanzi, E., Magné, B., n.d. Long-term economic growth projections in the Shared Socioeconomic Pathways. Glob. Environ. Change. 10.1016/j.gloenvcha.2015.06.004
12. Dietrich et al., 2014 J.P. Dietrich, C. Schmitz, H. Lotze-Campen, A. Popp, C. MüllerForecasting technological change in agriculture—an endogenous implementation in a global land use model Technol. Forecast. Soc. Change, 81 (2014), pp. 236-249, 10.1016/j.techfore.2013.02.003
13. Ebi et al., 2013 K.L. Ebi, S. Hallegatte, T. Kram, N.W. Arnell, T.R. Carter, J. Edmonds, E. Kriegler, R. Mathur, B.C. O'Neill, K. Riahi, H. Winkler, D.P.V. Vuuren, T. ZwickelA new scenario framework for climate change research: background, process, and future directions Clim. Change, 122 (2013), pp. 363-372, 10.1007/s10584-013-0912-3

14. Foley et al., 2005 J.A. Foley, R. DeFries, G.P. Asner, C. Barford, G. Bonan, S.R. Carpenter, F.S. Chapin, M.T.Coe, G.C. Daily, H.K. Gibbs, J.H. Helkowski, T. Holloway, E.A. Howard, C.J. Kucharik, C. Monfreda, J.A.Patz, I.C. Prentice, N. Ramankutty, P.K. SnyderGlobal consequences of land use Science, 309 (2005), pp. 570-574, 10.1126/science.1111772
15. Fricko et al., 2016 Fricko, O., Havlik, P., Rogelj, J., Klimont, Z., Gusti, M., Johnson, N., Kolp, P., Strubegger, M., Valin, H., Amann, M., Ermolieva, T., Forsell, N., Herrero, M., Heyes, C., Kindermann, G., Krey, V., Mccollum, D.L., Obersteiner, M., Pachauri, S., Rao, S., Schmid, E., Schoepp, W., Riahi, K., 2016. SSP2: A middle of the road scenario for the 21st century. Global Environ. Change, this Special Issue.
16. Fujimori et al., 2014 S. Fujimori, T. Hasegawa, T. Masui, K. TakahashiLand use representation in a global CGE model for long-term simulation: CET vs. logit functions Food Secur., 6 (2014), pp. 685-699, 10.1007/s12571-014-0375-z
17. Fujimori et al., 2016 Fujimori, S., Hasegawa, T., Masui, T., Takahashi, K., Silva Herran, H.Y.H., Dai, Y., Kainuma, M., et al., 2016. Global Environ. Change, SSP3: AIM Implementation of Shared Socioeconomic Pathways, this Special Issue.
18. Havlík et al., 2013 P. Havlík, H. Valin, A. Mosnier, M. Obersteiner, J.S. Baker, M. Herrero, M.C. Rufino, E.SchmidCrop productivity and the global livestock sector: implications for land use change and greenhouse gas emissions Am. J. Agric. Econ., 95 (2013), pp. 442-448
19. Havlík et al., 2014 P. Havlík, H. Valin, M. Herrero, M. Obersteiner, E. Schmid, M.C. Rufino, A. Mosnier, P.K.Thornton, H. Böttcher, R.T. Conant, S. Frank, S. Fritz, S. Fuss, F. Kraxner, A. NotenbaertClimate change mitigation through livestock system transitions Proc. Natl. Acad. Sci., 111 (2014), pp. 3709-3714, 10.1073/pnas.1308044111
20. Humpenöder et al., 2014 F. Humpenöder, A. Popp, J.P. Dietrich, D. Klein, H. Lotze-Campen, M. Bonsch, B.L.Bodirsky, I. Weindl, M. Stevanovic, C. MüllerInvestigating afforestation and bioenergy CCS as climate change mitigation strategies Environ. Res. Lett., 9 (2014), p. 64029, 10.1088/1748-9326/9/6/064029

21. Hurtt et al., 2011 G.C. Hurtt, L.P. Chini, S. Frolking, R.A. Betts, J. Feddema, G. Fischer, J.P. Fisk, K. Hibbard, R.A. Houghton, A. Janetos, C.D. Jones, G. Kindermann, T. Kinoshita, K.K. Goldewijk, K. Riahi, E.Shevliakova, S. Smith, E. Stehfest, A. Thomson, P. Thornton, D.P. Vuuren, van WangHarmonization of land-use scenarios for the period 1500-2100: 600 years of global gridded annual land-use transitions, wood harvest, and resulting secondary lands Clim. Change, 109 (2011), pp. 117-161, 10.1007/s10584-011-0153-2
22. Hussein et al., 2013 Z. Hussein, T. Hertel, A. GolubClimate change mitigation policies and poverty in developing countries Environ. Res. Lett., 8 (2013), p. 35009, 10.1088/1748-9326/8/3/035009
23. Jackson et al., 2008 R.B. Jackson, J.T. Randerson, J.G. Canadell, R.G. Anderson, R. Avissar, D.D. Baldocchi, G.B. Bonan, K. Caldeira, N.S. Diffenbaugh, C.B. Field, B.A. Hungate, E.G. Jobbágy, L.M. Kueppers, M.D.Nosetto, D.E. PatakiProtecting climate with forests Environ. Res. Lett., 3 (2008), p. 44006, 10.1088/1748-9326/3/4/044006
24. Jones et al., 2015 A.D. Jones, K.V. Calvin, W.D. Collins, J. EdmondsAccounting for radiative forcing from albedo change in future global land-use scenarios Clim. Change, 131 (2015), pp. 691-703, 10.1007/s10584-015-1411-5
25. KC and Lutz, 2016 KC, S., Lutz, W., n.d., The human core of the shared socioeconomic pathways: Population scenarios by age, sex and level of education for all countries to 2100. Glob. Environ. Change. 10.1016/j.gloenvcha.2014.06.004
26. Kareiva et al., 2007 P. Kareiva, S. Watts, R. McDonald, T. BoucherDomesticated nature: shaping landscapes and ecosystems for human welfare Science, 316 (2007), pp. 1866-1869, 10.1126/science.1140170
27. Kindermann et al., 2006 G.E. Kindermann, M. Obersteiner, E. Rametsteiner, I. McCallumPredicting the deforestation-trend under different carbon-prices Carbon Balance Manag., 1 (2006), p. 15, 10.1186/1750-0680-1-15
28. Kok et al., 2016 M.T.J. Kok, K. Kok, G.D. Peterson, R. Hill, J. Agard, S.R. CarpenterBiodiversity and ecosystem services require

IPBES to take novel approach to scenarios Sustain. Sci. (2016), pp. 1-5, 10.1007/s11625-016-0354-8

29. Kriegler et al., 2012 E. Kriegler, B.C. O'Neill, S. Hallegatte, T. Kram, R.J. Lempert, R.H. Moss, T. WilbanksThe need for and use of socio-economic scenarios for climate change analysis: a new approach based on shared socio-economic pathways Glob. Environ. Change, 22 (2012), pp. 807-822, 10.1016/j.gloenvcha.2012.05.005
30. Kriegler et al., 2014 E. Kriegler, J. Edmonds, S. Hallegatte, K.L. Ebi, T. Kram, K. Riahi, H. Winkler, D.P. van VuurenA new scenario framework for climate change research: the concept of shared climate policy assumptions Clim. Change, 122 (2014), pp. 401-414, 10.1007/s10584-013-0971-5
31. Kriegler et al., 2016 E. Kriegler, N. Bauer, A. Popp, F. Humpenöder, M.J.S. Leimbach, J. Streffler, L. Baumstark, B. Bodirsky, J. Hilaire, D. Klein, I. Mouratiadou, I. Weindl, C. Bertram, J.P. Dietrich, G. Luderer, M. Pehl, R. Pietzcker, F. Piontek, H. Lotze-Campen, A. Biewald, M. Bonsch, A. Giannousakis, U. Kreidenweis, C.Müller, S. Rolinski, A. Schultes, J. Schwanitz, M. Calvin, K. Emmerling, J. Fujimori, S. Edenhofer, M.StevanovicFossil-fueled development (SSP5): an energy and resource intensive scenario for the 21st century Glob. Environ. Change (2016), 10.1016/j.gloenvcha.2016.05.015 (this special issue)
32. Lambin and Meyfroidt, 2011 E.F. Lambin, P. MeyfroidtGlobal land use change, economic globalization, and the looming land scarcity Proc. Natl. Acad. Sci., 108 (2011), pp. 3465-3472, 10.1073/pnas.1100480108
33. Lawrence et al., 2016 Lawrence, D. M., Hurtt, A., Arneth, V., Brovkin, K. V. Calvin, Jones, C. D. Jones, P. J. Lawrence, N. de Noblet-Ducoudré, J. Pongratz, S. I. Seneviratne, E. Shevliakova, 2016. The Land Use Model Intercomparison Project (LUMIP): Rationale and experimental design. Geosci. Model Dev. Discuss. 2016: 1–42.
34. Luyssaert et al., 2014 S. Luyssaert, M. Jammet, P.C. Stoy, S. Estel, J. Pongratz, E. Ceschia, G. Churkina, A.Don, K. Erb, M. Ferlicoq, B. Gielen, T. Grünwald, R.A. Houghton, K. Klumpp, A. Knohl, T. Kolb, T.Kuemmerle, T. Laurila, A. Lohila, D. Loustau, M.J.

McGrath, P. Meyfroidt, E.J. Moors, K. Naudts, K.Novick, J. Otto, K. Pilegaard, C.A. Pio, S. Rambal, C. Rebmann, J. Ryder, A.E. Suyker, A. Varlagin, M.Wattenbach, A.J. DolmanLand management and land-cover change have impacts of similar magnitude on surface temperature Nat. Clim. Change, 4 (2014), pp. 389-393, 10.1038/nclimate2196

35. Mueller et al., 2012 N.D. Mueller, J.S. Gerber, M. Johnston, D.K. Ray, N. Ramankutty, J.A. FoleyClosing yield gaps through nutrient and water management Nature (2012), 10.1038/nature11420
36. Nakicenovic and Swart, 2000 N. Nakicenovic, R. SwartSpecial Report on Emissions Scenarios (SRES) Cambridge University Press, Cambridge (2000)
37. Nelson et al., 2014 Nelson, G.C., Valin, H., Sands, R.D., Havlík, P., Ahammad, H., Deryng, D., Elliott, J., Fujimori, S., Hasegawa, T., Heyhoe, E., Kyle, P., Lampe, M.V., Lotze-Campen, H., d'Croz, D.M., Meijl, H., van, Mensbrugghe, D. van der, Müller, C., Popp, A., Robertson, R., Robinson, S., Schmid, E., Schmitz, C., Tabeau, A., Willenbockel, D., 2014. Climate change effects on agriculture: Economic responses to biophysical shocks. Proc. Natl. Acad. Sci. 111, 3274–3279. 10.1073/pnas.1222465110.
38. O'Neill et al., 2013 B.C. O'Neill, E. Kriegler, K. Riahi, K.L. Ebi, S. Hallegatte, T.R. Carter, R. Mathur, D.P.vanVuurenA new scenario framework for climate change research: the concept of shared socioeconomic pathways Clim. Change, 122 (2013), pp. 387-400, 10.1007/s10584-013-0905-2
39. O'Neill et al., 2016 O'Neill, B.C., Kriegler, E., Ebi, K.L., Kemp-Benedict, E., Riahi, K., Rothman, D.S., van Ruijven, B.J., van Vuuren, D.P., Birkmann, J., Kok, K., Levy, M., Solecki, W., n.d. The roads ahead: Narratives for shared socioeconomic pathways describing world futures in the 21 st century. Glob. Environ. Change. 10.1016/j.gloenvcha.2015.01.004.
40. Obersteiner et al., 2001 Obersteiner, M., Azar, C., Kauppi, P., Möllersten, K., Moreira, J., Nilsson, S., Read, P., Riahi, K., Schlamadinger, B., Yamagata, Y., Yan, J., Ypersele, J.-P. van, 2001. Managing Climate Risk. Science 294, 786–787. 10.1126/science.294.5543.786b.

41. Pitman et al., 2009 A.J. Pitman, N. de Noblet-Ducoudré, F.T. Cruz, E.L. Davin, G.B. Bonan, V. Brovkin, M.Claussen, C. Delire, L. Ganzeveld, V. Gayler, B.J.J.M. van den Hurk, P.J. Lawrence, M.K. van der Molen, C. Müller, C.H. Reick, S.I. Seneviratne, B.J. Strengers, A. VoldoireUncertainties in climate responses to past land cover change: first results from the LUCID intercomparison study Geophys. Res. Lett., 36 (2009), p. L14814, 10.1029/2009GL039076

42. Popp et al., 2010 A.Popp, H. Lotze-Campen, B. BodirskyFood consumption, diet shifts and associated non-CO_2 greenhouse gases from agricultural production Glob. Environ. Change, 20 (2010), pp. 451-462, 10.1016/j.gloenvcha.2010.02.001

43. Popp et al., 2011 A. Popp, J.P. Dietrich, H. Lotze-Campen, D. Klein, N. Bauer, M. Krause, T. Beringer, D.Gerten, O. EdenhoferThe economic potential of bioenergy for climate change mitigation with special attention given to implications for the land system Environ. Res. Lett., 6 (2011), p. 34017, 10.1088/1748-9326/6/3/034017

44. Popp et al., 2013 A.Popp, S.K. Rose, K. Calvin, D.P.V. Vuuren, J.P. Dietrich, M. Wise, E. Stehfest, F.Humpenöder, P. Kyle, J.V. Vliet, N. Bauer, H. Lotze-Campen, D. Klein, E. KrieglerLand-use transition for bioenergy and climate stabilization: model comparison of drivers, impacts and interactions with other land use based mitigation options Clim. Change, 123 (2013), pp. 495-509, 10.1007/s10584-013-0926-x

45. Popp et al., 2014 A.Popp, F. Humpenöder, I. Weindl, B.L. Bodirsky, M. Bonsch, H. Lotze-Campen, C. Müller, A.Biewald, S. Rolinski, M. Stevanovic, J.P. DietrichLand-use protection for climate change mitigation Nat. Clim. Change, 4 (2014), pp. 1095-1098, 10.1038/nclimate2444

46. Preston et al., 2011 B.L. Preston, E.J. Yuen, R.M. WestawayPutting vulnerability to climate change on the map: a review of approaches, benefits, and risks Sustain. Sci., 6 (2011), pp. 177-202, 10.1007/s11625-011-0129-1

47. Riahi et al., 2012 Riahi, K., Dentener, F., Gielen, D., Grubler, A., Jewell, J., Klimont, Z., et al (2012). Global Energy Assessment Chapter 17: Energy Pathways for Sustainable Development. In Global Energy Assessment – Toward a Sustainable Future,

Cambridge University Press, Cambridge, UK and New York, NY, USA and the International Institute for Applied Systems Analysis, Laxenburg, Austria, pp. 1203–1306.

48. Riahi et al., 2016 K. Riahi, D.P. van Vuuren, E. Kriegler, J. Edmonds, B. O'Neill, S. Fujimori, N. Bauer, K. Calvin, R. Dellink, O. Fricko, W. Lutz, A. Popp, J. Crespo Cuaresma, K.C. Samir, M. Leimback, L. Jiang, T. Kram, S. Rao, J. Emmerling, K. Ebi, T. Hasegawa, P. Havlik, F. Humpenöder, L.A. Da Silva, S. Smith, E.Stehfest, V. Bosetti, J. Eom, D. Gernaat, T. Masui, J. Rogelj, J. Strefler, L. Drouet, V. Krey, G. Luderer, M.Harmsen, K. Takahashi, L. Baumstark, J. Doelman, M. Kainuma, Z. Klimont, G. Marangoni, H. Lotze-Campen, M. Obersteiner, A. Tabeau, M. TavoniShared socioeconomic pathways: an overview Global Environ. Change (2016), 10.1016/j.gloenvcha.2016.05.009

49. Robinson et al., 2014 S. Robinson, H. van Meijl, D. Willenbockel, H. Valin, S. Fujimori, T. Masui, R. Sands, M.Wise, K. Calvin, P. Havlik, D. Mason d'Croz, A. Tabeau, A. Kavallari, C. Schmitz, J.P. Dietrich, M. von Lampe Comparing supply-side specifications in models of global agriculture and the food system Agric. Econ., 45 (2014), pp. 21-35, 10.1111/agec.12087

50. Rudel et al., 2009 T.K. Rudel, L. Schneider, M. Uriarte, B.L. Turner, R. DeFries, D. Lawrence, J. Geoghegan, S.Hecht, A. Ickowitz, E.F. Lambin, T. Birkenholtz, S. Baptista, R. GrauAgricultural intensification and changes in cultivated areas, 1970–2005 Proc. Natl. Acad. Sci., 106 (2009), pp. 20675-20680, 10.1073/pnas.0812540106

51. Schmitz et al., 2012 C. Schmitz, A. Biewald, H. Lotze-Campen, A. Popp, J.P. Dietrich, B. Bodirsky, M. Krause, I.WeindlTrading more food: Implications for land use, greenhouse gas emissions, and the food system Glob. Environ. Change, 22 (2012), pp. 189-209, 10.1016/j.gloenvcha.2011.09.013

52. Schmitz et al., 2014 C. Schmitz, H. van Meijl, P. Kyle, G.C. Nelson, S. Fujimori, A. Gurgel, P. Havlik, E. Heyhoe, D.M. d'Croz, A. Popp, R. Sands, A. Tabeau, D. van der Mensbrugghe, M. von Lampe, M. Wise, E. Blanc, T. Hasegawa, A. Kavallari, H. ValinLand-use change trajectories up to 2050: insights from a global agro-economic model comparison Agric. Econ., 45 (2014), pp. 69-84, 10.1111/agec.12090

53. Smith et al., 2010 Smith, P., Gregory, P.J., Vuuren, D., van, Obersteiner, M., Havlík, P., Rounsevell, M., Woods, J., Stehfest, E., Bellarby, J., 2010. Competition for land. Philos. Trans. R. Soc. B Biol. Sci. 365, 2941–2957. 10.1098/rstb.2010.0127.
54. Smith et al., 2013 P. Smith, H. Haberl, A. Popp, K. Erb, C. Lauk, R. Harper, F.N. Tubiello, A. de Siqueira Pinto, M. Jafari, S. Sohi, O. Masera, H. Böttcher, G. Berndes, M. Bustamante, H. Ahammad, H. Clark, H. Dong, E.A. Elsiddig, C. Mbow, N.H. Ravindranath, C.W. Rice, C. Robledo Abad, A. Romanovskaya, F. Sperling, M. Herrero, J.I. House, S. RoseHow much land-based greenhouse gas mitigation can be achieved without compromising food security and environmental goals? Glob. Change Biol., 19 (2013), pp. 2285-2302, 10.1111/gcb.12160
55. Stehfest et al., 2009 E. Stehfest, L. Bouwman, D. van Vuuren, M. den Elzen, B. Eickhout, P. Kabat Climate benefits of changing diet Clim. Change, 95 (2009), pp. 83-102, 10.1007/s10584-008-9534-6
56. Stehfest et al., 2014 Stehfest, E., van Vuuren, D., Kram, T., Bouwman, L., Alkemade, R., Bakkenes, M., Biemans, H., Bouwman, A., den Elzen, M., Janse, J., Lucas, P., van Minnen, J., Müller, M., Prins, A. 2014. Integrated Assessment of Global Environmental Change with IMAGE 3.0. Model description and policy applications, The Hague: PBL Netherlands Environmental Assessment Agency.
57. Steinfeld and Gerber, 2010 H. Steinfeld, P. GerberLivestock production and the global environment: consume less or produce better? Proc. Natl. Acad. Sci., 107 (2010), pp. 18237-18238, 10.1073/pnas.1012541107
58. Tilman et al., 2011 D. Tilman, C. Balzer, J. Hill, B.L. BefortGlobal food demand and the sustainable intensification of agriculture Proc. Natl. Acad. Sci., 108 (2011), pp. 20260-20264, 10.1073/pnas.1116437108
59. Tubiello et al., 2015 Tubiello, F.N., Salvatore, M., Ferrara, A.F., House, J., Federici, S., Rossi, S., Biancalani, R., Condor Golec, R.D., Jacobs, H., Flammini, A., Prosperi, P., Cardenas-Galindo, P., Schmidhuber, J., Sanz Sanchez, M.J., Srivastava, N., Smith, P., 2015. The Contribution of Agriculture, Forestry and other Land Use activities to Global Warming, 1990–2012. Glob. Change Biol. n/a-

n/a. 10.1111/gcb.12865

60. Valdivia et al., 2015 R.O. Valdivia, J.M. Antle, C. Rosenzweig, A.C. Ruane, J. Vervoot, M. Ashfaq, I. Hathie, S.H.-K. Tui, R. Mulwa, C. Nhemachena, P. Ponnusamy, H. Rasnayaka, H. SinghRepresentative agricultural pathways and scenarios for regional integrated assessment of climate change impacts, vulnerability, and adaptation C. Rosenzweig, D. Hillel (Eds.), Handbook of Climate Change and Agroecosystems: The Agricultural Model Intercomparison and Improvement Project (AgMIP) Integrated Crop and Economic Assessments, Part 1. ICP Series on Climate Change Impacts, Adaptation, and Mitigation, vol. 3, Imperial College Press(2015), pp. 101-156, 10.1142/9781783265640_0005
61. Valin et al., 2014 H. Valin, R.D. Sands, D. van der Mensbrugghe, G.C. Nelson, H. Ahammad, E. Blanc, B.Bodirsky, S. Fujimori, T. Hasegawa, P. Havlik, E. Heyhoe, P. Kyle, D. Mason-D'Croz, S. Paltsev, S.Rolinski, A. Tabeau, H. van Meijl, M. von Lampe, D. WillenbockelThe future of food demand: understanding differences in global economic models Agric. Econ., 45 (2014), pp. 51-67, 10.1111/agec.12089
62. Verburg et al., 2009 R. Verburg, E. Stehfest, G. Woltjer, B. Eickhout The effect of agricultural trade liberalisation on land-use related greenhouse gas emissions Glob. Environ. Change, 19 (2009), pp. 434-446, 10.1016/j.gloenvcha.2009.06.004
63. Van Vuuren et al., 2011 Van Vuuren, D.P., van, Edmonds, J., Kainuma, M., Riahi, K., Thomson, A., Hibbard, K., Hurtt, G.C., Kram, T., Krey, V., Lamarque, J.-F., Masui, T., Meinshausen, M., Nakicenovic, N., Smith, S.J., Rose, S.K., 2011. The representative concentration pathways: an overview. Clim. Change 109, 5–31. 10.1007/s10584-011-0148-z
64. van Vuuren et al., 2012 D.P. van Vuuren, M.T.J. Kok, B. Girod, P.L. Lucas, B. de VriesScenarios in Global Environmental Assessments: Key characteristics and lessons for future use Glob. Environ. Change, 22 (2012), pp. 884-895, 10.1016/j.gloenvcha.2012.06.001
65. Van Vuuren et al., 2016 D. Van Vuuren, E. Stehfest, D.E.H.J. Gernaat, J. Doelman, M. van den Berg, M.Harmsen, H.-S. De Boer,

L. Bouwman, D. Vassilis, O.Y. Edelenbosch, B. Girod, T. Kram, L. Lassaletta, P.Lucas, H. van Meijl, C. Müller, B.J. van Ruijven, A. Tabeau Energy, land-use and greenhouse gas emissions trajectories under a green growth paradigm Global Environ. Change (2016) (this special issue)

66. von Lampe et al., 2014 von Lampe, M., Willenbockel, D., Ahammad, H., Blanc, E., Cai, Y., Calvin, K., Fujimori, S., Hasegawa, T., Havlik, P., Heyhoe, E., Kyle, P., Lotze-Campen, H., Mason d'Croz, D., Nelson, G.C., Sands, R.D., Schmitz, C., Tabeau, A., Valin, H., van der Mensbrugghe, D., van Meijl, H., 2014. Why do global long-term scenarios for agriculture differ? An overview of the AgMIP Global Economic Model Intercomparison. Agric. Econ. 45, 3–20. 10.1111/agec.12086.
67. Vuuren et al., 2013 Vuuren, D.P., van, Kriegler, E., O'Neill, B.C., Ebi, K.L., Riahi, K., Carter, T.R., Edmonds, J., Hallegatte, S., Kram, T., Mathur, R., Winkler, H., 2013. A new scenario framework for Climate Change Research: scenario matrix architecture. Clim. Change 122, 373–386. 10.1007/s10584-013-0906-1.
68. Weindl et al., 2015 Weindl, H. Lotze-Campen, A. Popp, C. Müller, P. Havlík, Mario Herrero, C. Schmitz, S.Rolinski Livestock in a changing climate: production system transitions as an adaptation strategy for agriculture Environ. Res. Lett., 10 (2015), p. 94021, 10.1088/1748-9326/10/9/094021
69. Wiebe et al., 2015 Wiebe, K., Lotze-Campen, H., Sands, R., Tabeau, A., Mensbrugghe, D. van der, Biewald, A., Bodirsky, B., Islam, S., Kavallari, A., Mason-D'Croz, D., Müller, C., Popp, A., Robertson, R., Robinson, S., Meijl, H., van, Willenbockel, D., 2015. Climate change impacts on agriculture in 2050 under a range of plausible socioeconomic and emissions scenarios. Environ. Res. Lett. 10, 85010. 10.1088/1748-9326/10/8/085010
70. Wilbanks and Ebi, 2013 T.J. Wilbanks, K.L. EbiSSPs from an impact and adaptation perspective Clim. Change, 122 (2013), pp. 473-479, 10.1007/s10584-013-0903-4

71. Wise et al., 2014 M. Wise, K.V. Calvin, G.P. Kyle, P. Luckow, J.A. EdmondsEconomic and physical modeling of land use in GCAM 3.0 and an application to agricultural productivity, land, and terrestrial carbon Clim. Change Econ., 5 (2014), p. 1450003, 10.1142/s2010007814500031
72. Worldbank, 2015 WORLDBANK. 2015. World Development Indicators. http://data.worldbank.org/data-catalog/world-development-indicators.

3 INTEGRATING SPATIAL VALUATION OF ECOSYSTEM SERVICES INTO REGIONAL PLANNING AND DEVELOPMENT

Ilpo Tammi[a], Kaisa Mustajärvi[b], and Jussi Rasinmäki[c]

[a]Council of Tampere Region, 33101 Tampere, Finland

[b]Ramboll Finland Oy, Tampere, Finland

[c]Simosol Oy, Riihimäki, Finland

ABSTRACT

The transition of the ecosystem service framework from academic discourse into practical land use management and policy guidance is in the making. Planners and decision makers seek spatial valuation data, comprehensive examples of which are few or hindered by sectoral research traditions. We present a case of linking land use to multimethod spatial ecosystem service valuation aiming at comprehensiveness and commensurability, based on a project run parallel to regional land-use planning in the Tampere region, Finland. A spectrum of ecosystem services was scrutinized, the annual value of which was estimated at €0.8–1B. Compared to land-use planning, core areas of ecological networks proved relatively poor in terms of valuation, but hot-spots of human–nature interaction such as recreational, groundwater and landscape areas immensely valuable. Strong urban-rural trends in ecosystem service value were found, emphasizing the importance of urban nature and the context-specificity of natural capital discourse. We argue that some mismatches exist between the ecosystem service framework and its practical applicability, and that the main problem is not necessarily the transferability of tools and indicators, but the transfer of valuation and the assumptions and choices behind it. Notwithstanding its problems, the applied framework proved valuable in evaluating and guiding future land use.

Keywords: Ecosystem services, Valuation, GIS, Land-use planning and Natural capital

INTRODUCTION

The past few decades have seen widespread adoption of the ecosystem approach as an overarching framework for environmental management discourse – at least in the academic sphere (Drakou et al., 2015; Polizzi et al., 2015). Its operationalization has been increasingly approached via the interrelated conceptual entity of ecosystem services (ES), and the valuation of these benefits obtained by humans from ecosystems and their functions (de Groot et al., 2012). Even though the number of studies concerning ES valuation has been constantly on the rise, the practical application of valuation has been criticized as somewhat superficial and its utility for policy guidance questionable (Primmer and Furman, 2012; Schägner et al., 2013). Perhaps due to some vagueness or unfamiliarity of the rapidly diversifying scope of ES discourse from the perspective of the "hands-on sphere", the field of ES has been increasingly approached via a more comprehensive and possibly often even more comprehensible concept of natural capital, its stocks, flows and their values (Costanza et al., 1997; Crossman et al., 2013).

Ecosystem services and natural capital are inherently spatial by nature (Boyd and Banzhaf, 2007; Schägner et al., 2013), although some services are unarguably easier to pinpoint on a map than others with the same accuracy, precision and resolution. Notwithstanding this place-bound essence, geospatial applications of ES valuation methodology have gained momentum only more recently (Maes et al., 2012a), not least due to a growing ubiquity of geographic information systems (GIS) in both study and practice (Schägner et al., 2013). However, while one of the main objectives of mapping and valuing ES is arguably visualization and communication of information into decision-making processes concerned with natural resources management (Jäppinen and Heliölä, 2015; Polizzi et al., 2015), the bridge between research and decision making is yet being built (Primmer and Furman, 2012; Bagstad et al., 2013). Current discourse on the status quo of practical ES applications has suggested a need for binding knowledge on key areas of green infrastructure and natural values with that of ES hotspots, thus enhancing the legitimacy of ES in land use related decision making (Jäppinen and Heliölä, 2015; Vierikko and Niemelä, 2016).

The recently published TEEB for Finland (The Economics of Ecosystems and Biodiversity) highlighted ES valuation as a tool for holistic land-use planning (Jäppinen and Heliölä, 2015), and the integration of spatial ES assessment data into planning, decision making and management was called for. In Finland, there is a wealth of ES-related studies, yet these often focus on mapping ES supplying structures or have had a dominantly service-specific scope, and no comprehensive and commensurate spatial valuation has been previously carried out. This may not be least due to enduring sectoral traditions in natural resources governance and research (Primmer and Furman, 2012), which may have also contributed to the relative scarcity of studies bound specifically to an ES framework (Seppelt et al., 2011). Traditions in thematically tightly-scoped research can also be seen in an imbalance of attention given to different ES – so far most emphasis has been on recreation and water ecosystems (Jäppinen and Heliölä, 2015).

In this paper, we present a regional case of spatially explicit mapping and valuation of ecosystem services in a Nordic context, in the Tampere region in Southwest Finland. The paper expands on outputs of a research project focusing on ecosystem services and natural capital, set in the context of land-use planning and regional development. The project was run alongside a comprehensive, strategic regional land-use planning process targeting the year 2040, thus establishing a connection to regional and local decision making. Besides green accounting, the project aimed to contribute to the evaluation and iteration of said regional plan draft (in the Finnish land-use planning system, plans have four phases: participation and assessment scheme, draft, proposal and approval), its land use policies and impacts, as well as aid in ES-related resource allocation. The paper aims to:

- Describe the use of novel GIS techniques in creating a uniform spatial framework for ES inventory.
- Derive commensurable monetary values for mapped ES and natural capital.
- Compare the spatial configuration of ES supply value to current and planned land use contexts.
- Evaluate the capability of the ES framework to answer practical needs in land-use management.

MATERIALS AND METHODS

Study setting

The study area covers the Tampere administrative region (Finnish: *Pirkanmaa*, Fig. 1) with an area of circa 14 600 km^2, of which 5% is urban, 11% agroecosystems, 69% forests or forestry lands, 1% wetland and 14% inland watercourses according to CORINE Land Cover (CLC) 2012 data (Fig. 2). The region and the spatial configuration of its ES supply are heavily characterized by being located in the intersection of different "landscape regions" – the region's southwestern–western parts being agricultural lowland, central and eastern parts a mosaic of forest, lakes and agricultural land, and the northern parts being dominantly forested and host to the region's most wetlands – due to being located on the highlands of a major drainage divide (*Suomenselkä*). Most of the region is situated on the western frontier of the so-called Finnish Lakeland – a geographical region characterized by a multitude of lakes and mosaic-like landscapes. Similarly pivotal geomorphological features from the ES perspective are the numerous eskers crisscrossing the region. Almost the entire study area is situated within the Kokemäenjoki (*Kokemäki River*) drainage basin (of which in turn most is located in the region), named after a major river traversing westwards through the region. As of 2015, the region, second-most populated in the country after the Helsinki region, had circa 503 500 inhabitants, the majority of whom live in the centrally located Tampere city region – the most populous inland city region in the Nordic countries. The adult population (15–74 y/o), used in multiple reference studies, was circa 370 800. According to the regional plan's population development scenario, the region will grow by an estimated 120 000 inhabitants by 2040 (+24%).

Figure 1: Location of the study area in the context of Northern Europe.

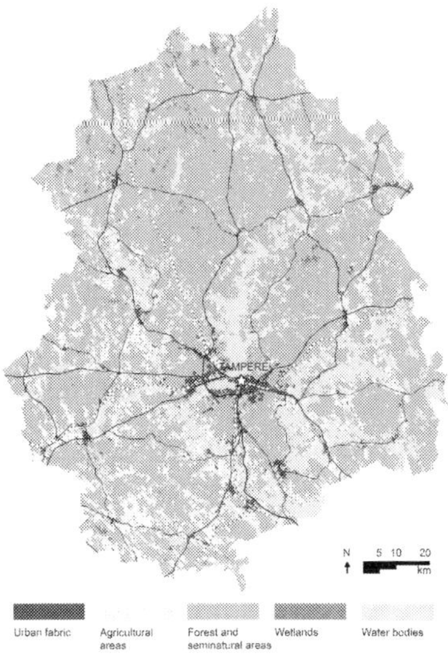

Figure 2: Land cover in the Tampere region (CLC2012 data), including main road and rail networks.

Ecosystem service mapping and valuation framework

A cross-section of ES was selected for scrutiny, from all different sections defined in the Common International Classification of Ecosystem Services (CICES 4, Haines-Young and Potschin, 2013). Alongside ES, some accompanying abiotic outputs from natural capital or natural processes were examined, in case these were strongly interconnected with biotic processes. These abiotic outputs mainly concerned energy provision. The selection of ES was based on the availability of (spatial) data and existing valuation methodology as well as the opinion of the study project's steering group consisting of 17 natural resource or land-use management experts and researchers (incl. corresponding author), but also unavoidably limited by the time frame set for the project (1 year). The mapping of ES was based dominantly on refining available spatial data, preferably open access data when applicable. Numerous spatial datasets were utilized in the study, mainly open data provided by various Finnish authorities. The main data layers included national forest inventory data, topographic databases, various spatial ES-related statistics, land-use planning data and several derivatives thereof. Several land use related aspects were examined based on data from the regional plan and background analyses thereof.

In order to enable examinations of accumulation and trade-offs between different ecosystem services, and to provide a spatial framework for commensurable valuation of ES, a spatial database was created for the region, consisting of approximately 5 900 hexagonal cells with a size of 260 ha each (roughly corresponding to the size of a 1.5 km grid cell) – the size being found here optimal for generalizing the variety of types and resolutions of the raw data as well as for "fuzzifying" some potentially sensitive spatial information regarding private livelihoods and real estate ownership. Although the coarseness of outputs from previous spatially explicit studies has been seen an obstacle for implementing an ES framework in decision making (Schägner et al., 2013), the selected spatial resolution was found appropriate for a regional-level context, given the heterogeneity of input data and underlying methodology. A hexagonal cell structure was chosen for the spatial database over a typical rectangular grid due to its better ability to represent and account for certain spatial continuities and connectivity in the rather complex landscape features typical to the region – especially the near-ubiquitous eskers and branching elongated lakes. The mapping was mainly carried out utilizing three interrelated open-source software: QGIS, PostGIS and R.

Valuation was aimed at combining use and non-use values to determine Total Economic Value (TEV) of ES (de Groot et al., 2010). ES were valued based on their annual flow or utilization in common monetary units, €/year, inflation-adjusted to 2015 euros. The valuation of ES was based on market price statistics mainly from within 2004–2015 or market cost approaches whenever applicable, and in other cases on value or benefit transfer from previous valuation studies. Most of the reference studies had utilized stated or revealed preference methods as the primary valuation approach. Reference case studies were dominantly Finnish and from regions with physical and socioeconomic settings as similar to the study area as possible – in order to minimize context-related value transfer errors (Nelson and Kennedy, 2008; Brander et al., 2012). Its difficulties and deficiencies acknowledged, especially in regard to regulation, maintenance and cultural ES, a monetary valuation approach was chosen for its commensurable nature and its strength in providing understanding of the relativeness and preferences in resource allocation decision making (de Groot et al., 2010, 2012; Primmer and Furman, 2012), and in enabling cost-benefit analysis based impact assessment and trade-offs in land-use planning (Jäppinen and Heliölä, 2015).

Ecosystem service supply determination

Provision ES

Annual crop outputs (food and plants for agricultural use) from agroecosystems were determined from municipal agricultural yield statistics and mean producer prices from 2004 to 2013 (Natural Resources Institute Finland, 2014, 2015b), combined with field parcel data and hectarage from the National Land Survey's (NLS) Topographic Database. For livestock a similar approach was used, but pen location, livestock type and headcount data was provided by ELY (Centre for Economic Development, Transport and the Environment). Main reared species were taken into consideration: cattle (meat and dairy), pork, poultry (meatand eggs) and sheep. Additionally, the technical potential for biogas energy provision was estimated from major agrobiomass by-products, i.e. straw, hay (e.g. from crop rotation) and manure, their estimated methane output and the energy content thereof (Salminen and Rintala, 2002; Mäenpää, 2005; Lehtomäki, 2006; Mäkinen et al., 2006; Lehtomäki et al., 2007; Seppälä et al., 2009). Valuation was based on Nord Pool power market mean spot prices for Finland (2011–2015).

Freedom to roam is pivotal for ES use in Finland. Especially picking wild berries and mushrooms is a popular pastime and supplementary livelihood (Jäppinen and Heliölä, 2015). The most extracted berry species (80–90% of total) and thus economically important are bilberry (Vaccinium myrtillus), lingonberry (*Vaccinium vitis-idaea*) and cloudberry (*Rubus chamaemorus*), the latter dominantly in northern Finland (Ihalainen et al., 2005; Turtiainen et al., 2007). The spatial distribution of bilberry and lingonberry yield potential was calculated with calibrated regional yield prediction expert models for boreal forests and peatlands (Ihalainen et al., 2005; Turtiainen et al., 2005, 2007), using Multi-source National Forest Inventory spatial data (MS-NFI, 20 m resolution, Natural Resources Institute Finland, 2015a). Similar models were not yet available for fungi, although such are in development (Miina et al., 2013). Therefore, their output was estimated non-spatially based on mushroom sales and national yield estimates. Unit values were based on sales price market surveys from 2012 to 2014 commissioned by the Finnish Agency for Rural Affairs.

Wildlife hunting is commonplace in the region, the main hunted species being elk (*Alces alces*) and white-tailed deer (*Odocoileus* virginianus). Hunting permits are given based on quotas to game management associations, whose spatial delimitations loosely correspond to municipal boundaries. Permits for elk and other deer are given to contiguous hunting grounds of 1000 and 500 ha minimum, respectively. Suitable areas were analyzed with Morphological Spatial Pattern Analysis or MSPA (JRC, 2015) from CLC2012 data (forest and semi-natural areas), based on the ecological network analysis of the regional plan. Data on game quotas and killings acquired from the RiistaWeb Game Information database was joined to contiguous areas within each association's extent (Finnish Wildlife Agency and LUKE, 2012), weighted by the number of intersecting *Cervidae* observations, routes and pasturing grounds (data from ELY). Small game was valued non-spatially, since data for mapping this type of hunting was insufficient. Game meat values were based on official hunting statistics (FGFRI, 2014a, 2014b).

Fishing is a popular, yet declining sport. Professional fishermen are very few in the region and fishing is mainly recreational (FGFRI, 2014b), targeting species like European perch (*Perca fluviatilis*), Northern pike (*Esox lucius*), zander (*Sander lucioperca*), common roach (*Rutilus rutilus*), common bream (*Abramis brama*) and signal crayfish (*Pacifastacus leniusculus*). Mean fish catch and price data was based on 2010–2012 statistics (FGFRI, 2014b). Regional figures were derived from a larger conglomerate administrative area (*Häme*) based on the location of lakes, population and holiday housing within this area. As majority of the catch value is recreational (c. 93%) and tightly connected to holiday housing, the spatial configuration of holiday housing (summer cottages) with shore access was used as a basis for mapping. The value of fisheries is closely tied to water quality – the value declines as relative phosphorus content increases (Marttunen et al., 2012). Thus, fishery values were mapped to water bodies weighted per water quality, area and number of onshore holiday housing. Water quality data was acquired from the national lake register containing lakes and river segments over one hectare in size (SYKE, 2015).

In Finland, water extraction for household, public and economic uses are typically based on the same water sources and infrastructure – i.e. there is mainly no division between water acquisition and delivery systems for human and other uses. Main water sources nationally are ground water, surface water and artificial ground water. Water extraction was mapped for lake basins (and their immediate drainage basins) used as surface water sources, as well as for utilized groundwater aquifers and wells. Completion of the ongoing change of the main water source for the city of Tampere was assumed in mapping. Average daily extraction rates per well from 2009 to 2011 were acquired from ELY. Valuation was based on municipal water use fee averages from 2013 to 2014 and the unit values were attached to extraction facilities within corresponding municipalities. Future extraction potential was estimated as of 2040 from water supply development plans conducted by the regional council and ELY.

The annual net change in growing biomass of trees and tree-to-soil biomass inputs was modelled using a forest growth simulator (Rasinmäki et al., 2009), based on growth models by Hynynen et al. (2002). The modelling approach included different management activities, including removal of growing and dead trees for different land use classes, mimicking common Finnish silviculture management practices. The maximum sustainable removal rate of saw timber and pulpwood (spruce, pine, birch and other broad-leaved trees), as defined in the National Forest Inventory, was used in the modelling to define maximum forestry yields. The intensity of current usage was estimated based on regional statistics on forest uses (Natural Resources Institute Finland, 2015c) and applied to the modelling results as a spatially uniform ratio of the maximum. As energy wood is mainly harvested as a by-product of the aforementioned, its harvest's spatial configuration was assumed to follow that of the removal of spruce and pine especially. Wood trunk and fiber were valued according to stumpage price averages from 2004 to 2014 and energy wood according to average forest chip fuel prices from 2013 to 2014 (Statistics Finland, 2015).

Regulation and maintenance ES

Initial carbon stock sizes and sequestration rates in growing biomass were primarily taken from the latest MS-NFI results available. The dataset includes biomass estimates based on satellite imagery and biomass models (Repola et al., 2007, Helmisaari and Hallbäcken, 1998; Muukkonen et al., 2006). Dry weight biomass estimates were converted to carbon estimates by multiplying by 0.5 (Liski et al., 2006). As the MS-NFI results cover only forest areas, agricultural areas as well as some forested urban areas are not covered by the data. For these areas land use class (CLC) average values were used as scaled averages compared to average forest biomass in the area. The initial carbon stock size in soil was modelled using Yasso07 soil carbon model (Tuomi et al., 2011). A retrospective 50 year simulation period with constant input of organic material per land use class was used to predict the current soil carbon stock. Data by Liski and Westman (1997) was used to get the initial value of organic soil carbon at the beginning of the simulation period. EU Emission Trading Scheme prices for carbon dioxide allowance units (2010–2014 average from the European Energy Exchange) were used as a market value proxy for annual carbon sequestration, although it should be noted that the emission trading schemes are practically devised for emissions from electricity production and other industries, while sequestration is basically a counterforce of these.

An operational, national scale nutrient loading model developed for Finnish watersheds (VEMALA) was used to analyze ES related to hydrological and nutrient cycles. The model simulates nutrient processes (e.g. load, retention and leaching) and transport on land and in water bodies. It includes two main sub-models, the WSFS hydrological model (Vehviläinen, 1994) and the VEMALA water quality model (Huttunen et al., 2016). The WSFS model enables tracking of retention to different compartments of watersheds and was used to calculate the relative role of each catchment in retention of phosphorus and nitrogen. Finest available resolution of the model, water bodies, was utilized, based on the national lake register (SYKE). Nutrient retention was valued per water body with the replacement cost method, based on water purification rates and costs for phosphorus and nitrogen in constructed wetlands (Majoinen, 2005; Maes et al., 2012a, 2012b). The value of pollinationservices was mapped as pollination demand according to the spatial configuration of crop and berry yields (Crossman et al., 2013), their monetary values, and their dependency rate on biotic pollination – as for insects generally and bees specifically (Robinson et al., 1989; Lehtonen, 2012).

Cultural ES

A major component of cultural ES is close-to-home recreation (85% of recreation in the region happens within 5 km from home). Data on recreational use of nature studied in the LVVI2 inventory (2009–2010) was used for recreation frequencies (Sievänen and Neuvonen, 2011). Accessibility to close-to-home recreation was mapped for green spaces at least 1.5 ha in size (Söderman et al., 2012). Population data (250 m resolution) from SYKE was allocated to green spaces defined in municipal master plans and regional plans using a probabilistic gravity model (Huff, 1964) and expert-opinion based multi-criteria evaluation (MCE), evaluating several criteria for green space attractiveness (see e.g. de Groot et al., 2010; Paracchini et al., 2014): proximity to water, density of recreational path networks, degree of naturalness, landscape values based on important landscape area inventories, density of supporting sport facilities (University of Jyväskylä, 2014), surrounding population density, and intensity of aesthetic ES (see below). Valuation was based on previous willingness to pay (WTP) studies (Tyrväinen, 2001; Lankia et al., 2013). Other nature trips were mapped for major state-owned protection areas with visitor surveys in place: the Seitseminen and Helvetinjärvi national parks and Vehoniemenharju and Siikaneva protection sites. Statistics and money generation model based data on visitor money usage (2009–2013) for the sites or similar reference sites in Southern Finland were used in valuation, alongside with survey-based estimates of the financial value of health and well-being benefits from nature (Kaikkonen et al., 2014; Vähäsarja, 2014; Metsähallitus, 2015).

Recreational valuation of gathering berries and fungi was based on LVVI2 and the value of everyman's rights based trips (Maes et al., 2012b) – spatial distribution of recreational values was assumed to follow that of the yields since the spatial configuration of utilization is unknown. As of hunting, the recreational value was estimated at 60% of total hunting value (Aarnio and Härkönen, 2007). For fishing, WTP-based value transfer of recreational values was used (Toivonen et al., 2004). A commonplace characteristic of recreation in the Nordic region is the near-ubiquitous presence of holiday housing and the environmental benefits therefrom – swimming, other uses of water, enjoyment of scenery, being in the nature, etc. (Jäppinen and Heliölä, 2015). Holiday house locations were acquired from the Topographic Database (NLS). Two types of WTP were used in valuation – econometric estimation of travel cost (value transfer) and hedonic pricing according to the VIRVA model (Lankia, 2011; Seppälä et al., 2014). Holiday housing real estate prices were used in hedonic valuation, acquired for both cottages with and without shore access (NLS, 2014). Mean interest rates of the Bank of Finland (2000–2014) were used in deriving annual price amortization rates. In both models water quality affected the unit value.

Bequest was examined as willingness to pay for preserving natural environments. Nature protection areas (SYKE) and planned protection sites were mapped. Two distinct approaches were used to value bequest. Firstly, in Finland there is a market-based system for encouraging voluntary nature protection, especially for habitats suitable for endangered species – the Forest Biodiversity Programme METSO, 2008–2025 (METSO, 2008). METSO allows for temporary or permanent nature protection, the latter of which was used here as a reference. METSO payments to landowners are based on lost forestry-based income and the "ground value" based on biogeological forest site types. Gross wood and ground values were calculated for all nature protection areas according to METSO criteria. Where water bodies were present, these were valued according to mean water area real estate prices (NLS) from 1990 to 2012. In order to derive an annual value for the ES "flow", the value was divided by 70 years, a typical length of a silvicultural cycle in the region. Secondly, a contingent valuation approach was taken, based on WTP studies concerning the continuation of nature conservation in Eastern Finland (Kniivilä, 2004). WTP values for adult population were assigned to sites using METSO criteria as MCE weights roughly indicating site quality.

A supplementary, non-monetary mapping of aesthetic or experiential human-nature interactions was conducted. As a proxy for these services often insufficiently integrated into ES inventories (Daniel et al., 2012), geolocated nature-related imagery was "mined" from social media, using popular services Panoramio (Google) and Instagram (Facebook) (see e.g. Di Minin et al., 2015). Data was acquired from corresponding application program interfaces (API), from 2005 to 2015 (Panoramio) and 2014 (Instagram) – the different time spans due to different service ages, data quantities and API processing limits. As the services are not exclusively built for nature photography, raw data was filtered by tags corresponding to circa 200 nature-related keywords. In order to avoid a selection bias and to distinguish areas of high aesthetic interaction on average, the number of individual nature (or in-nature) photographers per grid cell was calculated. This resulted in a nature photography activity or aesthetic ES index.

Abiotic provision outputs

Potential outputs of wind power (current wind power production is virtually non-existent in the region) were based on the regional plan draft, in which potential major wind power production sites had been defined via a rigorous site exclusion and impact assessment procedure. Maximum number of wind power plants per area, nominal power outputs, annual wind information and estimated annual operating hours were derived from accompanying analyses. Feed-in tariffs were not accounted for, since these are currently practically halted for new wind power plants in Finland. Hydropower usage was based on regional electricity production statistics and the locations and nominal powers of hydroelectric power plants (>1 MW). Off-site effects were accounted for two major dams located nearby outside the region, whose reservoirs are located in the region. Since no readily suitable model for the valuation of the hydropower "production chain" was available, expert opinion was utilized. The value of power provision was divided between dam locations and reservoirs, the latter of which was split again between the lakes and their immediate catchments. Most of the major lakes in the region are regulated, and a "cascading-like" approach was used to allocate the value along the chain of lakes and watercourses, the relative importance of which for power production was estimated based on their volume and flow. Hydroelectricity and potential wind power outputs were valued according to power market prices (2011–2015 mean).

Peat extraction is a major component in the region's energy production (especially in fuel mixes with wood in combined heat and power generation) and peat is also used in large quantities for agricultural purposes. Mapping major peat extraction sites was also based on the regional plan. The estimated volume of annually extractable peat for each area was derived from per-site total estimates, divided by a typical 20-year site life cycle. An energy content of 0.9 MWh per loose cubic meter of milled peat was used as a reference (Alakangas, 2000). Valuation was based on mean energy (heat) production price statistics from 2009 to 2013 (Statistics Finland, 2015). Iivonen (2008) was used for valuing major non-energy uses of peat.

Linking ES into land use and management contexts

In order to gain an understanding of the linkages between land use, its management and ES, the results of spatial ES valuation were compared to different environment-related land use categories of the regional plan 2040 draft (Fig. 3), as well as to current land use according to CLC2012 level 1 data. The land use types examined from the regional plan were: core areas of ecological networks (areas important for ES and core areas of biodiversity), all nature protection areas (>1 ha), areas belonging to the Natura 2000 network, recreational areas, groundwater areas, nationally or regionally valuable landscape areas and valuable cultural landscapes. The core areas of ecological networks were based on an analysis made for the regional plan on regionally important green core areas and the connections between them. The analysis had been conducted using MSPA, spatial MCE of qualitative ecological spatial data, and methods of participatory mapping (see e.g. Tammi and Kalliola, 2014). ES were aggregated into spatial delineations of current and planned and use based on whether or not a specific ES is supplied at specific land use type (e.g. wild berries are supplied basically only at forest or wetlands) or is legally utilizable at specific areas defined in the land use plan (e.g. forestry is practically forbidden in protection areas).

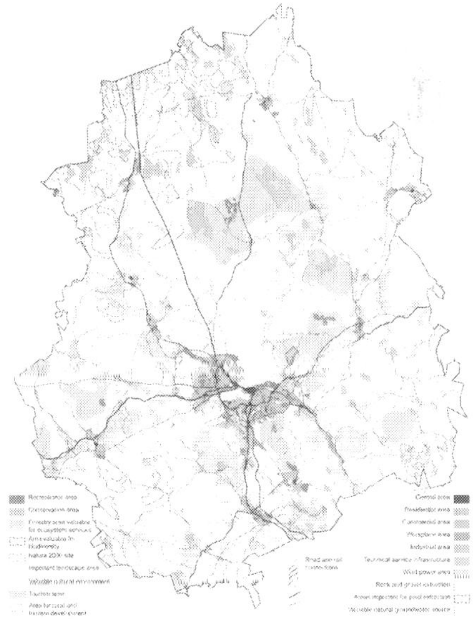

Figure 3: A generalization of the Tampere regional plan 2040 draft. Numerous plan elements have been omitted in order to enable readability. Main urban structure and environment related elements are included. Council of Tampere Region.

A municipal level aggregation of ES values was originally carried out in order to aid in future local ES implementations, but was found to lack explanatory power because of numerous municipal mergers, due to which the municipalities often comprise functionally vastly different areas – from urban cores to wilderness. Therefore an alternative approach independent of administrative boundaries was chosen. ES values were compared against a multi-source functional urban–rural typology created at SYKE as a tool for zonal planning and development independent of administrative contexts (Fig. 4, Helminen et al., 2014). The typology, ranging from inner urban cores to sparsely populated rural areas, combines socioeconomic, demographic, traffic and commuting data as well as urban structure and land use information and indicators derived from these. Both absolute and relative ES statistics were calculated for the different typology classes.

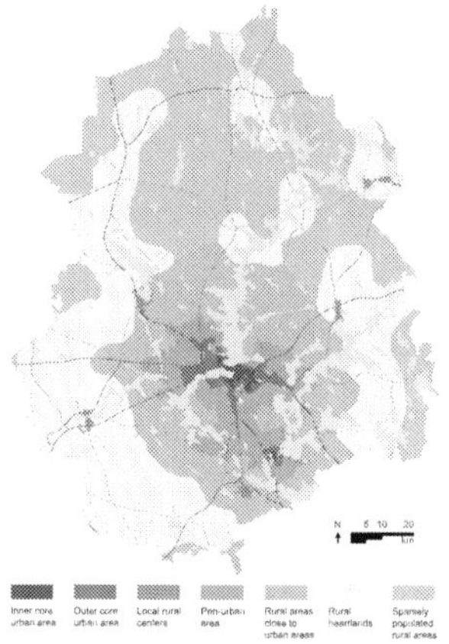

Figure 4: Urban–rural spatial typology in the Tampere region. Major lakes and road networks have been added. Helminen et al., 2014 (modified)

RESULTS

Spatial valuation of ecosystem services

The annual value of ES in the Tampere region scrutinized here is circa 760–910 million euros, or 835 million euros on average (Table 1). Including abiotic outputs the value reaches approximately 810–960 million euros per annum (average 885 M€). When considering the potential rates examined here in addition, the total annual value reaches circa 0.9–1.2 billion euros. However, these values should be considered a lower boundary of ES TEV (see e.g. Costanza et al., 1997), since neither the selection of ES nor valuation were exhaustive. Spatially the utilization of ES is heterogeneous (Fig. 5a, b). Logically many ES follow the landscape settings (e.g. SW crop yields and pollination, S–SE forest growth and carbon sequestration or NW abiotic outputs and potentials from peat and wind power), or the combination of these and settlement patterns (e.g. fishing in largest lakes surrounded by most holiday housing, water usage from lakes and eskers around the city region or recreation dominantly around the urban core and holiday housing clusters). Some ES show more complex patterns rooted in for example historical ES utilization or mixtures of biogeographic and anthropogenic factors (e.g. animal rearing or nutrient retention in lakes) and some have a clear linkage to ES user movement or human-made decisions (e.g. aesthetic interactions or bequest).

Table 1: Current and potential annual rates and values for ES and abiotic natural outputs in the Tampere region. The potential value is mainly used for comparative purposes, since changes in the supply and demand balance due to realization of said potentials and their probable impact on unit values have not been accounted for. NS=non-spatially valued ES

Ecosystem service	**Current rate p.a.**	**Value (M€/a)**	**Potential rate p.a.**	**Potential value (M€/a)**	**Mean unit value**
Provision	**NA**	**380**	**NA**	**515**	**NA**
Cultivated crops	0.6–37 t/ha	48.5 + 22.3 NS	NA	NA	130–590 €/t
Reared animals	28 Mkg meat, 0.12 Mm^3 milk, 1.5 Mkg eggs	93.5	NA	NA	1.4–2.7 €/kg (meat), 0.4 €/l (milk), 1 €/kg (eggs)
Wild plants for nutrition	0.5–1.5 Mkg	0.6–2.0+1.1 NS	15 Mkg (berries), 12 Mkg (fungi)	19.8+35 NS	1.3€/kg (berries), 2.9e/kg (fungi)
Wild animals for nutrition	7 100 (*Cervidae*)+2.3 Mkg (fish, 15% freed)	2.7 (deer)+1.5 (other game) NS+5.7 (fish)	8 000 (*Cervidae*)+2.3 Mkg (fish)	3.1 (deer)+1.5 small game (NS)+7.0 (fish)	6.5 €/kg (deer), 2.7 €/kg (fish)
Surface water	16.8 Mm^3	16.8	21.2 Mm^3	21.2	1.0 €/m^3
Ground water	18.7 Mm^3	22.5	23.7 Mm^3	28.4	1.2 €/m^3

Fibers from wood	3.7 Mm^3	110	5.2 Mm^3	159	17.1 €/m^3 (fiber), 53.3 €/m^3 (saw wood)
Bioenergy (wood)	1.3 Mm^3/2 600 GWh	54	1.8 Mm^3/3 600 GWh	76	21 €/MWh
Bioenergy (agro-)	NA	NA	496 GWh (plant), 136 GWh (animal)	18.9 (plant), 5.20 (animal)	38 €/MWh
Regulation and maintenance	**NA**	**79–96**	**NA**	**79–96** (+83 incl. all CO_2sequestration)	**NA**
CO_2sequestration	3.4	31	12.7	31 (114)	9.0 €/t CO_2
Nutrient retention	54–61 t P, 0.9–1.2 Mt N	43–54	NA	NA	410 €/kg P, 25 €/kg N
Pollination (demand-based)	NA	4.1–9.9 (incl. domestic needs)+1.1 (honey)	NA	NA	310 €/ha (crop yield contribution), 6.8 €/ha (berries).
Cultural	**NA**	**306–434**	**NA**	**306–434**	**NA**
Close-to-home recreation and nature trips)	Close-to-home trips 33.6 M+2.8 M NS, nature trips 0.09 M	118–144+9.7–12 NS (close-to-home), 9.3–17 (nature trips)			3.5–4.3 €/close-to-home trip, 103–194 €/nature trip

Recreational values from holiday housing and waters (e.g. boating)	38.4 d/a or 18 trips/user/a (cottages). 20.8 trips/pers, 4.2 M trips (boating etc.)	83–120 (holiday housing)+42–58 (boating etc.)	NA	NA	111–203 €/trip (WTP), 2 780–5 050 €/cottage (hedonic, 80% of plot & 30% of building value)
Recreational value from hunting-gathering	10 000 hunters, 23.5 hunting days p.a. 168 000 fishers, 15.3 fishing days p.a.	6.3 (hunting)+12–33 (fishing)+12.9 (berries)+9.5 (fungi)	NA	NA	9.3 €/trip (everyman's rights), 60% of total hunting/fishing value, 69.4 €/a (fishing, WTP)
Bequest	NA	2.9 (METSO)–21.5 (WTP)	NA	NA	39 €/m³(wood)+ 150–580 €/ha (ground), 58 €/a/pers. (WTP)
(Aesthetic) nature interactions	3.2 photos/user, 18 000 photos/a, 0.3 users/km²	NA	NA	NA	NA
Abiotic (non-ES)	**NA**	**44**	**NA**	**117**	**NA**
Hydropower	150 MW/548 GWh	21–24	NA	NA	38 €/MW h
Wind power	~ 0 GW h	0	1.7 TWh	63	38 €/MW h
Peat extraction	1.1 Mm³/1.0 TWh (energy)+0.3 Mm³(other)	23	1.3 Mm³/1.2 TWh (energy)+0.5 Mm³ (other)	30	19 €/MW h, 16 €/m³

Total	**NA**	**760–910**	**NA**	**900–1 130 (incl. current use)**	**NA**
Total incl. abiotic	**NA**	**810–960**	**NA**	**940–1 180 (incl. current use)**	**NA**

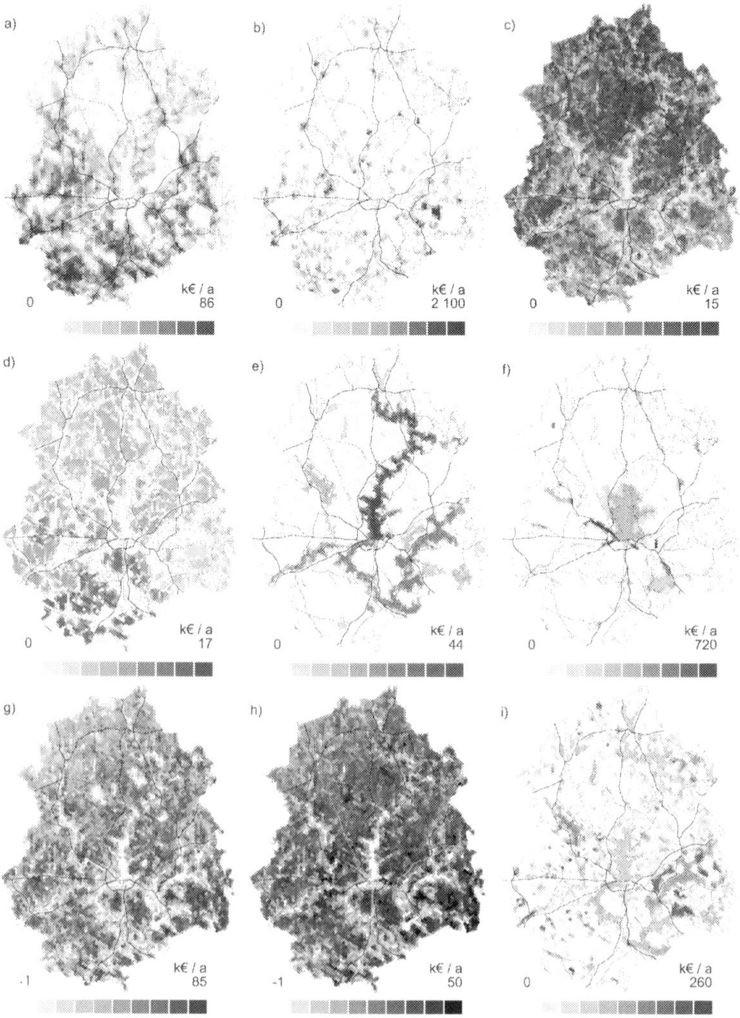

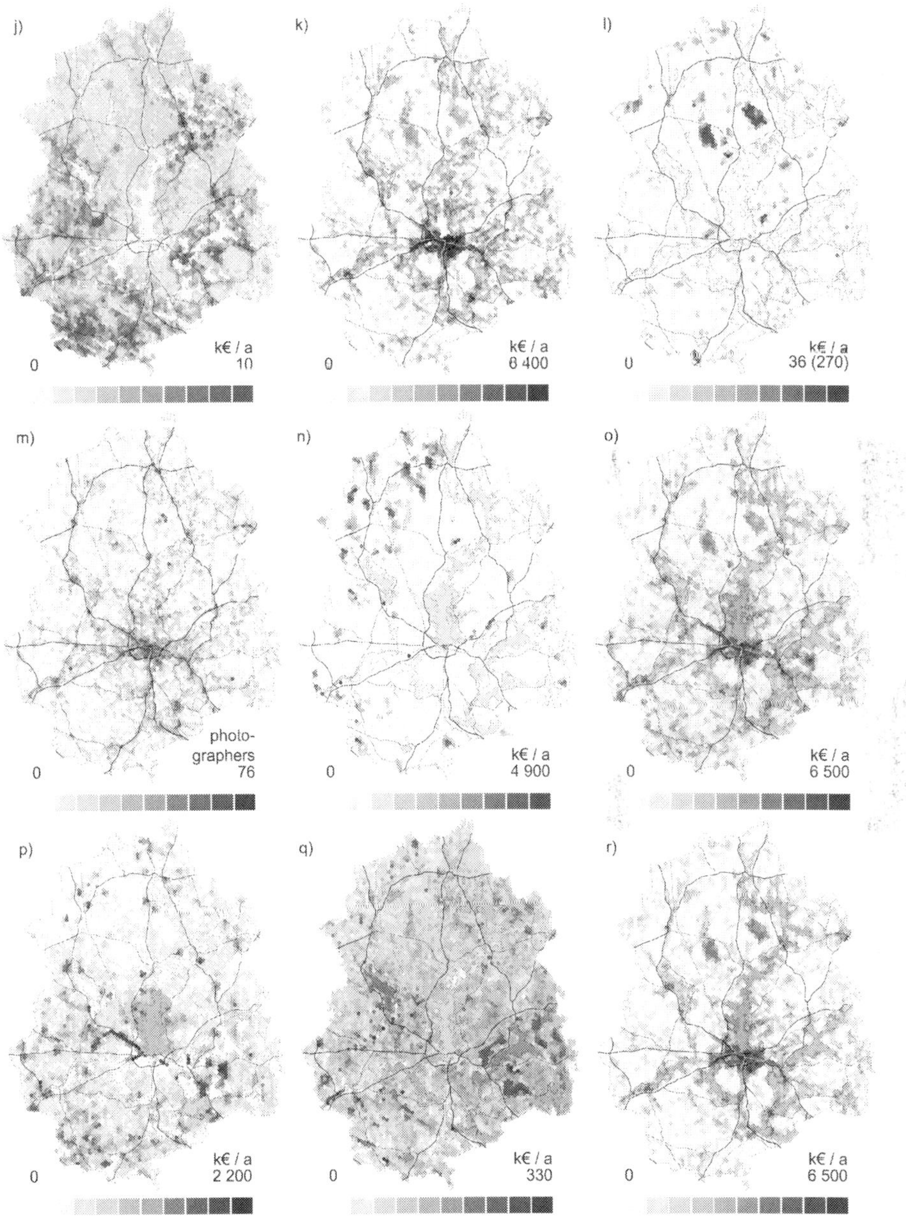

Figure 5: **a.** Annual ES TEV supply in the Tampere region: a) cultivated crops incl. plant biomass based energy potential, b) reared animals incl. manure based energy potential, c) yield potential for bilberry and lingonberry incl. recreational value from current utilization, d) *Cervidae* hunting incl. recreational value, e) professional and recreational fishing incl. meat and recreational value, f)

ground and surface water usage, g) supply of saw timber, pulpwood and energy wood, h) carbon sequestration, i) retention of phosphorus and nitrogen in water bodies. (width: 2 columns). 5b. Annual ES value (TEV) supply in the Tampere region: j) insect pollination of crops and wild berries, k) close-to-home recreation, nature trips and recreational values from holiday housing incl. uses of waters, l) bequest, mapped by market-cost pricing and WTP (in brackets), m) aesthetic nature interactions (non-monetary), described by the number of individual nature photographers per grid cell, n) abiotic provision of hydropower, peat (energy and other uses) and wind power (potential), o) ES total (excl. abiotic outputs), p) total provision ES, q) total regulation and maintenance ES, r) total cultural ES. (width:2 columns).

Comparing ES with current and planned land use and spatial typologies

The linkage between ES and land use reveals clear patterns between extensive and intensive forms of land use in relation to ES value supply and utilization (Table 2), with highest relative rates in areas of high anthropogenic influence, while total values follow the overall landscape composition and its extensive forms of land use, especially forestry and agriculture. When comparing ES values to land-use planning, similar trends highlighting the human–nature interface arise (Table 3). Areas with high use frequencies, such as recreational or groundwater areas, seem by far the most valuable in terms of relative annual ES flow, followed by valuable landscape areas, which often encompass a mixture of agricultural, forestry and recreational uses. An ES value divide exists between protection sites (incl. Natura 2000) and core areas of the ecological network (areas important for biodiversity and areas important for ecosystem services) – mainly due to a higher presence of bequest and recreational uses, as mapped here. The anthropocentrism of the ES framework becomes evident when comparing ES values to the urban–rural typology (Fig. 6). Simply put, the higher the population density, the stronger the ES demand, and in low population density areas, the per capita ES value rises dramatically. One should note that ES supply and demand go often strongly hand-in-hand in our case area, since there is little scarcity of supplying ecosystem structures due to the relative ubiquity of nature in Finnish cities – but this is not always nor everywhere the case.

Table 2: Comparison of ES values with current land cover as described by CLC2012 level 1 data. ES-specific values were calculated as averages in case a range of values was present or several valuation methodologies were utilized. Immediately utilizable ES potentials or services, whose current utilization rate is unknown were included in the analysis, arriving at a total ES value of c. 916 M€/a. It should be noted that due to the resolution of the data and its underlying modelling methodology, artificial surfaces include some urban nature and semi-natural areas, which are especially valuable from the perspective of cultural ES. Similarly, many ground water areas consist usually of a combination of artificial, semi-natural and natural surfaces, and in agricultural areas animal rearing often intersects with artificial surfaces from the land cover perspective

CLC2012 class	Total area (km^2)	Provision ES (M€/a)	Reg. and maintenance ES (M€/a)	Cultural ES (M€/a)	Total ES (M€/a)	Total ES (€/a/ha)
Artificial surfaces	710	28.2	1.68	54.5	84	1 197
Agricultural areas	1 700	148	3.81	23.5	175	1 060
Forest and semi-natural areas	10 030	226	29.6	173	430	428
Wetlands	120	2.33	0.32	4.12	6.8	549
Water bodies	2 100	22.8	44.7	153	220	1 049
Total	14 660	427	80	408	916	625

Table 3: Comparison of (currently utilized) ES values with land use types from the regional plan 2040 draft. Values of ES that were subject to more than one type of valuation methodology were calculated as averages of these. Note that some of the different area types can be overlapping, e.g. areas belonging to the Natura 2000 network often – but not always – include nature protection areas. For recreational areas two different delineation approaches were used, excluding and including adjacent water areas, since the water element played a major role in the valuation function for recreational areas

Land use	Area (km^2)	Provision ES (M€/a)	Reg. and maintenance ES (M€/a)	Cultural ES (M€/a)	Total ES (M€/a)	Total ES value incl. abiotic outputs (M€/a)	ES €/a/ha	ES incl. abiotic €/a/ha
Areas important for biodiversity	698	12	3.9	11	27	27	389	394
Areas important for ecosystem services	980	21	4.7	12	37	37	380	382
Protection areas	372	1.3	2.1	29	32	33	866	875
Natura 2000 areas	348	1.0	2.1	25	29	29	820	830
Recreational areas	167	4.2	0.8	17–27	22–32	23–33	1 380–1 980	1 440–2 040
Valuable landscapes	1 431	56	11	55	121	129	847	900
Cultural landscapes	911	44	5.6	20	70	72	765	788
Groundwater areas	337	30	1.4	13	44	45	1 310	1 340

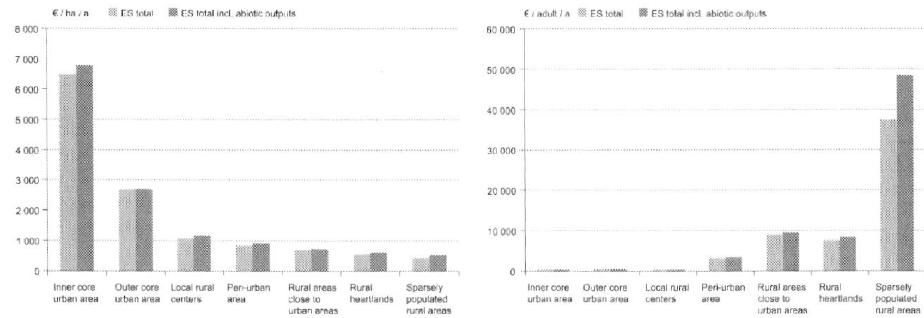

Figure 6: Comparison of ES value mapping with an urban–rural spatial typology (Helminen et al., 2014). The graphs describe ES value (€/a) both including and excluding abiotic outputs a) per hectare and b) per person (15–74 y/o).

DISCUSSION AND CONCLUSIONS

Assessing uncertainty

Capturing ecosystem functions in monetary values triggered skepticism early on (Vatn and Bromley, 1994), and valuation remains tricky. We formed a moderate estimate for the lower boundary of ES value supplied (and utilized) by the region. However, since ES often form dynamically in complex chains of ecosystem functions at different spatial levels, a relatively fixed-scale approach aimed at producing localizable, practically applicable data, will miss some of the ES dynamics, especially regulation services typically relevant and observable at larger scales (de Groot et al., 2010). Converting these services to monetary values remains difficult partly due to problems in identifying the providing ecosystem structures. This is partly evident in our results, which emphasize provision and cultural services that can be pinpointed and monetized in a relatively straightforward manner. Care should be taken when considering the relatively low value of regulation and maintenance ES and their potentially providing structures (major forests and wetlands). Then again, the values that can be most easily monetized may also be most relevant and influenceable in practical land-use planning. The regulation and maintenance ES that may be easiest valued in future research, could be local services such as visual screening or microclimate and air quality regulation provided by vegetation, which again would increase the relative value of urban and semi-natural environments.

The maximum simultaneous TEV remains fuzzy. Care was taken in avoiding double counting of ES values (Costanza, 1997), but take for instance carbon sequestration and forest growth, which correlate strongly. Sequestration rates were estimated against the "unutilized" part of forest growth, but unleashing the potential TEV of sequestration would require significant reductions in forestry. The Pareto optimality of the ES use spectrum remains unclear and is likely to fluctuate according to societal changes. Critical interpretation should be also practiced when comparing current and potential ES flows, which are partly dependent on realized land use plans, partly "immediately utilizable". Although newest available valuation references were used, the availability of price time series used in value transfer varied between different ES. A temporal misalignment may have widened the error margin, since prices can fluctuate notably due to e.g. climatic, political and macroeconomic changes. Aggregating results to different spatial delineations from a single-resolution ES inventory creates uncertainty in quantitative comparisons. Parts of the spatial database located on the region's borders lacked some ES TEV due to edge effects caused by the grid delineation, which was softened by partial extrapolation. The spatial delineation of ES scrutiny is also a somewhat philosophical question, especially with WTP and off-site effects (see e.g. Seppelt et al., 2011). Here we estimated that the region's inhabitants are willing to pay for, say, bequest ES, but there surely are people outside the region willing to pay for its nature protection. Choices made on defining the ES beneficiaries inevitably affect the valuation, and narrowing down the uncertainty would require in-depth ES user base analysis.

The generalizability and transferability of ES assessments are context-dependent – the danger of generalization errors exists in value transfer (Nelson and Kennedy, 2008; Brander et al., 2012). The results of this study could be satisfactorily transferred to at least boreal-hemiboreal settings with similar societal conditions, but the further one moves away from circumstantial similarity, the stronger the uncertainty. Many of the distinguished ES hot-spots were connected to locations special due to their biogeomorphology or cultural-historical significance that would be nearly impossible to identify using for example generalized land cover based models. There, on the other hand, lies the strength of multimethod ES assessment frameworks. Here we combined ad hoc approaches with different types of existing spatial modelling processes, aiming at the golden mean between sufficient customizability and practical usability. Concerning questions of the generality of ES tools, indicators, and the feasibility of their implementation into practical land-use management (Primmer and Furman, 2012), we argue that problems are faced especially when dealing with values, since these are strongly rooted in societal contexts – yet seem to be among the most sought-after information. The tools, approaches or indicators tested here are however not strictly context-specific, but a lot depends on the quality and quantity of ES-related data at hand for the planner.

ES from a land-use planning and policy perspective

Before case study onset, the regional planning process had been surrounded by discourse on interactions between urban and rural areas, where land-use planning was criticized of concerning mainly the former, and comprehensive consideration of natural capital in planning and development was called for. This call came especially from declining rural areas where possibly a feeling of neglect towards their potential was felt. Our results underline a few aspects to this discourse. Firstly, a rather linear decline in ES/ha value from urban to rural settings was seen – prime ES hot-spots tended to be located in urban settings. This is perhaps not surprising considering ES are benefits gained by humans from ecosystems – and need people to enjoy these benefits. Also, historically population centers tended to form around favorable environmental settings – and the demand for cultural ES only seems to emphasize this relationship. Next, one should consider intensity versus extensiveness (see e.g. de Groot et al., 2010, 2012). In absolute figures, ES value declined from rural towards urban areas. Of course, rural areas were tenfold compared to urban ones. ES such as food and forestry outputs are highly valuable, but require vast spaces and "flow" slowly – compare for example hourly water usage or daily recreation to a 70-year boreal silviculture cycle. These slowly regenerated natural capital stocks are highly valuable (e.g. forest stocks were valued at c. €4B at current prices, or carbon storages of c. 700 $MtCO_2$ at €6.3B using emission trading prices), but the flows are experienced gradually over time and space.

Low population density in extensive rural areas elevated the per-capita ES value tremendously. This may bring an interesting viewpoint into land use discourse. A common housing preference of (young) Finnish adults has been a detached house at the urban–rural intersect, at a commutable distance to the city yet close to nature for restorative purposes (Korpela et al., 2001). This nature-bound housing preference could be possibly translated loosely into a preference of ES abundance over having to share the ES "commons"; the most rural areas generated 120–150 times more ES flow value per capita compared to their most urban counterparts. At the same time, the relative importance of abiotic provision outputs versus ES decreased along the rural–urban gradient, mainly due to the fact that many abiotic outputs cannot be generated in urban areas. However, in the discourse surrounding the study, stakeholders from rural settings seemed to lean more on traditional natural resource viewpoints, while their urban counterparts dominantly promoted an ES framework. Potential linkages between such place-based land use discourse differences and ES versus natural capital viewpoints would require further verification. We recommend that the focus in implementing an ES framework in land-use planning should be placed at urban and peri-urban green (and blue) spaces. In our case region, we expect the demand for ES to grow hand in hand with population growth. Due to e.g. infill development, the increasingly valuable urban ES refuges may be threatened and should be put under the magnifying glass in land-use planning.

Our results indicate that there is indeed a need to bind knowledge on "key green and blue areas" with that of ES hot-spots (Jäppinen and Heliölä, 2015), since these may often be relatively separate. Large forest and wetland areas, considered core areas of ecological networks in the regional plan, did not exhibit particularly strong ES value. There are probably a few reasons behind this. The ES framework approaches valuation from a human perspective, yet in the regional plan ecological networks had been considered mainly from the perspective of fauna and flora. It is thus not necessarily surprising that ES value doesn't seem very high in remote parts of the ecological network, often infrequently utilized by humans. Although, their value is often related to biodiversity or regulation and maintenance ES – which were not exhaustively covered by our analysis. As for recreation, the lack of value in these areas was partly caused by the fact that little spatial information exists on their recreational use. However, only an estimated 7% of recreational trips (excl. holiday housing) are made to these areas (valued here non-spatially). The value difference is partly technical by nature; due to no-overlap rules in the land use plan, protection sites within core green areas are technically enclaves and thus didn't contribute to the value of the core areas. Nevertheless, results from the CLC comparison support the observation of low values. A smaller, similar value difference was observed between Natura 2000 areas and other protection sites. The former are often used exclusively for conservation, while the set of protection areas as a whole contains many areas used simultaneously for both conservation and recreation, thus exhibiting higher value here. Regardless of large forests and wetlands seeming relatively weak from the perspective of annual ES flow, they contain massive natural capital stocks, as pointed out above. Future research should aim towards devising a reliable means of commensurating natural capital stocks and ecosystem services, as these are often brought up simultaneously in planning discourse, yet may differ notably even spatiotemporally. To put it otherwise, how should slow and fast ecosystem services really be compared?

Putting ES into future land-use management

Regional level ES assessments seem most common so far (Seppelt et al., 2011; Crossman et al., 2013) – but why? According to our experience there may be a few key reasons behind this. Firstly, statistics and reference studies suitable for ES mapping and valuing as well as spatial data are often produced at a regional administrative level. Much related at least in Finland is the fact that people's everyday movement is increasingly regional by extent, not limited to local settings. As ES values are often heavily connected to the location of the beneficiaries (Paracchini et al., 2014), the people, local-scale implementations may be unable to account for the big picture of ES user mobility. Thirdly, a regional approach offers possibly a suitable level of generalization between local and national levels, which may often prove too precise with the current knowledge or too coarse to be useful in practical implementations, respectively. A useful nuance of the Tampere region is the fact that it is mostly part of the same major drainage basin. Indeed, a drainage basin based spatial delimitation for ES approaches has been previously found valuable in mapping ES supply, especially when hydrological and nutrient cycles and other regulation and maintenance services are concerned (Valtanen et al., 2015; Vierikko and Niemelä, 2016).

Finally, a reason brought up during the project for the lack of local-scale ES implementations was simply a lack of resources at the municipal level. There are indeed a few obstacles to satisfactory ES framework implementation in practical land-use management, and often these can be tracked down to time and money, specifically the lack thereof. Decision-making processes are often criticized for their short-sightedness and desire for benefits in the short-term over the long-term. This is reflected in land-use planning processes, and if an ES assessment is made to support these to begin with, its comprehensiveness and rigor are often compromised by insufficient resourcing, which prevents the planner from properly familiarizing with ES indicators, data, models and tools. This in turn calls for an era of open information. For instance Schägner et al. (2013) have noted that mapping and storing ES values in spatial databases has the advantage of making site-specific valuation data readily available for decision makers and policy evaluation. We argue that ES data should be readily available for everyone (see e.g. de Groot et al., 2010), in order to promote the contributions of ES to human well-being as well as to encourage further ES implementations and data refinement. The ecosystem approach challenges planners and decision makers to use best available scientific knowledge, and the open data principle can only help in overcoming this challenge.

We wanted to uproot sectoral modelling traditions and plant them into the CICES framework (Haines-Young and Potschin, 2013). From the perspective of practical implementation the framework needed bending, especially in the case of regulation, maintenance and cultural ES. The problem was often a seemingly artificial separation of abiotic and biotic services from an everyday perspective. We find this problematic since ecosystems themselves consist of biotic–abiotic interactions – an issue already discussed in CICES itself and recently by van der Meulen et al. (2016). For instance, recreational or aesthetic experiences can be derived from both living organisms and the ground below them, nutrient retention in lakes is a combination of biotic and abiotic structures and processes (e.g. sedimentation and erosion), and visual screening is achieved by combining physical and biological structures. Similar examples are trying to find their place in CICES. It is often the case that planners and decision makers want information on both biotic and abiotic aspects, and to examine these separately can be even dangerous since there are strong feedbacks from one to the other. For example the utilization of hydropower in our case region has created obstacles to fish migration (and fishing-related ES) and nutrient flows, and rock extraction has had major effects on recreation, water balances and the overlying land cover – basically removing the ES supplying structures. The production boundary between ecosystem services and goods remains somewhat diffuse in practical applications, which here seemed to drift towards a mixture of CICES and the SEAA Central Framework or the concept of "natural flows", already highlighted by Haines-Young and Potschin (2013). The big picture of ES-related flows is complex, and proper cost-benefit analyses would need to consider non-ecological flows or inputs (Boyd and Banzhaf, 2007), the surface of which was merely scratched here. Similarly, the concepts of ES provision potential, supply, potential and sustainable supply, demand and potential demand can have several, often ambiguous interrelated interpretations (Jäppinen and Heliölä, 2015), and their mixed use is often unavoidable in practice due to the contextual frameworks from which the analyses stem from. Which of these concepts is most relevant in planning practice, remains yet unclear.

Notwithstanding its problems, the ES framework provided a workable foundation for spatial ES valuation. The results of the study already affected the Tampere regional plan 2040 proposal, altering the plan towards a more comprehensive guidance solution for ecosystem service hot-spots. The key issue here was a spatially explicit inventory and valuation framework, which could be directly compared with land-use planning data and the plan itself and thus be used as a foundation for delineating plan features and reformulating the accompanying regulations. A common characteristic of strategic level land-use planning is the time it takes for a plan to be realized, thus monitoring and assessing the aftermath of implementing an ES framework remains to be seen. However, the task of regional planning has partly been seen as delivering information further down the planning hierarchy and raising awareness on land use -related issues under discussion. Our approach succeeded in generating stakeholder interest for locally utilizing the gained information in future development projects. The study raised notable media interest, and has affected and aided already several large and small scale planning processes from impact assessments to bio- and circular economy projects to participatory landscape planning, not only within the region but in different parts of the country. In some other regions work is currently planned or underway to assess ecosystem services mimicking the methodology and analytic processes described here. From what we have observed so far, most interest has been generated by three aspects: relativity of ES values, comprehensiveness of their scrutiny, as well as mapping and valuing non-market services – especially cultural and experiential ES. When it comes to indicators of the latter, we anticipate a growing interest towards data generated by the ES beneficiaries themselves, that is, "crowdsourced" ES-related information, be it stated or revealed preferences obtained via social media or other means of participatory data collection.

Drawing on our experience, we conclude that there is a place for ES-based approaches and tools in land-use planning and development, but the practical planning sphere may require clarification and synthesis of the widening scope of ES, natural capital and environmental economics discourse. In practice, there is seldom a single framework to go forward with, but the suitable pieces of the ecosystem service puzzle need to be assembled. The same applies to ES tools and indicators – it seems unrealistic that any single data source or tool would fulfill the information needs of those initiating an ES planning scheme. Our work described here consisted largely of sewing existing or refined pieces of information together. This would suggest that in such contexts where for instance authorities or research institutes already have multiple ES-related data collection, reporting or sharing mechanisms in place, the main focus of future work enabling the implementation of ecosystem service frameworks into sustainable land-use management, should be put into integrating existing, separate sources of information in a manner that would readily combine the spatial and value components of ES-related information – this again would allow for more effortless and even semi-automated ecosystem service monitoring and assessment processes. The utility of ES assessments to policy guidance had been previously questioned (Primmer and Furman, 2012). Nevertheless, we argue that by comprehensively incorporating both spatial and value dimensions to ES analyses and communicating the information via visualizations and data openness, immediate changes in land use may not appear, but awareness of the issue will emerge and gradually reach different levels of planning – and hopefully the implementation of the plans, eventually.

ACKNOWLEDGEMENT

The ecosystem service project in which the authors of this paper participated, "Pirkanmaan ekosysteemipalvelut", was supported by the European Regional Development Fund programme for Western Finland 2007–2013 (Project A32791). We would like to thank Anne Mäkynen, Miina Vainio, Riina Känkänen and Thomas Banafa for their support.

REFERENCES

1. Aarnio and Härkönen, 2007 J. Aarnio, S. HärkönenHirvestä hyötyjä ja kustannuksia Metsätieteen aikakauskirja (2007) 2/2007
2. Alakangas, 2000 E. AlakangasProperties of Fuels Used in Finland Technical Research Center of Finland, VTT Energy, Jyväskylä, Finland (2000) (Research Notes 2045)
3. Bagstad et al., 2013 K.J. Bagstad, D.J. Semmens, S. Waage, R. WinthropA comparative assessment of decision-support tools for ecosystem services quantification and valuation Ecosyst. Serv., 5 (2013), pp. e27-e39, 10.1016/j.ecoser.2013.07.004
4. Boyd and Banzhaf, 2007 J. Boyd, S. BanzhafWhat are ecosystem services? The need for standardized environmental accounting units Ecol. Econ., 63 (2007), pp. 616-626, 10.1016/j.ecolecon.2007.01.002
5. Brander et al., 2012 L.M. Brander, I. Bräuer, H. Gerdes, A. Ghermandi, O. Kuik, A. Markandya, S. Navrud, P.A.L.D. Nunes, M. Schaafsma, H. Vos, A. WagtendonkUsing meta-analysis and GIS for value transfer and scaling up: valuing climate change induced losses of European Wetlands Environ. Resour. Econ., 52 (2012), pp. 395-413, 10.1007/s10640-011-9535-1
6. Costanza et al., 1997 R. Costanza, R. d'Arge, R. de Groot, S. Farberk, M. Grasso, B. Hannon, K. Limburg, S.Naeem, R.V. O'Neill, J. Paruelo, R.G. Raskin, P. Sutton, M. van den BeltThe value of the world's ecosystem services and natural capital Nature, 387 (1997), pp. 253-260, 10.1038/387253a0
7. Crossman et al., 2013 N.D. Crossman, B. Burkhard, S. Nedokv, L. Willemen, K. Petz, I. Palomo, E.G. Drakou, B.Martín-Lopez, T. McPhearson, K. Boyanova, R. Alkemade, B. Egoh, M.B. Dunbar, J. MaesA blueprint for mapping and modelling ecosystem services

Ecosyst. Serv., 4 (2013), pp. 4-14, 10.1016/j.ecoser.2013.02.001

8. Daniel et al., 2012 T.C. Daniel, A. Muhar, A. Arnberger, O. Aznar, J.W. Boyd, K.M.A. Chan, R. Costanza, T.Elmqvist, C.G. Flint, P.H. Gobster, A. Grêt-Regamey, R. Lave, S. Muhar, M. Penker, R.G. Ribe, T.Schauppenlehner, T. Sikor, I. Soloviy, M. Spierenburg, K. Taczanowska, J. Tam, A. von der DunkContributions of cultural services to the ecosystem services agenda Proc. Natl. Acad. Sci., 109 (2012), pp. 8812-8819, 10.1073/pnas.1114773109
9. Di Minin et al., 2015 E. Di Minin, H. Tenkanen, T. ToivonenProspected and challenges for social media data in conservation science Front. Environ. Sci., 3 (2015), p. 63, 10.3389/fenvs.2015.00063
10. Drakou et al., 2015 E.G. Drakou, N.D. Crossman, L. Willemen, B. Burkhard, I. Palomo, J. Maes, S. PeedellA visualization and data-sharing tool for ecosystem service maps: lessons learnt, challenges and the way forward Ecosyst. Serv., 13 (2015), pp. 134-140, 10.1016/j.ecoser.2014.12.002
11. Finnish Environment Institute (SYKE), 2015 Finnish Environment Institute (SYKE), 2015. Avoin tieto. (accessed 2.4.15.), (accessed version of website obsolete). 〈http://www.syke.fi/avointieto〉 .
12. Finnish Game and Fisheries Research Institute (FGFRI), 2014a2014 Finnish Game and Fisheries Research Institute (FGFRI), 2014a. Hunting 2013. Riista- ja kalatalous – Tilastoja. Official Statistics of Finland – Agriculture, Forestry and Fishery.
13. Finnish Game andFisheries Research Institute (FGFRI), 2014b Finnish Game andFisheries Research Institute (FGFRI), 2014b. Statistics. (acessed 15.5.15.). 〈http://www.rktl.fi/julkaisut/c/4/〉 .
14. Finnish Wildlife Agency and LUKE, 2012 Finnish Wildlife Agency and LUKE, 2012. RiistaWeb. Game information. (accessed 14.5.15.). 〈https://riistaweb.riista.fi/index.mhtml?Lang=en〉 .
15. de Groot et al., 2010 R. de Groot, R. Alkemade, L. Braat, L. Hein, L. WillemenChallenges in integrating the concept of ecosystem services and values in landscape planning, management and decision making Ecol. Complex., 7 (2010), pp. 260-272, 10.1016/j.ecocom.2009.10.006
16. de Groot et al., 2012 R. de Groot, L. Brander, S. van der Ploeg, R. Costanza, F. Bernand, L. Braat, M. Christie, N. Crossman, A.

Ghermandi, L. Hein, S. Hussain, P. Kumar, A. McVittie, R. Portela, L.C. Rodriquez, P.ten Brink, P. van BeukeringGlobal estimates of the value of ecosystems and their services in monetary units Ecosyst. Serv., 1 (2012), pp. 50-61, 10.1016/j.ecoser.2012.07.005

17. Haines-Young and Potschin, 2013 Haines-Young, R., Potschin, M., 2013. Common International Classification of Ecosystem Services (CICES): Consultation on Version 4, August-December 2012. EEA Framework Contract No EEA/IEA/09/003.
18. Helmisaari and Hallbäcken, 1998 H.-S. Helmisaari, L. Hallbäcken Tree biomass belowground F. Andersson, F.H. Brække, L. Hallbäcken (Eds.), Nutrition and Growth of Norway Spruce Forests in a Nordic Climatic and Deposition Gradient. Tema Nord, 566, Nordic Council of Ministers, Copenhagen(1998), pp. 80-90
19. Helminen et al., 2014 Helminen, V., Nurmio, K., Rehunen, A., Ristimäki, M., Oinonen, K., Tiitu, M., Kotavaara, O., Antikainen, H., Rusanen, J., 2014. Urban–rural classification. Reports of the Finnish Environment Institute 25/2014.
20. Huff, 1964 D.L. HuffDefining and estimating a trading area J. Mark., 28 (1964), pp. 34-38
21. Huttunen et al., 2016 I.Huttunen, M. Huttunen, V. Piirainen, M. Korppoo, A. Lepistö, A. Räike, S. Tattari, B.VehviläinenA national scale nutrient loading model for Finnish watersheds – VEMALA Environ. Model. Assess., 21 (2016), pp. 83-109, 10.1007/s10666-015-9470-6
22. Hynynen et al., 2002 J. Hynynen, R. Ojansuu, H. Hökkä, J. Siipilehto, H. Salminen, P. HaapalaModels for Predicting Stand Development in MELA System Vantaa Research Center, Vantaa, Finland (2002)
23. Ihalainen et al., 2005 M. Ihalainen, T. Pukkala, O. SaastamoinenRegional expert models for bilberry and cowberry yields in Finland Boreal Environ. Res., 10 (2005), pp. 145-158
24. Iivonen, 2008 Iivonen, S., 2008. Ympäristöturpeet ja niiden käyttö. Ruralia-instituutin raportteja 32.
25. JRC, 2015 JRC, 2015. (accessed 19.12.15.). Morphological Spatial Pattern Analysis. 〈http://forest.jrc.ec.europa.eu/download/software/guidos/mspa/〉 .
26. Jäppinen and Heliölä, 2015 J.-P. Jäppinen, J. HeliöläTowards a

sustainable and genuinely green economy. The value and social significance of ecosystem services in Finland (TEEB for Finland) J.-P. Jäppinen, J. Heliölä (Eds.), Synthesis and roadmap. The Finnish Environment 1en/2015., The Finnish Ministry of Environment, Helsinki (2015)

27. Kaikkonen et al., 2014 H. Kaikkonen, V. Virkkunen, L. Kajala, J. Erkkonen, M. Aarnio, R. KorpelainenHealth and well-being from Finnish national parks – a study on benefits perceived by visitors Nat. Prot. Publ. Metsähall. Ser. A (2014), p. 210
28. Kniivilä, 2004 M. KniiviläContingent Valuation and Cost-benefit Analysis of Nature Conservation: A Case Study in North Karelia, Finland. Academic dissertation Joensuu University Press, Joensuu, Finland (2004)
29. Korpela et al., 2001 K. Korpela, T. Hartig, F.G. Kaiser, U. FuhrerRestorative experience and self-regulation in favorite places Environ. Behav., 33 (2001), pp. 572-589, 10.1177/00139160121973133
30. Lankia, 2011 T. LankiaRecreation Value of A Summer House Visit – An Application of the Travel Cost Method (MSc thesis) Department of Economics and Management, University of Helsinki, Helsinki, Finland (2011)
31. Lankia et al., 2013 Lankia, T., Kopperoinen, L., Pouta, E., Neuvonen, M., 2013. Valuing recreational ecosystem service flow in Finland. European Association of Environmental and Resource Economists. In: Proceedings of the 20th Annual Conference at Toulouse, France, 26–29 June 2013.
32. Lehtomäki, 2006 A.LehtomäkiBiogas Production from Energy Crops and Crop Residues. Jyväskylä Studies in Biological and Environmental Science 163 University of Jyväskylä (2006)
33. Lehtomäki et al., 2007 A.Lehtomäki, S. Huttunen, J.A. RintalaLaboratory investigations on co-digestion of energy crops and crop residues with cow manure for methane production: effect of crop to manure ratio Resour. Conserv. Recycl., 51 (2007), pp. 591-609, 10.1016/j.resconrec.2006.11.004
34. Lehtonen, 2012 T. LehtonenThe Economic Value of Honey Bee Pollination in Finland among the Most Important (MSc thesis)

Department of Agricultural Sciences, University of Helsinki, Helsinki, Finland (2012)

35. Liski et al., 2006 J. Liski, A. Lehtonen, T. Palosuo, T. Peltoniemi, T. Eggers, P. Muukkonen, R. MäkipääCarbon accumulation in Finland's forests 1922–2004 – an estimate obtained by combination of forest inventory data with modelling of biomass, litter and soil Ann. For. Sci., 63 (2006), pp. 687-697
36. Liski and Westman, 1997 J. Liski, C.J. WestmanCarbon storage in forest soil of Finland Biogeochemistry, 36 (1997), pp. 261-274, 10.1023/A:1005742523056
37. Maes et al., 2012a J. Maes, B. Egoh, L. Willemen, C. Liquete, P. Vihervaara, J.P. Schägner, B. Grizzetti, E.G.Drakou, A. La Notte, G. Zulian, F. Bouraoui, M.L. Paracchini, L. Braat, G. BidoglioMapping ecosystem services for policy support and decision making in the European Union Ecosyst. Serv., 1 (2012), pp. 31-39, 10.1016/j.ecoser.2012.06.004
38. Maes et al., 2012b J. Maes, J. Hauck, M.L. Paracchini, O. Ratamäki, M. Termansen, M. Perez-Soba, L.Kopperoinen, K. Rankinen, J.P. Schägner, P. Henrys, I. Cisowska, M. Zandersen, K. Jax, A. La Notte, N.Leikola, E. Pouta, S. Smart, B. Hasler, T. Lankia, H.E. Andersen, C. Lavalle, T. Vermaas, M.H. Alemu, P.Scholefield, F. Batista, R. Pywell, M. Hutchins, M. Blemmer, A. Fonnesbech-Wulff, A.J. Vanbergen, B.Münier, C. Baranzelli, D. Roy, V. Thieu, G. Zulian, M. Kuussaari, H. Thodsen, E.-L. Alanen, B. Egoh, P.B.Sørensen, L. Braat, G. BidoglioA spatial assessment of ecosystem services in Europe: methods, case studies and policy analysis - phase 2. PEER Report No 4 Partnership for European Environmental Research, Ispra (2012), 10.2788/41831
39. Majoinen, 2005 L. MajoinenConstructed Wetlands (MSc Thesis), Helsinki University of Technology, Espoo, Finland (2005)
40. Marttunen et al., 2012 M. Marttunen, M. Dufva, K. Martinmäki, I. Sammalkorpi, T. Hjerppe, I. Huttunen, V.Lehtoranta, E. Joensuu, E. Seppälä, M. Partanen-HertellInteractive and comprehensive river basin management planning – a summary of the results of the project Integrating scenario approach into the river basin management planning in the Karvianjoki river basin Finn. Environ., 15 (2012)

41. METSO, 2008 METSO. (2008). Government Decision in Principle on an Action Programme to Protect Biodiversity in Forests in Southern Finland 2008-2016 (METSO-Programme). 27 March.
42. Metsähallitus, 2015 Metsähallitus, 2015. Käyntimääriä suojelu- ja retkeilyalueilla sekä asiakaspalvelupisteissä. (accessed 12.2.15.). 〈http://www.metsa.fi/kayntimaarat〉 .
43. van der Meulen et al., 2016 E.S. van der Meulen, L.C. Braat, J.M. BrilsAbiotic flows should be inherent part of ecosystem services classification Ecosyst. Serv., 19 (2016), pp. 1-5, 10.1016/j.ecoser.2016.03.007
44. Miina et al., 2013 J. Miina, M. Kurttila, K. SaloKauppasienisadot itäsuomalaisissa kuusikoissa – koealaverkosto ja tuloksia vuosilta 2010–2012 Working Papers of the Finnish Forest Research Institute, Vantaa, Finland (2013), p. 266
45. Muukkonen and Mäkipää, 2006 P. Muukkonen, R. MäkipääEmpirical biomass models of understorey vegetation in boreal forests according to stand and site attributes Boreal Environ. Res., 11 (2006), pp. 355-369
46. Mäenpää, 2005 Mäenpää, I., 2005. Kansantalouden ainevirtatilinpito. Laskentamenetelmät ja käsitteet. Suomen ainetaseet 1999. Statistics Finland and Thule Institute, University of Oulu. Helsinki.
47. Mäkinen et al., 2006 Mäkinen, T., Soimakallio, S., Paappanen, T., Pahkala, K., Mikkola, H., 2006. Greenhouse gas balances and new business opportunities for biomass-based transportation fuels and agrobiomass in Finland. VTT Research Notes 2357.
48. National Land Survey of Finland (NLS), 2014 National Land Survey of Finland (NLS)Real Estate Purchase Price Statistics 2013 Official Statistics of Finland (2014) (Prices 2014:1)
49. Natural Resources Institute Finland, 2014 Natural Resources Institute Finland, 2014. MTT Economy doctor. (accessed 7.9.14.). 〈https://portal.mtt.fi/portal/page/portal/economydoctor〉 .
50. Natural Resources Institute Finland, 2015a Natural Resources Institute Finland, 2015a. National Forest Inventory (NFI). (accessed 14.4.15.). 〈http://www.metla.fi/ohjelma/vmi/info-en.htm〉 .
51. Natural Resources Institute Finland, 2015b Natural Resources Institute Finland, 2015b. Producer prices of agricultural products.

(acessed 15.4.15.). 〈http://stat.luke.fi/en/producer-prices-of-agricultural-products〉.

52. Natural Resources Institute Finland, 2015c Natural Resources Institute Finland, 2015c. Report generator. NFI11 (data collected2009–2013). (accessed 2.5.15.). 〈http://mela2.metla.fi/mela/tupa/index-en.php〉.
53. Nelson and Kennedy, 2008 J.P. Nelson, P.E. KennedyThe Use (and Abuse) of meta-analysis in environmental and natural resource economics: an assessment Environ. Res. Econ., 42 (2008), pp. 345-377, 10.1007/s10640-008-9253-5
54. Paracchini et al., 2014 M.L. Paracchini, G. Zulian, L. Kopperoinen, J. Maes, J.P. Schägner, M. Termansen, M.Zandersen, M. Perez-Soba, P.A. Scholefield, G. BidoglioMapping cultural ecosystem services: a framework to assess the potential for outdoor recreation across the EU Ecol. Indic., 45 (2014), pp. 371-385, 10.1016/j.ecolind.2014.04.018
55. Polizzi et al., 2015 C. Polizzi, M. Simonetto, A. Barausse, N. Chaniotou, R. Känkänen, S. Keränen, A. Manzardo, K. Mustajärvi, L. Palmeri, A. ScipioniIs ecosystem restoration worth the effort? The rehabilitation of a Finnish river affects recreational ecosystem services Ecosyst. Serv., 14 (2015), pp. 158-169, 10.1016/j.ecoser.2015.01.001
56. Primmer and Furman, Primmer, E., Furman, E. Operationalising ecosystem service approaches for governance: Do measuring, mapping and valuating integrate sector-specific knowledge systems? Ecosystem Services 1, 85–92. 〈http://dx.doi.org/10.1016/j.ecoser.2012.07.008〉.
57. Rasinmäki et al., 2009 J. Rasinmäki, A. Mäkinen, J. KalliovirtaSIMO: an adaptable simulation framework for multiscale forest resource data Comput. Electron. Agric., 66 (2009), pp. 76-84, 10.1016/j.compag.2008.12.007
58. Repola et al., 2007 J. Repola, R. Ojansuu, M. KukkolaBiomass functions for Scots pine, Norway spruce and birch in Finland Work. Pap. Finn. For. Res. Inst. (2007), p. 53
59. Robinson et al., 1989 W.S. Robinson, R. Nowogrodski, R.A. MorseThe value of honey bees as pollinators of U.S. crops Am. Bee

J., 129 (1989), pp. 477-478

60. Salminen and Rintala, 2002 E. Salminen, J. RintalaAnaerobic digestion of organic solid poultry slaughterhouse waste - a review (Review paper) Bioresour. Technol., 83 (2002), pp. 13-26
61. Schägner et al., 2013 J.P. Schägner, L. Brander, J. Maes, V. HartjeMapping ecosystem services' values: Current practice and future prospects Ecosyst. Serv., 4 (2013), pp. 33-46, 10.1016/j.ecoser.2013.02.003
62. Seppelt et al., 2011 R. Seppelt, C.F. Dorman, F.V. Eppink, S. Lautenbach, S. SchmidtA quantitative review of ecosystem service studies: approaches, shortcomings and the road ahead J. Appl. Ecol., 48 (2011), pp. 630-636, 10.1111/j.1365-2664.2010.01952.x
63. Seppälä et al., 2014 E. Seppälä, T. Hjerppe, M. MarttunenThe benefits of water pollution control in the Gulf of Finland – Water quality impacts on recreational use of water bodies estimated with VIRVA model Rep. Finn. Inst. (2014) (2014), p. 66
64. Seppälä et al., 2009 M. Seppälä, T. Paavola, A. Lehtomäki, J. RintalaBiogas production from boreal herbaceou grasses - specific methane yield and methane yield per hectare Bioresour. Technol., 100 (2009), pp. 2952-2958, 10.1016/j.biortech.2009.01.044
65. Sievänen and Neuvonen, 2011 Sievänen, T., Neuvonen, M., (Eds.), (2011). Luonnon virkistyskäyttö 2010. Working Papers of the Finnish Forest Research Institute, 212.
66. Statistics Finland, 2015 Statistics Finland, 2015. Consumer Prices of Domestic Fuels inEnergy Production. (accessed 2.2.15.). 〈http://pxnet2.stat.fi/PXWeb/pxweb/en/StatFin/StatFin__ene__ehi/020_ehi_tau_102_en.px/?Rxid=b6b5fc27-85e3-44a5-8762-d9de6a317100〉 .
67. Söderman et al., 2012 T. Söderman, L. Kopperoinen, P. Shemeikka, V. Yli-PelkonenEcosystem service criteria for sustainable development in urban regions J. Environ. Assess. Policy Manag., 14 (2012), pp. 1250008-1-125008-48, 10.1142/S1464333212500081
68. Tammi and Kalliola, 2014 I.Tammi, R. KalliolaSpatial MCDA

in marine spatial planning: experiences from the Mediterranean and Baltic seas Mar. Policy, 9 (2014), pp. 73-83, 10.1016/j.marpol.2014.03.015

69. Toivonen et al., 2004 A.L. Toivonen, E. Roth, S. Navrud, G. Gudbergsson, H. Appelblad, B. Bengtsson, P.TuunainenThe economic value of recreational fisheries in Nordic countries Fish. Manag. Ecol., 1 (2004), pp. 1-14, 10.1046/j.1365-2400.2003.00376.x
70. Tuomi et al., 2011 M. Tuomi, J. Rasinmäki, A. Repo, P. Vanhala, J. LiskiSoil carbon model Yasso07 graphical user interface Environ. Model. Softw., 26 (2011), pp. 1358-1362, 10.1016/j.envsoft.2011.05.009
71. Turtiainen et al., 2005 M. Turtiainen, K. Salo, O. SaastamoinenModel-based estimates of regional and national bilberry and lingonberry yields on mineral soils in Finland (Research Notes 167) University of Joensuu, Faculty of Forestry, Helsinki, Finland (2005)
72. Turtiainenetal.,2007M.Turtiainen,K.Salo,O.SaastamoinenNational and regional estimates of blueberry (Vaccinium myrtillus L.) and lingonberry (V. vitis-idaea L.) yields on peatlands in Finland Finn. Peatl. Soc., 58 (2007), pp. 87-98
73. Tyrväinen, 2001 L. TyrväinenEconomic valuation of urban forest benefits in Finland J. Environ. Manag., 62 (2001), pp. 75-92, 10.1006/jema.2001.0421
74. University of Jyväskylä, 2014 University of Jyväskylä, 2014. Geographic database for sport facilities. 〈https://www.jyu.fi/sport/laitokset/liikunta/liikuntapaikat/sport%20facility.en〉 .
75. Valtanen et al., 2015 M. Valtanen, N. Sillanpää, H. SeteläKey factors affecting urban runoff pollution under cold climatic conditions J. Hydrol., 529 (2015), pp. 1578-1589, 10.1016/j.jhydrol.2015.08.026
76. Vatn and Bromley, 1994 A.Vatn, D. BromleyChoices without prices without apologies J. Environ. Econ. Manag., 26 (1994), pp. 129-148, 10.1006/jeem.1994.1008
77. Vehviläinen, 1994 B. VehviläinenThe Watershed Simulation and Forecasting System in the National Board of Waters and the

Environment Publications of the Water and Environment Research Institute. National Board of Waters and the Environment, Finland (1994) (No. 17)

78. Vierikko and Niemelä, 2016 Vierikko, K., Niemelä, J., 2016. Bottom-up Thinking – Identifying Socio-cultural Values of Ecosystem Services in Local Blue-green Infrastructure Planning in Helsinki, Finland. 〈http://dx.doi.org/10.1016/j.landusepol.2015.09.031〉.

79. Vähäsarja, 2014 V. VähäsarjaAssessment of the financial value of the health and well-being benefits of natural environments Nat. Prot. Publ. Metsähall. Ser. A (2014), p. 210

4 MODELLING THE FUTURE IMPACTS OF CLIMATE AND LAND-USE CHANGE ON SUSPENDED SEDIMENT TRANSPORT IN THE RIVER THAMES (UK)

Gianbattista Bussi [a], Simon J. Dadson [a] , Christel Prudhomme [b,c] , and Paul G. Whitehead [a]

[a]School of Geography and the Environment, University of Oxford, South Parks Road, OX1 3QY Oxford, UK

[b]Centre for Ecology and Hydrology (CEH), Benson Lane, Crowmarsh Gifford, OX10 8BB Wallingford, UK

[c]Department of Geography, Loughborough University, Loughborough LE11 3TU, UK

ABSTRACT

The effects of climate change and variability on river flows have been widely studied. However the impacts of such changes on sediment transport have received comparatively little attention. In part this is because modelling sediment production and transport processes introduces additional uncertainty, but it also results from the fact that, alongside the climate change signal, there have been and are projected to be significant changes in land cover which strongly affect sediment-related processes. Here we assess the impact of a range of climatic variations and land covers on the River Thames catchment (UK). We first calculate a response of the system to climatic stressors (average precipitation, average temperature and increase in extreme precipitation) and land-cover stressors (change in the extent of arable land). To do this we use an ensemble of INCA hydrological and sediment behavioural models. The resulting system response, which reveals the nature of interactions between the driving factors, is then compared with climate projections originating from the UKCP09 assessment (UK Climate Projections 2009) to evaluate the likelihood of the range of projected outcomes. The results show that climate and land cover each exert an individual control on sediment transport. Their effects vary depending on the land use and on the level of projected climate change. The suspended sediment yield of the River Thames in its lowermost reach is expected to change by −4% (−16% to +13%, confidence interval, $p = 0.95$) under the A1FI emission scenario for the 2030s, although these figures could be substantially altered by an increase in extreme precipitation, which could raise the suspended sediment yield up to an additional +10%. A 70% increase in the extension of the arable land is projected to increase sediment yield by around 12% in the lowland reaches. A 50% reduction is projected to decrease sediment yield by around 13%.

Keywords: Sediment transport, Climate change, Land-use change, INCA model, River Thames and Scenario-neutral approach

INTRODUCTION

Climate change is expected to alter soil erosion and sediment transport processes, although the extent and magnitude of these variations are poorly understood (Peizhen et al., 2001; Pruski and Nearing, 2002). According to Nearing et al. (2004), changes in precipitation and temperature and their interactions with land use and vegetation cover are the main climate change-related stressors that are likely to affect sediment transport in the future. These factors are expected to alter sediment production and soil loss, as well as in-channel mobilisation of sediment, phosphorus and contaminants (Mullan et al., 2012). For example, sediment transport is strongly affected by extreme precipitation and river discharge, owing to the non-linear relation between water discharge and sediment transport rate (e.g., Julien, 2010). In many catchments, short and intense precipitation events are responsible for a large part of the total sediment transport (González-Hidalgo et al., 2010, 2013). Climate models predict a change in the behaviour of precipitation extremes. For the UK, extreme precipitation is forecast to increase in the next decades (Fowler et al., 2010; Palmer and Räisänen, 2002). Its impact on soil erosion has been assessed, for example, by Boardman (2015) and Burt et al. (2015), who detected an upward trend in average rainfall per rain day in southern England which could increase soil erosion.

Land cover changes also impact soil erosion and sediment mobilisation processes (Blöschl et al., 2007; Turner et al., 2007). Land use and land management are themselves also changing, mainly due to human interventions (Buendia et al., 2015; Evans et al., 2015; Rodríguez-Lloveras et al., 2016, 2015), but also as a result of the indirect effects of climate and other human-induced environmental changes (Mullan et al., 2012). For example, soil erosion in South Eastern England has been largely affected by the shift from grassland to arable land in the second half of the 20th century, due to mechanisation and intensification of agriculture (Boardman, 2003; Howden et al., 2013).

While the impacts of climate change on sediment transport are increasingly reported in the literature (e.g., Cerdà, 1998; Foster et al., 2012; Zhao et al., 2013), only a few studies have also considered simultaneous changes in land-use and land management (de Vente et al., 2013). This has been done in the literature by a scenario-type analysis (Mullan et al., 2012), where plausible future land-use scenarios were hypothesised and used to alter the model parameterisation under different climatic conditions (see for example Nunes et al., 2009). Furthermore, there is large uncertainty about the forecasted effect of climate change on sediment transport, given that previous studies have demonstrated that different emission scenarios can lead to opposite results (Bussi et al., 2014a). The effects of both climate change and land-use change on soil erosion and sediment transport should be analysed simultaneously, as they can have important synergistic or antagonistic effects. Eventual mitigation measures based on land management and land-use change, such as reduction of arable land, extension of forested areas and introduction of better agricultural practices (Haines-Young et al., 2014), must also be evaluated under the framework of climate change to assess their effectiveness and cost.

In order to analyse the non-linear interactions between climate-driven processes that affect sediment transport at the catchment scale, hydrological and sediment models have typically been used along with climate projections from global circulation models and regional climate models (e.g., Nearing et al., 2005; Ito, 2007; and Mouri, 2015). In these approaches, climate model outputs (mainly precipitation and temperature series) obtained under specific greenhouse gas emissions scenarios are used to drive regional or catchment-scale mathematical models, which in turn provide predictions of the variable of interest (e.g., water or sediment discharge) under the climatic scenarios considered. In the field of soil erosion and sediment transport research, this approach was used recently in Nunes et al. (2009), Coulthard et al. (2012), Bangash et al. (2013), Mullan (2013), Bussi et al. (2014), Routschek et al. (2014), Francipane et al. (2015), Paroissien et al. (2015) and Simonneaux et al. (2015). As mentioned before, some of these studies also incorporated an analysis of the system response to different land-cover scenarios. The majority of these studies found a strong dependence on the climatic scenario concerned, with the different sensitivity of the response depending on combinations of climate and land-cover (Mullan, 2013; Nunes et al., 2009). Some of them also found that climate change-induced land-use change and soil management exert a larger control on soil erosion rates than climate variability (Routschek et al., 2014).

Here we employ an alternative, scenario-neutral approach, which is based on the definition of relevant climatic stressors that affect the variable of interest to quantify the joint effect of climate change and land-use change whilst at the same time evaluating uncertainty associated with the choice of climate scenario. These climatic stressors – potentially including, for example, average temperature and precipitation, precipitation intensity, seasonality and the occurrence of extremes – are perturbed within a Monte Carlo framework to establish the sensitivity of the model's outcomes to their variation. The framework is designed explicitly to quantify interactions between climatic variables and land use. The model results then form a response surface which can be compared with changes predicted using climate models. Recent applications of this scenario neutral method within hydrology, water resources and water quality research are reported by Bastola et al. (2011), Fronzek et al. (2011), Wetterhall et al. (2011), Brown et al. (2012), Brown and Wilby (2012), Prudhomme et al. (2013), Poff et al. (2015), Prudhomme et al. (2015) and Bussi et al. (2016). The scenario-neutral methodology has some important advantages for sediment-oriented studies. For example, it allows exploration of the system's resilience to the full range of possible climate scenarios independently from individual climate modelling results. This can be very important in sediment transport studies as it enables to highlight the climate drivers associated with critical thresholds in the system, due to the non-linearity of the processes. Such thresholds might not appear when using a conventional top-down approach if the drivers are outside the range of available climate change projections. The response surface also acts as a tool for decision-makers that can be used to explore a wide range of possible sensitivities within the system to guide low- or no-regrets adaptation measures.

In this paper we assess the effects of climate change and land-use change on the sediment transport of the River Thames (UK), which has tangible value for human water consumption, for the ecosystem and for conservation. We present an extension of the scenario-neutral methodology which accounts for the joint effects of different climatic stressors and land-use scenario. Using the hydrological and water quality model INCA (Whitehead et al., 1998a; Lázár et al., 2010) we quantify the response of the specific suspended sediment yield (SSY; i.e. the average mass of suspended sediment transported over a year per unit area) to changes in annual average precipitation, extreme precipitation and annual average temperature. This analysis is repeated under four different scenarios of land use: current land use, a future scenario describing an expansion in arable land (Castellazzi et al., 2010; Crossman et al., 2013), a future scenario considering a substantial reduction in the arable land (−50%) and a theoretical scenario of total agriculture abandonment. The resulting response surfaces are compared with future climatic scenarios (Murphy et al., 2007) to establish the likelihood of changes of different magnitudes and to test the hypothesis that climate and land-cover changes exert a joint control on soil erosion and sediment transport.

In this study, we intend to propose a new methodological approach based on the scenario-neutral methodology, which can be used to quantify the impact of land-use change under different climatic scenarios, taking into account the model parametric uncertainty. Using this novel tool, we address the following research questions:

- What climatic variable(s) exert the strongest control on soil erosion and sediment transport in the River Thames catchment?
- What are the interactions and feedbacks between different climatic variables and land use and what is their effect on sediment transport?
- What is the role and extent of land-use change in affecting sediment transport under a changing climate and how can it be used to contrast sediment transport?

STUDY AREA

The Thames catchment

The study area of this paper is the River Thames (Figs. 1 and 9, 927 km^2), located in southern England and draining toward the city of London. Its water is used for freshwater supply to fourteen million people (Whitehead et al., 2013) and its non-tidal section (upstream of Teddington Weir) receives treated wastewater from approximately three million inhabitants (Kinniburgh and Barnett, 2009). The climate of the River Thames catchment is temperate with both Atlantic and continental influences. The annual precipitation is 730 mm per year (calculated between 1985 and 2014, with a minimum of 545 mm in the year 1996 and a maximum of 950 mm in the year 2000) and the average temperature is 10.7 °C (between 1985 and 2014, minimum: 9.3 °C in 1986, maximum 12.1 °C in 2014), with a difference of around 2 °C between the uplands and the lowlands. The average flow is 67 $m^3\ s^{-1}$ (20–115 $m^3\ s^{-1}$), with a Q95 of 206 $m^3\ s^{-1}$. High flows usually occur in winter to early spring and low flows in summer to late autumn. The geology of the catchment is dominated by a chalk strip that crosses the catchment in its central part from east to west. The headwaters are composed predominantly of limestone, and clay/mudstone and sandstone are also present both upstream and downstream of the chalk strip (Bloomfield et al., 2011). The catchment is characterised by arable land use in its upper part (around 80% of the catchment draining to reach 4 in Fig. 1 is dedicated to agriculture and grassland), with little urban land in the headwaters but with intensively urbanised areas in the lowlands. A non-trivial fraction of the catchment is covered by forest (13% of the catchment drained by reach 19 in Fig. 1).

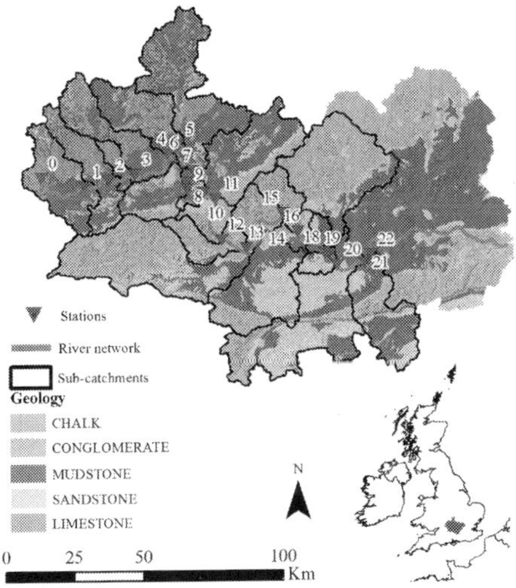

Figure 1: Location and characteristics of the River Thames (UK). The geological map is a simplified version of the 1:50,000 geology map of England by the British Geological Survey (Smith, 2013).

Data

Meteorological inputs for the hydrological and water quality model, consisting of daily precipitation and temperature time series, were obtained from the UK Met Office (Met Office, 2012). The daily precipitation, minimum temperature and maximum temperature from all the available stations within the Thames catchment were interpolated on a 5 × 5 km grid using the Thiessen polygon method, and then the daily average precipitation and temperature series were computed and used as model input. Land cover data were obtained from Fuller et al. (2002). For the hydrological sub-model model calibration and validation, records of continuous daily water discharge at the downstream section of three of the four INCA reaches were obtained from the National River Flow Archive (NRFA, ceh.ac.uk/data/nrfa/). The sediment sub-model was calibrated using different datasets, including weekly observations of suspended sediment concentration from the Thames Initiative research platform dataset (Bowes et al., 2012), collected by the UK Centre of Ecology and Hydrology (CEH). The suspended concentration was measured by collecting a single water sample for each measurement. The samples were collected using a suspended sediment sampler, positioning the sampler intake around the 60% of the river depth (when possible). The samples were then stored in plastic bottles and subsequently taken to the laboratory. The laboratory analysis was carried out by filtering the sample through microfibre filters. The papers were then dried, weighed, and suspended solid concentrations were calculated. The suspended sediment sample were collected with a weekly frequency, from March 2009 until May 2014, in six stations along the main stem of the River Thames, corresponding with the stations 1, 3, 4, 11, 13 and 19. The samples were collected regardless of the stage, flow or season.

METHODOLOGY

The INCA model

In this study, the INCA hydrological and water quality model was employed to reproduce the water and sediment dynamics of the River Thames (UK). The INCA model was initially developed as a nitrogen (Whitehead et al., 1998a) and phosphorus (Wade et al., 2002b) model, although several other sub-models were added later, such as a soil erosion and sediment transport sub-model (Lázár et al., 2010). The hydrological and water quality sub-models of INCA have been applied to several basins across the UK and Europe, and, in particular, to the River Thames catchment (Bussi et al., 2016; Crossman et al., 2013; Jin et al., 2012; Lu et al., 2016; Whitehead et al., 2015b, 2013). INCA is a semi-distributed process-based model which simulates the transformation of rainfall into runoff and the propagation of water through a river network (Wade et al., 2002a). Its inputs are daily time series of precipitation, temperature, hydrologically effective rainfall, and soil moisture deficit. The latter two are estimated using another semi-distributed hydrological model, called PERSiST (Futter et al., 2014). PERSiST is a semi-distributed catchment-scale rainfall-runoff model which is specifically designed to provide input series for the INCA family of models. It is based on a user-specified number of linear reservoirs which can be used to represent different hydrological processes, such as snow melting, direct runoff generation, soil storage, aquifer storage and stream network movement.

The description of its application to the river Thames can be found in Futter et al. (2014).

The sediment sub-model of INCA has been used in several studies with focus on soil erosion and sediment transport (Farkas et al., 2013; Jarritt and Lawrence, 2007, 2006; Lázár et al., 2010; Rankinen et al., 2010; Whitehead et al., 2010). It is also a component of the phosphorus, carbon, pathogen and organic contaminant versions of the INCA model, due to absorption processes and interaction with bed sediments (Crossman et al., 2013; Futter et al., 2007; Lu et al., 2016; Nizzetto et al., 2016; Wade et al., 2002b; Whitehead et al., 2015b). A sensitivity/uncertainty analysis of its structure can be found in Jackson-Blake and Starrfelt (2015). This model was selected because of its simple structure and relatively low data demands, but also because of the large number of previous applications in the study area and in other catchments. It is structured in two parts: the land-phase model, which simulates the soil erosion and sediment production and transport processes on the hillslopes, and the in-stream compartment, which reproduces the processes and storage within the river reaches. In the land-phase model (Jarritt and Lawrence, 2007), the sediment production is caused by splash detachment, sheet erosion and rill erosion. Soil is eroded from an unlimited pool of parental material and driven to a pool of readily available sediment. The splash detachment (S_{SP}, kg km^{-2} day^{-1}) is estimated as in Eq. (1):

$$S_{SP} = c_{x1} p (E_{SP})^{\left(\frac{10}{10-v}\right)} \cdot 8.64 \cdot 10^{10} \tag{1}$$

S_{SP} is function of precipitation (p, mm), a soil-specific erosion potential parameter (E_{SP}, kg m^{-2} s^{-1}), a vegetation cover index (v, –) and a scaling parameter (c_{x1}). Analogously, the net detachment due to flow erosion (sheet and rill erosion) is calculated depending on another soil-specific erosion potential parameter (E_{FL}, kg m^{-2} s^{-1}), following a similar formulation but depending on direct runoff (q_{DR}, m^3 s^{-1} km^{-2}) rather than precipitation. A function K is defined as follows (Eq. (2)):

$$K = a_1 E_{FL} \left(\frac{A q_{DR}}{L} - a_2 \right)^{a_3} \cdot 86,400 \tag{2}$$

where a_1, a_2 and a_3 are calibration parameters (s m^{-1}, m^2 s^{-1} and non-dimensional), A the catchment area (km^2) and L the reach length (m). q_{DR} is used to simulate rill formation due to intense rainfall and runoff and can be set to a different value depending on the land use, in order to activate rill erosion where it is important (arable lands) and deactivated where the vegetation cover does not allow rill formation.

Then the flow erosion (S_{FL}, kg km^{-2} day^{-1}) is defined as follows (Eq. (3)):

$$S_{FL} = \frac{K(S_{TC} - S_{SP})}{(S_{TC} + K)} \tag{3}$$

where S_{TC} (kg km^{-2} day^{-1}) is the maximum overland flow sediment transport capacity (i.e. the maximum amount of material that can be transported from the land-phase to the stream). This is estimated depending on the catchment characteristics and the direct runoff (Eq. (4)):

$$S_{TC} = a_4 \left(\frac{Aq_{DR}}{L} - a_5 \right)^{a_6} \cdot 86,400 \tag{4}$$

where a_4, a_5 and a_6 are calibration parameters (kg m^{-2} km^{-2}, m^2 s^{-1} and non-dimensional, respectively). The mass of sediment transported to the reach depends on the availability of readily available sediment and the overland flow transport capacity. Erosion by other sources, such as farm tracks, is not explicitly considered, although it is reasonable to assume that it is included in the parameterisation of arable areas, if the model is correctly calibrated. Soil crusting is also not included in the INCA model conceptualisation.

The suspended sediment then enters the stream via surface runoff. The transport capacity of the flow is used to separate this material into deposited and suspended material and to route the suspended part towards downstream. Most of the existing sediment models are based on similar transport capacity/material availability concepts (Aksoy and Kavvas, 2005; Bussi et al., 2013). Depending on the size of the sediment and the turbulence of the river, sediment begins to settle to the river bed and accumulate. At the same time, in-stream flows entrain settled sediment, depending on the availability and the transport capacity. Five categories of grain size are considered: clay (0–0.002 mm), silt (0.002–0.06 mm), fine sand (0.06–0.2 mm), medium sand (0.2–0.6 mm), and coarse sand (0.6–2 mm). The river shear velocity, computed depending on water depth, slope and a calibration parameter (a_7, –), is used to determine the maximum entrainable grain size. This maximum grain size is used to estimate the proportion of each size class that can be potentially entrained from the bed sediment mass. Effective entrainment of each sediment class is calculated using Bagnold (1966) stream power equation (Eq. (5)), which is widely known and used in similar applications:

$$\omega = \rho_f g d v s = \rho_f g \frac{Q}{W} s \tag{5}$$

where ω is he stream power per unit area of bed (J s^{-1} m^{-2}), ρ_f is the density of water (kg m^{-3}), g the gravitational acceleration (m s^{-2}), v is the flow velocity (m s^{-1}), s is the channel slope (–), Q is the water discharge (m^3 s^{-1}) and W is the channel width (m). The effective entrainment (m_{ent}, kg m^{-2} s^{-1}) is calculated as follows (Eq. (6)):

$$m_{ent} = a_8 m_{bed} M_{PROP} \omega f \cdot 86,400 \tag{6}$$

where m_{bed} is the bed mass per unit area (kg m^{-2}), M_{PROP} is the proportion of grain size class that can be entrained (–), f a friction factor (–) calculated as the ratio of the actual hydraulic radius and the maximum hydraulic radius of a channel with the same width and a_8 a calibration parameter (entrainment coefficient, s^2 kg^{-1}). In the river channel, sediment settles to the river bed and accumulates, depending on the size of the sediment and the turbulence of the river. At the same time, in-stream flows mobilise settled sediment. These equations allow taking into account in-channel processes such as deposition, storage and re-entrainment. Background release of material from gullies and channels is also taken into account as a non-linear function of the river discharge depending on a calibrated coefficient (release scaling factor, a_9, kg m^{-1} m^{-3}) and a calibrated exponent (release non-linear coefficient, a_{10}, non-dimensional).

The model can be calibrated by adjusting its parameters (Lázár et al., 2010). Some of the most influential hydrological model parameters are the direct runoff, soil water and groundwater residence times, which control the hydrological response of the catchment (i.e., the overland flow, sub-superficial flow and base flow velocities), the maximum soil moisture deficit (see Whitehead et al., 1998b) and the flow routing parameters (a, coefficient and b, exponent, used to calculate flow velocity in the river channel, see Whitehead et al., 1998b). The model also has several sediment parameters. For example, the sediment production is controlled by the splash and flow erosion potential parameters (E_{SP} in Eq. (1) and E_{FL} in Eq. (2)), the splash erosion scaling parameter (c_{x1} in Eq. (1)) and the flow erosion calibration parameters (a_1, a_2 and a_3 in Eq. (2)). The transport of material from the hillslope to the channel network is controlled by the transport capacity calibration coefficients (a_4, a_5 and $a_6$in Eq. (3)). The sediment transport and deposition in the river channel is controlled by the shear velocity coefficient (a_7), the entrainment coefficient (a_8 in Eq. (6)) and two background release calibration coefficients (a_9 and a_{10}).

The INCA model has already been applied to the River Thames catchment (Crossman et al., 2013; Jin et al., 2012; Lu et al., 2016; Whitehead et al., 2015b, 2013). In this study, the same model structure is proposed, where the catchment is divided into 22 sub-catchments and the river into 22 corresponding reaches. For each of them, different parameters are considered. For example, topography is considered through the average slope of the sub-catchment and the slope of the channels, sub-catchment shape is considered through the use of the ratio between area and length, soil texture is considered as an input parameter, variable depending on the sub-catchment and on the land use (and it is then used to separate the material routed to the channel network into different size classes) and geology is taken into account by employing different base flow index values depending on the sub-catchment. The following land-use categories were considered: urban, arable, grassland and pasture, wetlands and forest land.

Model general sensitivity analysis

A general sensitivity analysis was applied to the INCA model of the River Thames (Spear and Hornberger, 1980; Whitehead et al., 2015a). Following a preliminary sensitivity analysis, and based on the modeller's knowledge, the following parameters were selected as the most influential and the sensitivity of the model results to them was analysed: direct runoff, soil water and ground water residence times, rainfall excess proportion, maximum infiltration rate, flow-velocity coefficient, flow threshold for saturation excess direct runoff, flow erosion direct runoff threshold (q_{DR}), splash detachment soil erodibility parameter (E_{SP}), flow erosion soil erodibility parameter (E_{FL}), transport capacity scaling factor (a_1), transport capacity non-linear coefficient (a_3), channel entrainment coefficient (a_8), release scaling factor (a_9) and release non-linear coefficient (a_{10}). E_{SP}, E_{FL} and q_{DR} are land-use-specific, although some expert knowledge-based rules were set to constrain their values, such as for example that E_{SP} and E_{FL} for arable land must be greater than E_{SP} and E_{FL} for grassland. The ranges of variation of the model parameters were also based on the modeller's knowledge previous studies (Lázár et al., 2010; Whitehead et al., 2010), although they were kept reasonably broad. The feasible space of model parameters was sampled randomly, and 10,000 different parameter sets were generated. Subsequently, the INCA model was run with each of these parameter sets, and its performance was assessed based on observed values of flow and sediment at two stations (reach 4 and reach 19), using data from 2010 to 2014. The metric used for model assessment was the Nash and Sutcliffe Efficiency (NSE - Nash and Sutcliffe, 1970). Thresholds of NSE values were used to split the 10,000 parameter sets into behavioural and non-behavioural (Spear and Hornberger, 1980). In particular, a threshold of 0.6 for the flow and a threshold of 0.1 for the suspended sediment concentration were used. Following the model evaluation guidelines of Moriasi et al. (2007), the flow threshold corresponds to a "good" model performance. The selected behavioural models were used in the rest of the study, providing ensemble results of flow and suspended sediment concentration.

Scenario-neutral methodology for climate change analysis

In contrast with top-down approaches to climate change studies, which use climate model outputs to drive hydrological and environmental models, the scenario-neutral method takes a bottom-up approach in which vulnerability ranges of a given hydrological or environmental indicator are defined. A response surface is then produced which depicts the changes in the relevant indicator subject to the range of climatic and other environmental changes under consideration (Singh et al., 2014). The likelihood of these changes is assessed by integrating information about future climate into the results of this methodology (Prudhomme et al., 2010). A schematic diagram showing the method used in this study is given in Fig. 2. First, the climatic stressors most likely to impact SSY were identified. Plausible changes in these climatic stressors, described in Section 3.4, were then applied to the current climatic observed series of daily precipitation and temperature from 1999 to 2015. This allowed the creation of a large set of perturbed input time series (precipitation and temperature) which were used to drive the INCA model. The INCA model, driven with the altered time series, produced a set of time series of daily water discharge and suspended sediment concentration, corresponding to each perturbed input series from which a corresponding SSY value was calculated. This procedure was repeated using four land use and land management scenarios (baseline, increase in arable land, decrease in arable land and agriculture abandonment) described in Section 3.5.

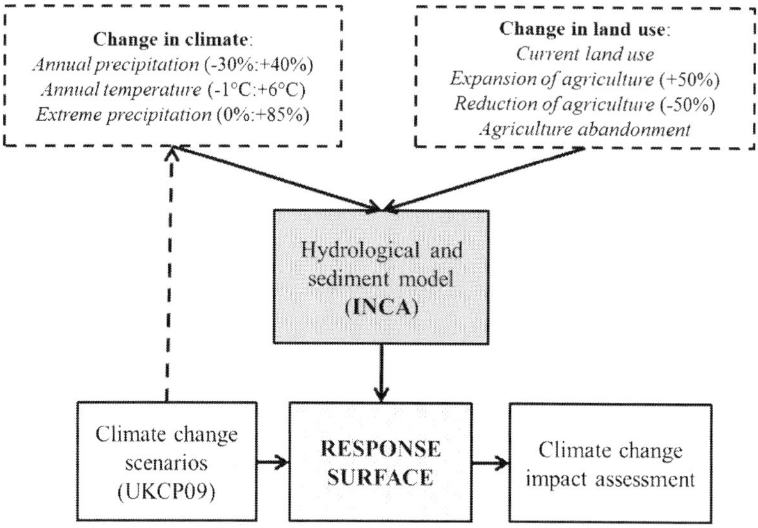

Figure 2: Conceptual scheme of the scenario-neutral methodology.

Climatic alterations

In order to produce useful results, the choice of climatic alterations used to construct the response surface should be restricted to the main climatic stressors that affect the variable of interest, but they must also sample the full range of possible climate futures effectively (Prudhomme et al., 2010). The climatic variables which exert the strongest controls on river flow are precipitation and temperature (which governs the spatial and temporal variations of soil moisture and evapotranspiration). Suspended sediment yield is most strongly affected by precipitation and river flow, which control soil erosion and in-channel processes of sediment mobilisation and deposition (Julien, 2010). However, it is widely known that suspended sediment entrainment and transport occurs disproportionately during precipitation and flow extremes (Boardman, 2015), due to the non-linear relationship between flow and sediment transport by water (Julien and Simons, 1985) and so we also consider the effect of changes in extremes.

The projected changes in total precipitation and average temperature following the UKCP09 projection are shown in Fig. 3a for the study region. They are in the range of −20% to +20% for the annual precipitation and 0 °C to +2 °C for the temperature. However, we considered a broader set of changes (not shown in Fig. 3a): changes in average temperature between −1 °C and +6 °C and changes in precipitation between −30% and +40%. In each case the range of possible changes was divided uniformly into fifteen divisions. The resulting set of 225 altered precipitation and temperature time-series were generated by applying a uniform "delta change" transformation (Hay et al., 2000) to observed daily precipitation and temperature values accordingly.

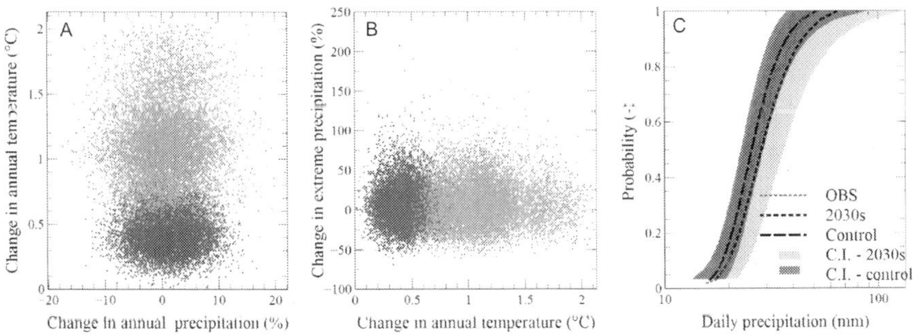

Figure 3: (a) Change in annual temperature and annual precipitation, following UKCP09 (blue: 2030s, orange: 2050s). (b) Change in annual temperature and extreme precipitation (90% percentile of the annual maximum daily precipitation distribution function), following UKCP09 (blue: 2030s, orange: 2050s). (c) Empirical cumulative distribution functions of the annual maximum daily precipitation (C.I.: 95% confidence interval from UKCP09 projections). Control: data from UKCP09 projections for the period 1961–1990, 2030s: data from UKCP09 projections for the period 2020–2049, OBS: data from the observed precipitation and temperature series for the period 1999–2015. For visualisation purposes, the UKCP09 data were taken from 10,000 transient stochastic daily series of precipitation and temperature produced by Glenis et al. (2015). (For interpretation of the references to color in this figure legend, the reader is referred to the web version of this article.)

The distribution of future changes projected by the UKCP09 probabilistic sample is given in Fig. 2b and c for the study region. As variations in extreme precipitations are not a standard product of UKCP09, these were obtained by analysing 10,000 transient stochastic daily series of precipitation and temperature produced by Glenis et al. (2015) and also used by Borgomeo et al. (2014). These are daily time series of precipitation and temperature from 1950 to 2060. Therefore, it was possible to estimate the maximum annual precipitation values for the control period (1960–1990) and for the future period (2030s and 2050s), compute an empirical cumulative distribution function for both time periods and calculate the difference. In Fig. 2b, the variations of temperature and extreme precipitation (expressed as the 90% percentile of the annual maximum daily precipitation distribution function) are depicted, following the UKCP09 over two different temporal horizons: 2030s and 2050s. In Fig. 2c, the empirical cumulative distribution function of the annual maxima of daily precipitation (CDF) is represented, both for the control period (1961–1990, median: broken line, 95% confidence interval: grey shaded area) and for the future period (2020–2049, median: dotted line, 95% confidence interval: violet shaded area), calculated from the 10,000 climate projections from the UKCP09 as described above. A shift towards higher values of annual maximum daily precipitation is projected.

Owing to the importance of hydrological extremes events for erosion processes (Wolman and Miller, 1960; González-Hidalgo et al., 2010), changes in extreme precipitation are also considered in this study. Specifically, extreme precipitation events are defined as events with daily rainfall above 15.7 mm, which is the minimum of the annual maxima of observed daily precipitation from 1960 to 2015. The changes were implemented by altering the baseline daily precipitation time series following a transformation function based on the empirical quantile mapping approach (Déqué, 2007). In order to explore a reasonable range of alteration in extreme precipitation values, two transformation functions were used. The two transformation functions used in this study were based on changes in extreme precipitation forecasted by the UKCP09. The first alteration, or transformation function, corresponds to a small but likely increase in extreme precipitation. Specifically, the median change forecasted by the UKCP09 was selected (e.g. a change larger than the change forecasted by 5000 out of 10,000 UKCP09 scenarios, i.e. probability = 50%). The second alteration corresponds to a larger but more unlikely shift in extreme precipitation than the first one. Specifically, a change larger than the change forecasted by 9750 out of 10,000 UKCP09 scenarios was selected (i.e., probability = 2.5%). For the sake of readability, these two extreme precipitation scenarios were called "small" and "large" increase in extreme precipitation, respectively. The two transformation functions are showed in Fig. 4 and Table 1. They are consistent with previous studies, including Fowler and Ekström (2009) who reported a change of −10% to +20% in the summer 10-day 5-year return period precipitation in South East England, and a range of 0 to +20% in winter. These transformation functions were applied on the precipitation days larger than the threshold specified above, and therefore they also alter slightly the total precipitation, by 0.9% and 4.5% respectively.

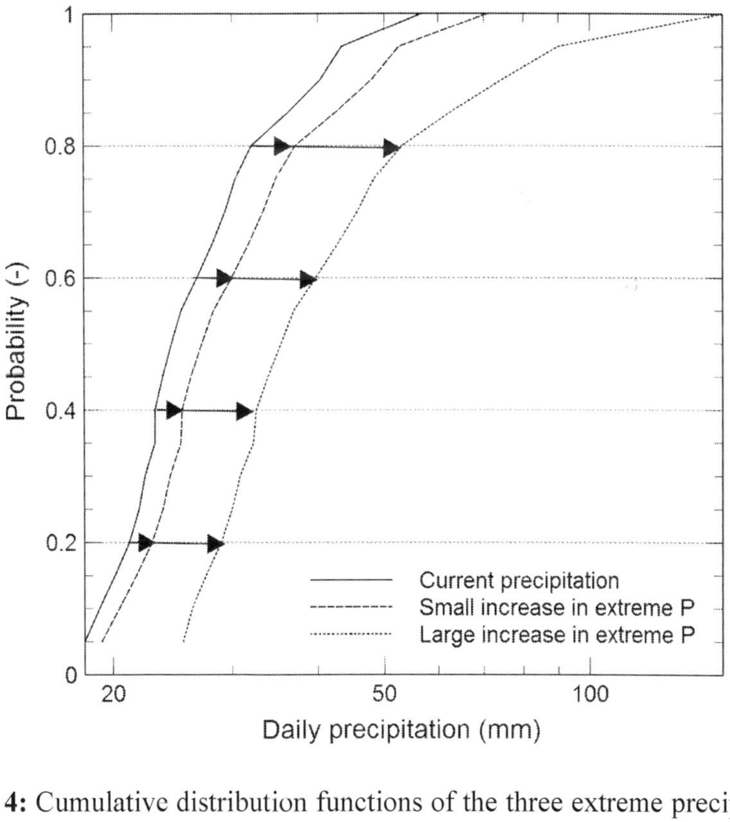

Figure 4: Cumulative distribution functions of the three extreme precipitation scenarios considered in this study.

Table 1: Extreme precipitation transformation function. The quantile is referred to the probability of the empirical distribution function of the observed annual maximum of daily precipitation

Quantiles (%)	Current precipitation (mm day^{-1})	Small increase of extreme precipitation (%)	Large increase of extreme precipitation (%)
10	20.0	+6.9	+37.2
20	21.9	+8.0	+36.9
30	23.1	+9.0	+38.2
40	24.8	+9.9	+41.2
50	26.9	+11.0	+44.8
60	29.2	+12.2	+48.9
70	31.1	+13.7	+55.9
80	35.5	+15.8	+65.9
90	40.9	+18.9	+84.6
95	43.5	+21.3	+107.7

One of the limitations of the scenario-neutral methodology is the reduction of number of stressors considered for a best readability and easier interpretation (here: 2 stressors only). This means that not all climate-related stressor that affect sediment transport have been considered. Other climatic stressors exist that might vary in the future and could have an impact on SSY. For example, a change in precipitation seasonality could change the wetting and drying cycle of the soil and therefore change soil properties such as infiltration and erodibility (Cerdà, 1999, 1997), in-channel sediment storage (Bussi et al., 2014b; Collins and Walling, 2007; Duijsings, 1986; López-Tarazón et al., 2011; Piqué et al., 2014) as well as affecting river flow (Prudhomme et al., 2010). A variation in the number of rainy days also could have an effect on sediment transport, in the case of both an increase or a decrease in the number of days of precipitation (Nearing et al., 2004). However, one of the advantages of this methodology is that it allows isolating the effects of single stressors leading to conclusions that can drive decision making. Because changes in the stressors mentioned here are expected to have a secondary role compared with the effect exerted by changes in average precipitation and temperature and changes in extreme precipitation, and because the effects of drying and wetting cycles on soil crusting and soil properties are not accounted for in the modelling of sediment delivery and transport model, they were not considered in this study.

Land-use variations

In this study we considered four land cover scenarios: (i) present day land use (called "baseline" scenario); (ii) arable land expansion; (iii) arable land reduction; and (iv) agriculture abandonment. Present day land use (Fig. 5) was obtained from the UK Land Cover Map 2007 (Smith et al., 2007). The catchment headwaters are dominated by arable land, while urban and forest land uses assume more importance in the lowlands. The arable land expansion scenario was defined according to the land cover model LandSFACTS (Castellazzi et al., 2010), which focuses on crop arrangement under increasing population, considering food security as a dominant driving force for land use change. Arable conversion from other land uses was permitted only on prime land and only from areas previously identified as grasslands (Crossman et al., 2013), and resulted in an increase from 3526 $km^2$to 5988 km^2 (over 9927 km^2 catchment area) of arable land (70% increase). This means that the portion of arable land increases from 36% to 60% for the Thames at Teddington (reach 22) (Fig. 5). This scenario was introduced into the INCA model by altering the fractions of the catchment assigned to each land use. The INCA model allows runoff production, soil erosion and sediment delivery parameters to be changed depending on the land use, thus representing the effect of changing land use on sediment transport. The arable land reduction scenario was set up by reducing the arable land by 50%, increasing forest land by 20% and assigning the remaining land to grassland. This was done in order to analyse the effect of reducing the arable land as a strategy to reduce sediment export by the River Thames. The agricultural abandonment scenario was implemented by setting arable land to 0%, increasing forest land by 20% and assigning the remaining land to grassland. This theoretical scenario was considered with the aim of analysing the hypothetical sediment transport response of the catchment if it was to be returned to a more natural status. This scenario, although highly unlikely in the foreseeable future for the Thames catchment, has already taken places in other parts of the world, such as the Spanish Pyrenees (Gallart and Llorens, 2004).

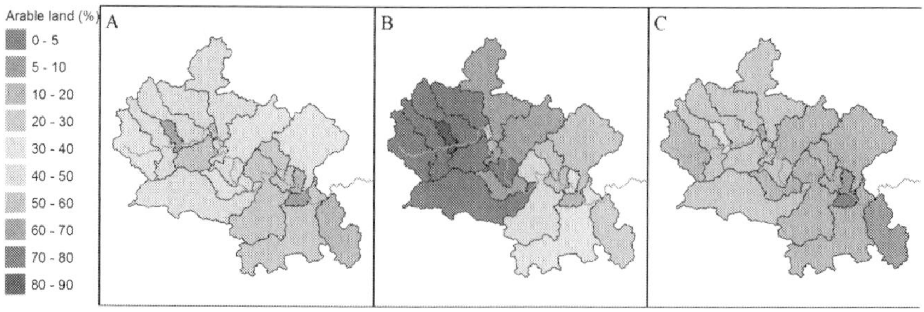

Figure 5: Proportion of arable land: (a) baseline, (b) arable land expansion, (c) arable land reduction. The agriculture abandonment scenario is not represented.

The 675 simulations described above (i.e. simulations driven by all possible combinations of 15 altered precipitation time series, 15 altered temperature time series and 3 alterations of the extreme precipitation) were repeated four times, one for each land-use scenario, and their results were compared to assess the net impact of land-use change on sediment transport under a changing climate.

Likelihood of predicted changes

The UKCP09 probabilistic change factor scenarios were developed by the UK Met Office to provide climate change projections of climate change over the UK with greater spatial and temporal detail than previous climate scenarios, but accounting for important uncertainties in Global Climate Models. These projections are based on the results of the HadCM3 coupled ocean-atmosphere Global Circulation model (Gordon et al., 2000), which was run as a perturbed physics ensemble to sample model and parameter uncertainties (Murphy et al., 2007). HadCM3 projections were downscaled on a 25 km grid over seven overlapping 30-yr time periods based on an ensemble of 11 variants of the regional climate model HadRM3, and a statistical procedure was applied to build local-scale distributions of changes for various climate variables. UKCP09 gives projections for each of three of the IPCC's Special Report on Emissions Scenarios (SRES) scenarios (A1FI - called High in UKCP09, A1B - Medium and B1 - Low). Among the available outputs, expected changes in average precipitation and temperature following the different emission scenarios are given (change factors). In the present study, we assess the risk of changes in SSY by comparison with climatic properties taken from a set of 10,000 change factors (Murphy et al., 2007) under the A1FI emission scenario for two temporal horizons: 2030s and 2050s. The A1FI scenario was chosen as it is the most severe scenario available, but one of the strength of the scenario-neutral methodology is that the scenario could be easily replaced without having to re-run all the simulations.

RESULTS

Model implementation

The Monte Carlo general sensitivity analysis resulted in 21 behavioural models. Fig. 6 shows their results for the River Thames at the reach 4, both in terms of flow and suspended sediment concentration. The results of the sediment sub-model are also shown in Fig. 7, where the distribution functions of the observed values of suspended sediment concentration are compared to the distribution functions of the modelled values. Note that the modelled values were resampled with the same time frequency as the observed values (e.g. weekly) for consistency. The model results indicate a SSY of 0.030 (0.020–0.063) Mg ha^{-1} yr^{-1} for reach 19 and 0.033 (0.016–0.053) Mg ha^{-1} yr^{-1} for reach 4 for the period 2010–2014.

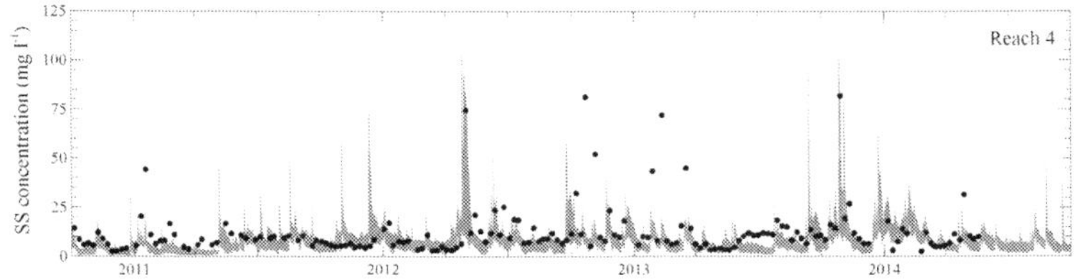

Figure 6: Water discharge model results (comparison with water discharge records from NRFA) and suspended sediment concentration model results (comparison with suspended sediment concentration values from the Thames Initiative dataset). The black broken line is the observed water discharge, the black dots the measured suspended sediment concentration values and the grey areas the model results.

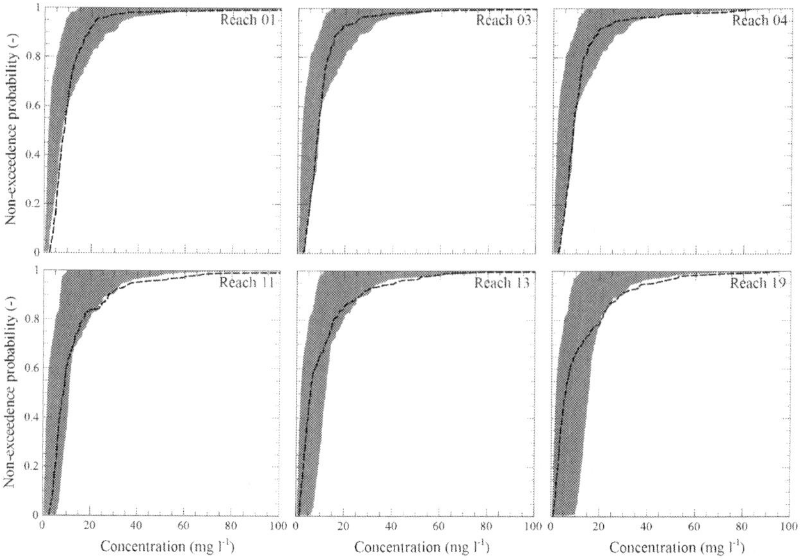

Figure 7: Cumulative distribution functions of suspended sediment concentration. The black broken lines are the observed suspended sediment concentration distribution functions, based on the available samples, the grey areas the modelled suspended sediment concentration distribution functions.

Fig. 7 also shows the spatial validation of the model. It can be observed that the results of the model at reach 1, 3, 11 and 13 (i.e. the reaches not used for behavioural model selection) are good in terms of reproduction of the observed distribution function of suspended sediment concentration. Spatial validation showed model biases of 0–27% for reach 1 and 4–65% for reach 13 regarding suspended sediment concentration.

Climate change impact under current land use

The precipitation and temperature changes considered in this study largely affect river flows, causing alterations in the average water discharge of the River Thames at reach 19 from −50% (caused by the combination of 30% reduction in precipitation and 6 °C increase in temperature) to +83% (caused by the combination of 40% increase in precipitation and 1 °C decrease in temperature). The increase in extreme precipitation considered causes a further increase of water discharge (around +1% in the case of a small increase in extreme precipitation, around +5% in the case of a large increase in extreme precipitation). The simulated SSY varies from 0.010 to 0.148 Mg ha^{-1} yr^{-1} for the Thames at reach 19, under uniform changes of precipitation and temperature and no increase in extreme precipitation. The small increase in extreme precipitation scenario considered in this study is responsible for an additional average increase of SSY of 2%, given the same condition of uniform precipitation and temperature change, while the large increase in extreme precipitation scenario causes an average additional SSY increase of 11%. Note that here we do not suggest that any of these changes is likely, neither that they are meteorologically plausible; we simply calculate them to be the catchment's response to such a combination of changes were they ever to occur.

Fig. 8 shows the results of the scenario-neutral methodology in terms of SSY response to climatic variations for station 4 under current land use. The top plots represent the median change in SSY and the bottom plots show the standard deviation of the change in SSY. The left-hand side plots represent the response of the system to changes in annual precipitation and temperature, with no changes in extreme precipitation, the central plots represent the response of the system to changes in annual precipitation and extreme precipitation, with no changes in annual temperature, and the right-hand side plots the response of the system to changes in annual temperature and extreme precipitation, with no changes in annual precipitation.

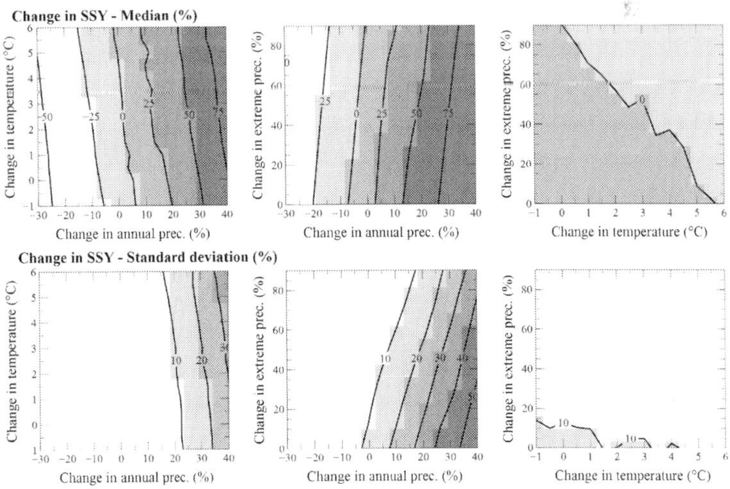

Figure 8: Response of the system to climatic alterations. Top plots: median change in SSY change, bottom plots: standard deviation of the change in SSY.

Climate change impact under different land use scenarios

The land-use change scenarios considered in this study have little effect on mean river flows compared to the impact of the climatic stressors, with a slight reduction in water discharge (estimated to be −0.2% by the INCA model), owing to an increase in evapotranspiration. By contrast, the impact of land-cover change on suspended sediment transport is considerable. Land-use change does not alter the pattern of the system response to changes in precipitation and temperature, but does affect the magnitude of SSY values. In Table 2, the net contribution of land-use change to the variations of SSY is shown for reach 4 and 19, i.e. the increase or decrease in SSY caused by a change in land use under the same climatic conditions. For example, for reach 4, the arable land expansion scenario, causes an average SSY increase of 41%, while under the arable land reduction scenario SSY decreases on average by −30%, and under the agriculture abandonment scenario −59%.

Table 2: Net contribution of the land use change to the variation of SSY.

Reach	**Extreme precipitation increase**	**Arable land expansion**	**Arable land reduction**	**Agriculture abandonment**
Reach 4	No increase in extreme precipitation	41 (41–41)	−30 (−30 to 30)	−59 (−59 to 59)
	Small increase in extreme precipitation	43 (43–43)	−31 (−31 to 31)	−60 (−60 to 60)
	Large increase in extreme precipitation	47 (47–47)	−33 (−33 to 33)	−64 (−64 to 64)
Reach 19	No increase in extreme precipitation	20 (4–40)	−18 (−28 to 4)	−32 (−40 to 20)
	Small increase in extreme precipitation	22 (6–44)	−17 (−27 to 1)	−30 (−39 to 18)
	Large increase in extreme precipitation	31 (15–55)	−10 (−21 to 8)	−25 (−34 to 12)

In Fig. 9, the joint effect of climatic change and land-use change on SSY is shown for the Thames at reach 4 (left) and reach 19 (right). These histograms describe the range and probability distribution of sediment response to a range of climate and land-use scenarios. Due to the use of a model ensemble rather than a single model, the histograms are represented as an envelop curve rather than a single line. In general, the land-use change impact is small compared with the climate change uncertainty range, but certain land-use changes can exacerbate or reduce the impact of climate change systematically. It is very important to note that there is a clear difference between the different land use management options.

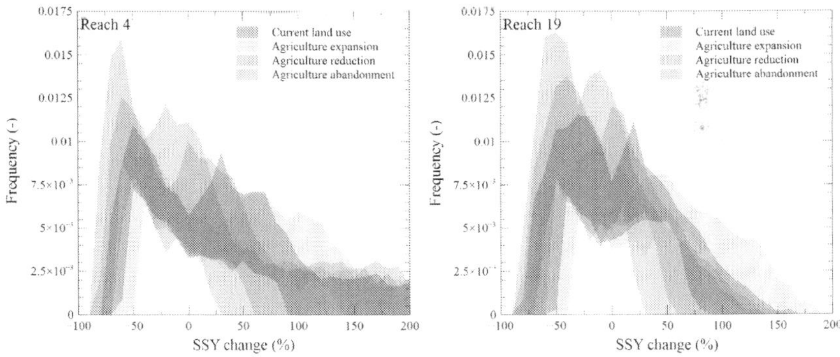

Figure 9: Histograms (represented as empirical probability distribution functions) of the change in suspended sediment yield (SSY) at reach 4 (left) and reach 19 (right) under different land use and land management scenarios under large increase in extreme precipitation.

Likelihood of predicted changes

The response surface plots shown above can be used for exploring combinations of climate change and land use or land management options. Nevertheless, they do not convey information about the plausibility or probability of the expected climatic changes, unless they are compared with projections from climate models (Prudhomme et al., 2010). In this section, we analyse the response of the system to changes in average precipitation, average temperature and extreme precipitation projected by the UKCP09. The values of change in SSY depending on the land use and extreme precipitation scenarios are reported in Table 3, where the median changes and their confidence interval (computed according to the variations in average temperature and precipitation given by UKCP09 for two different future periods: 2030s and 2050s) are shown for each combination of land use and extreme precipitation change.

Table 3: Change in SSY median value and p = 0.95 confidence intervals (%) corresponding to precipitation and temperature changes projected by UKCP09, under different scenarios of land use and extreme precipitation increase

Reach	**Future time slice**	**Extreme precipitation increase**	**Baseline**	**Arable land expansion**	**Arable land reduction**	**Agriculture abandonment**
Reach 4	2030s	No increase in extreme precipitation	−6 (−22 to 15)	16 (−3 to 42)	−22 (−35 to 4)	−37 (−48 to 24)
		Small increase in extreme precipitation	−4 (−19 to 20)	19 (−1 to 46)	−20 (−33 to 1)	−36 (−46 to 21)
		Large increase in extreme precipitation	7 (−10 to 32)	32 (11–62)	−11 (−25 to 9)	−29 (−40 to 13)
	2050s	No increase in extreme precipitation	−8 (−26 to 16)	13 (−9 to 42)	−23 (−38 to 4)	−39 (−50 to 24)
		Small increase in extreme precipitation	−6 (−23 to 19)	16 (−6 to 46)	−22 (−36 to 1)	−37 (−49 to 22)
		Large increase in extreme precipitation	5 (−14 to 31)	30 (6–61)	−13 (−28 to 9)	−31 (−43 to 14)

Reach 19	2030s	No increase in extreme precipitation	−4 (−16 to 13)	20 (4–40)	−18 (−28 to 4)	−32 (−40 to 20)
		Small increase in extreme precipitation	−2 (−14 to 16)	22 (6–44)	−17 (−27 to 1)	−30 (−39 to 18)
		Large increase in extreme precipitation	6 (−8 to 27)	31 (15–55)	−10 (−21 to 8)	−25 (−34 to 12)
	2050s	No increase in extreme precipitation	−6 (−18 to 13)	18 (0–40)	−19 (−30 to 4)	−32 (−42 to 20)
		Small increase in extreme precipitation	−4 (−17 to 16)	20 (2–44)	−18 (−29 to 1)	−31 (−41 to 18)
		Large increase in extreme precipitation	4 (−11 to 26)	30 (11–55)	−11 (−24 to 8)	−26 (−36 to 12)

Effectiveness of arable land reduction

In Fig. 10, the effectiveness of the arable land reduction as a sediment transport reduction mitigation measure is assessed. In this diagram, the difference between the SSY under the baseline scenario and arable land reduction scenario is represented (in %) with different shades of colour (red[1]-brown: smaller decrease in SSY, white-yellow: larger decrease in SSY). The SSY decrease is a proxy measure of the average effect of reducing arable land. Each pixel in this figure represents the effectiveness for reach 4 and reach 19 under a specific combination of change in average precipitation (in the x-axis) and increase in extreme precipitation (in the y-axis, quantified as the increase in the 90% percentile of the CDF), given a fixed increase in average temperature (+2 °C). The space of combinations of changes in average precipitation vs changes in extreme precipitation following the UKCP09 is also depicted as black dots. The larger plots represent the results obtained with the median of the model ensemble, while the smaller plots are the minimum and the maximum respectively. The standard deviation of the change in SSY is represented in the bottom plots, as derived from the ensemble of models employed in this study. These two plots provide an estimation of the uncertainty affecting the land management effectiveness assessment (i.e., large standard deviation: large uncertainty, and vice versa).

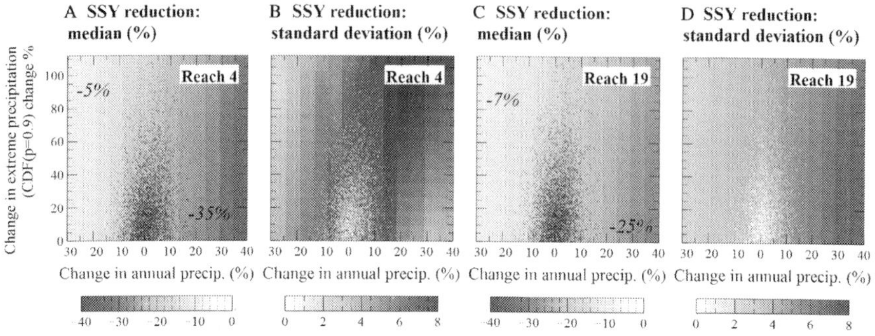

Figure 10: Effectiveness of land management scenario (% difference between SSY change under the baseline scenario and the arable land reduction scenario). (a) Median effectiveness, reach 4; (b) standard deviation of the effectiveness, reach 4; (c) median effectiveness, reach 19; (d) standard deviation of the effectiveness, reach 19. The black dots represent UKCP09 projections for the 2050s. For visualisation purposes, the UKCP09 data were taken from 10,000 transient stochastic daily series of precipitation and temperature produced by Glenis et al. (2015).

DISCUSSION

Model implementation

The model results, though satisfactory in terms of reproduction of the water and sediment dynamics of the River Thames catchments, reveal the presence of uncertainty in the model predictions. From Fig. 6, it can be seen that some of the highest values of suspended sediment concentration are underestimated by the model. This could be due to processes that are not well reproduced by the model, such as localised river bank failure or erosion from farm tracks. This appears not to be a concern in the upper reaches, as shown in Fig. 7, while in the lower reaches the model tends to slightly underestimate low probability concentrations. From the perspective of this study, this is likely to affect the model representation response of the lower Thames to changes in extreme precipitation, leading to small underestimations. Nevertheless, this study is focused to estimating the impact of different stressors on suspended sediment load, which is the product of flow and suspended sediment concentration. The thresholds chosen for model selection for flow and suspended sediment concentration (NSE = 0.65 and 0.1 respectively) are well within the ranges provided by Moriasi et al. (2007), and therefore the model estimates of suspended load can be considered reliable. The SSY obtained by the model is close to estimates of 0.068 Mg Ha^{-1} y^{-1} for the Thames at Days Lock (3500 km^2) by Neal et al. (2006), and lower compared with other large catchments in the UK (Worrall et al., 2013). Furthermore, the sensitivity analysis was carried out using a variety of climatic conditions, from a dry year (2010) to a very wet winter (2013–14), thus ensuring that the model is able to respond to a wide range of climatic conditions.

Climate change impact under current land use

A broad range of climatic variations was explored, to understand the effect of climatic variations not necessarily included by the available climate model projections. From the response surfaces, it can be seen that SSY increases with precipitation, due to larger soil erosion (given by both lower infiltration and increased splash erosion) and to larger river channel flow. SSY is also affected by temperature: higher temperatures increase evapotranspiration and thus decrease flow and SSY. Extreme precipitation also affects SSY, given that an increase in extreme precipitation triggers an increase in SSY. Concerning the uncertainty (standard deviation of the change in SSY, bottom plots in Fig. 8), this seems to be larger when large increase in annual precipitation are considered, while it is smaller when decrease in annual precipitation are considered. In other words, the ensemble of behavioural models used in this study tends to converge to similar results when a decrease in annual precipitation is considered, while it tends to diverge when a large increase in annual precipitation is taken into account.

Fig. 8 shows the magnitude of SSY change in response to climatic alteration and the uncertainty that affects these estimates, but it can also provide a measure of the sensitivity of the system to climatic changes. This can be interpreted by looking at the gradient of the values in the plots. For example, it can be seen that the gradient corresponding to changes in annual precipitation (x-axis of the left and central plots) is much larger than the gradient of the other two climatic variables (annual temperature and extreme precipitation). This means that, within the range of the climatic alterations considered in this study, the most influential, or dominant, is average precipitation, suggesting that the River Thames catchment has a much larger sensitivity to changes in average precipitation than to changes in extreme precipitation or temperature. There is more variability within simulations with the same temperature change than between them (and the same can be said about extreme precipitation), meaning that temperature changes and extreme precipitation changes have second order effects compared to the effects of changes in average precipitation.

These results substantially agree with previous studies on the impact of climate change on sediment transport (e.g. Nearing et al., 2005; Pruski and Nearing, 2002), although the extent of the role played by changes in extreme precipitation is usually high uncertain. In this study, this was quantified in detail for the River Thames. It must be noted that this effect is expected to be highly local, i.e. the impact of changes in extreme precipitation is expected to vary depending on climate and location of the catchment. Even within the same catchment, different sub-catchments respond differently to climatic alterations, due to different land-use configurations. For example, a scenario with an increase of 3 °C and a decrease of 10% in precipitation returns a decrease of 29% in SSY for reach 4 and 20% for reach 19, and a scenario with a decrease of 1 °C and an increase of 20% in precipitation returns an increase of 44% SSY at reach 4 and 39% at reach 19. Similarly, a large increase in extreme precipitation causes an average increase in SSY of 14% at reach 4 and of 11% at reach 19. This is because the uplands of the River Thames are more sensitive to changes in precipitation than the lowlands, due to the larger extension of arable land in the upper sub-catchments.

Climate change impact under different land use scenarios

The impact of land-use change on water discharge is very limited and consistent with that reported by Crooks and Davies (2001). No catchment-scale studies were available regarding the impact of land-use change on the sediment transport of the River Thames, to the authors' knowledge.

Land-use change appears to be a key driver of SSY alterations, although it is important to remember that both land-use change and climate change scenarios were chosen by the authors and do not indicate the likelihood of changes. It is very important to note that, despite modelling uncertainty, a strong signal of change between the different land-use options can be seen, i.e. all the behavioural models lead to the same conclusions in terms of impact of land-use change on SSY. This means that the approach proposed in the study was able to identify the effect of land use management on SSY taking into account the model uncertainty, leading to robust conclusions.

In terms of spatial variations of the effects of land-use change, Fig. 9 shows that there is a slightly different response of reach 4 (located in the upper part of the catchment) and reach 19 (located in the lower part). This is related mainly to the fraction of catchment dedicated to arable land use, and the proportion of arable land versus grassland, but it is also connected to other phenomena, such as the balance between sediment availability and sediment transport capacity of a river reach, which can be altered by changes in climate and land use. The use of a mathematical model, such as the INCA model, allows us understanding and extrapolating the extent of these non-linear interactions between different processes under conditions that have not been observed yet.

It is difficult to compare the results of this paper with previous studies, given that just a few modelling studies have been published so far about the joint impact of climate change and land-use change on sediment transport at the catchment scale. However, a few recent studies also noted that land-use change effects can be as relevant as climate change effects, especially in human-impacted catchments and agricultural areas. For example, Paroissien et al. (2015) found that the effect of land-use change was much more significant than the effect of precipitation change in an agricultural catchment in Southern France. Serpa et al. (2015) also pointed out the role of land-use changes in minimizing the indirect effects of climate changes for two catchments in Portugal and stressed the importance of an integrated approach combining the effects of climate and land-cover change for a realistic evaluation of the future state of natural resources. Similarly, Simonneaux et al. (2015) noted that climate changes alone might be of minor importance compared to changes in land use for an arid catchment in Morocco, especially regarding the evolution of badlands, which are closely conditioned by human actions. Rodríguez-Lloveras et al. (2016) showed that land–use change can counterbalance climate change for a catchment in Southern Spain. Nevertheless, these studies were conducted in more erosion-prone areas, and they might not be comparable with the River Thames. Routschek et al. (2014) presented a study on a temperate catchment in Germany, which led to conclude that land-use change and soil management induced played a more relevant role than climate change alone, and similar conclusions were drawn by Mullan et al. (2012). In the present study we show that for the River Thames the extent of climate and land-use change effect is variable depending on the sub-catchment and the river reach, and, although the climate change impact appears to be predominant at the catchment scale, the amount of arable land also controls an important part of the total sediment production and sediment transport.

Finally, it is worth mentioning that the INCA model land use and vegetation biomass parameters are static, and are not affected by intra- or inter-annual climatic variability. Therefore, the effect of climate change acting on vegetation and land use, which in turn affects sediment production, was not taken into account. Over relatively short time-scales in a heavily managed setting like southern Britain, where land cover is often not natural, the effects of this feedback were considered to be minimal. This is acknowledged as a priority for future regional-scale sediment model development (de Vente et al., 2013), although it is considered for example in the sediment transport model PESERA (Kirkby et al., 2008).

Likelihood of predicted changes

The small increase in extreme precipitation considered in this study has limited effect. On the other hand, the large increase in extreme precipitation has a very clear effect at both stations, increasing the SSY by around 13% at reach 4 and around 10% at reach 19, for the 2030s (+10% and +10% for the 2050s). These figures changes under different land-use changes scenarios: for example, under the arable land expansion scenario they are +16% and +11% respectively, for the 2030s, while under the arable land reduction scenarios they are +11% and +12%. While exploring a broad range of climatic combination may help understanding the system response under global change, the incorporation of climate model forecasts like the UKCP09 in the scenario-neutral methodology provides policy-makers with a clear figure of what the expected changes will be. It has to be acknowledge that the uncertainty of the final results is extremely large, but the methodology described in this paper can at least account for the climate model uncertainty, by using a large set of climate model outcome like the UKCP09 product, and the hydrological model parametric uncertainty, by employing an ensemble of equifinal behavioural models rather than a single model.

Effectiveness of arable land reduction

Fig. 10 shows that the reduction of arable land as a measure to reduce SSY in the Thames is effective under all the climate change scenarios considered for the Thames, with reductions of SSY ranging from 5% to −35% for reach 4 and between 7% and 25% for reach 19. The effectiveness of the arable land reduction within the area defined by the UKCP09 projections ranges between −20% and 30%. On the other hand, this plot also shows that the effectiveness of the arable land reduction may vary depending on the climate scenario. For both sub-catchments, the proposed arable land reduction is effective on a scenario of precipitation reduction (left part of the plot), but they are also effective in the case of increase in extreme precipitation. In particular, the SSY reduction is more sensitive to changes in annual precipitation rather than changes in extreme precipitation (i.e., the vertical gradient in the response surfaces is minimal compared to the horizontal gradient). Fig. 10 also shows the modelling uncertainty of the results. For example, it can be see that for reach 4 the standard deviation of the SSY reduction is larger than for reach 19 in the central area of the plot, i.e., the area of climatic outcomes defined as plausible by the UKCP09. Knowing the modelling uncertainty in reproducing the system response is of paramount importance for catchment managers, as it provides an estimate of the likelihood of a given land management measure to obtain the expected results.

This plot also shows the potential of the scenario-neutral methodology for assessing a soil erosion mitigation strategy under changing climate and changing land use. Owing to the particular approach, which analyses several different combinations of climate and land use, it was possible to assess whether this strategy was robust and effective under different climatic and land-use conditions. This provides decision makers and land managers with a simple tool that might inform climate change adaptation policy. However, this was not considered in this study because the representation of soil erosion mitigation measures into the model parameterisation requires a more detailed investigation that was out of the scopes of this paper.

CONCLUSIONS

This paper investigated the joint control exerted by climate change and land-cover change on suspended sediment discharge in the River Thames catchment (UK), through the use of a scenario-neutral method and the UKCP09 projections (Murphy et al., 2009). The results showed that UKCP09 changes in average precipitation and average temperature are likely to cause a median reduction in the suspended sediment yield of the River Thames by 6% in the uplands and by 4% in the lowlands, although the confidence interval ($p = 0.95$) is very broad (−22% to +15% and −16% to +13% respectively), owing to the high variability in expected future precipitation and temperature. The UKCP09 projections also project an increase in extreme precipitation, which is likely to increase suspended sediment yield by up to 13% in the uplands, potentially compensating the reduction due to changes in average precipitation and temperature (subject to the actual increase of extreme precipitation).

The main findings of this study are listed as follows:

(i) This paper has shown a methodological approach to assess the joint impact of climate and land-use change, taking into account the climate and sediment model uncertainties and leading to robust conclusions. If used along with a climate model, this methodology can also offer a measure of the plausibility of expected changes.

(ii) The control exerted on the soil erosion and sediment transport of the River Thames catchment by a change in average precipitation is larger than the effect of other stressors considered in this study.

(iii) Climate change and land use change exert a joint control on sediment transport, with interactions that cannot be neglected. The extent and magnitude of land-use and land-management impacts also vary depending on the location on the river and on the sub-catchment considered and must be assessed locally.

(iv) The proposed methodology allowed assessing the robustness of arable land reduction as a measure to control sediment transport. This measure appears effective under different climatic conditions, although with different effectiveness. This study also pointed out that their effectiveness may vary depending on the future climate outcomes, providing a quantification of how it varies across the spectrum of future climatic changes.

ACKNOWLEDGEMENTS

This study forms part of the POLL-CURB project (Changes in urbanisation and its effect on water quality and quantity from local to regional scale) which is funded by the Natural Environment Research Council under the Changing Water Cycle Programme (Grant NE/K002309/1) and the MaRIUS project (Managing the Risks, Impacts and Uncertainties of droughts and water Scarcity), which is funded by the Natural Environment Research Council (NERC) under the UK Droughts and Water Scarcity Programme (Grant NE/L010364/1). The meteorological data (precipitation and temperature) were provided by the UK Met Office. The river flow data were provided by the National River Flow Archive. The suspended sediment concentration data was provided by the Environment Agency of England and Wales and by the Centre of Ecology and Hydrology's Thames Initiative. We would like to thank two anonymous reviewers for their very valuable contributions.

REFERENCES

1. H. Aksoy, M.L. KavvasA review of hillslope and watershed scale erosion and sediment transport models Catena, 64 (2005), pp. 247-271, 10.1016/j.catena.2005.08.008
2. Bagnold, 1966 R.A. BagnoldAn approach to the sediment transport problem from general physics Geological Survey Professional Paper 422-I., U.S. Government Printing Office, Washington, DC (1966)
3. Bangash et al., 2013 R.F. Bangash, A. Passuello, M. Sánchez-

Canales, M. Terrado, A. López, F.J. Elorza, G.Ziv, V. Acuña, M. SchuhmacherEcosystem services in Mediterranean river basin: climate change impact on water provisioning and erosion control. Sci. Total Environ., 458–460C (2013), pp. 246-255, 10.1016/j.scitotenv.2013.04.025

4. Bastola et al., 2011 S. Bastola, C. Murphy, J. SweeneyThe sensitivity of fluvial flood risk in Irish catchments to the range of IPCC AR4 climate change scenarios Sci. Total Environ., 409 (2011), pp. 5403-5415, 10.1016/j.scitotenv.2011.08.042
5. Bloomfield et al., 2011 J.P. Bloomfield, S.H. Bricker, A.J. NewellSome relationships between lithology, basin form and hydrology: a case study from the Thames basin, UK Hydrol. Process., 25 (2011), pp. 2518-2530, 10.1002/hyp.8024
6. Blöschl et al., 2007 G. Blöschl, S. Ardoin-Bardin, M. Bonell, M. Dorninger, D. Goodrich, D. Gutknecht, D.Matamoros, B. Merz, P. Shand, J. SzolgayAt what scales do climate variability and land cover change impact on flooding and low flows? Hydrol. Process., 21 (2007), pp. 1241-1247, 10.1002/hyp.6669
7. Boardman, 2015 J. BoardmanExtreme rainfall and its impact on cultivated landscapes with particular reference to Britain Earth Surf. Process. Landforms (2015), 10.1002/esp.3792
8. Boardman, 2003 J. BoardmanSoil erosion and flooding on the eastern South Downs, southern England, 1976–2001 Trans. Inst. Brit. Geogr., 28 (2003), pp. 176-196, 10.1111/1475-5661.00086
9. Borgomeo et al., 2014 E. Borgomeo, J.W. Hall, F. Fung, G. Watts, K. Colquhoun, C. LambertRisk-based water resources planning: Incorporating probabilistic nonstationary climate uncertainties Water Resour. Res., 50 (2014), pp. 6850-6873, 10.1002/2014WR015558
10. Bowes et al., 2012 M.J. Bowes, E. Gozzard, A.C. Johnson, P.M. Scarlett, C. Roberts, D.S. Read, L.K.Armstrong, S.A. Harman, H.D. WickhamSpatial and temporal changes in chlorophyll-a concentrations in the River Thames basin, UK: are phosphorus concentrations beginning to limit phytoplankton biomass? Sci. Total Environ., 426 (2012), pp. 45-55, 10.1016/j.scitotenv.2012.02.056
11. Brown et al., 2012 C. Brown, Y. Ghile, M. Laverty, K. LiDecision scaling: linking bottom-up vulnerability analysis with climate

projections in the water sector Water Resour. Res., 48 (2012), 10.1029/2011WR011212

12. Brown and Wilby, 2012 C. Brown, R.L. WilbyAn alternate approach to assessing climate risks. Eos Trans. Am. Geophys. Union, 93 (2012), p. 401, 10.1029/2012EO410001
13. Buendia et al., 2015 C. Buendia, G. Bussi, J. Tuset, D. Vericat, S. Sabater, A. Palau, R.J. BatallaEffects of afforestation on runoff and sediment load in an upland Mediterranean catchment Sci. Total Environ., 540 (2015), pp. 144-157, 10.1016/j.scitotenv.2015.07.005
14. Burt et al., 2015 T. Burt, J. Boardman, I.D.L. Foster, N. HowdenMorc rain less soil: long-term changes in rainfall intensity with climate change Earth Surf. Process. Landforms (2015), 10.1002/esp.3868
15. Bussi et al., 2014a G. Bussi, F. Francés, E. Horel, J.A. López-Tarazón, R.J. BatallaModelling the impact of climate change on sediment yield in a highly erodible Mediterranean catchment J. Soils Sediments, 14 (2014), pp. 1921-1937, 10.1007/s11368-014-0956-7
16. Bussi et al., 2014b G. Bussi, F. Francés, J.J. Montoya, P.Y. JulienDistributed sediment yield modelling: importance of initial sediment conditions Environ. Model. Softw., 58 (2014), pp. 58-70, 10.1016/j.envsoft.2014.04.010
17. Bussi et al., 2013 G. Bussi, X. Rodríguez-Lloveras, F. Francés, G. Benito, Y. Sánchez-Moya, A. SopeñaSediment yield model implementation based on check dam infill stratigraphy in a semiarid Mediterranean catchment Hydrol. Earth Syst. Sci., 17 (2013), pp. 3339-3354, 10.5194/hess-17-3339-2013
18. Bussi et al., 2016 G. Bussi, P.G. Whitehead, M.J. Bowes, D.S. Read, C. Prudhomme, S.J. DadsonImpacts of climate change, land-use change and phosphorus reduction on phytoplankton in the River Thames (UK) Sci. Total Environ. (2016), 10.1016/j.scitotenv.2016.02.109
19. Castellazzi et al., 2010 M.S. Castellazzi, J. Matthews, F. Angevin, C. Sausse, G.A. Wood, P.J. Burgess, I.Brown, K.F. Conrad, J.N. PerrySimulation scenarios of spatio-temporal arrangement of crops at the landscape scale Environ. Model. Softw., 25 (2010), pp. 1881-1889, 10.1016/j.envsoft.2010.04.006

20. Cerdà, 1999 A CerdàSeasonal and spatial variations in infiltration rates in badland surfaces under Mediterranean climatic conditions Water Resour. Res., 35 (1999), pp. 319-328, 10.1029/98WR01659
21. Cerdà, 1998 A. CerdàEffect of climate on surface flow along a climatological gradient in Israel: a field rainfall simulation approach J. Arid Environ., 38 (1998), pp. 145-159, 10.1006/jare.1997.0342
22. Cerdà, 1997 A.CerdàSeasonal changes of the infiltration rates in a Mediterranean scrubland on limestone J. Hydrol., 198 (1997), pp. 209-225, 10.1016/S0022-1694(96)03295-7
23. Collins and Walling, 2007 A.L. Collins, D.E. WallingFine-grained bed sediment storage within the main channel systems of the frome and piddle catchments, Dorset, UK Hydrol. Process., 21 (2007), pp. 1448-1459, 10.1002/hyp.6269
24. Coulthard et al., 2012 T.J. Coulthard, J. Ramírez, H.J. Fowler, V. GlenisUsing the UKCP09 probabilistic scenarios to model the amplified impact of climate change on drainage basin sediment yield Hydrol. Earth Syst. Sci., 16 (2012), pp. 4401-4416, 10.5194/hess-16-4401-2012
25. Crooks and Davies, 2001 S. Crooks, H. DaviesAssessment of land use change in the Thames catchment and its effect on the flood regime of the river Phys. Chem. Earth Part B Hydrol. Ocean. Atmos., 26 (2001), pp. 583-591, 10.1016/S1464-1909(01)00053-3
26. Crossman et al., 2013 J. Crossman, P.G. Whitehead, M.N. Futter, L. Jin, M. Shahgedanova, M.S. Castellazzi, A.J. WadeThe interactive responses of water quality and hydrology to changes in multiple stressors, and implications for the long-term effective management of phosphorus Sci. Total Environ., 454–455 (2013), pp. 230-244, 10.1016/j.scitotenv.2013.02.033
27. De Vente et al., 2013 J. De Vente, J. Poesen, G. Verstraeten, G. Govers, M. Vanmaercke, A. Van Rompaey, M.Arabkhedri, C. Boix-FayosPredicting soil erosion and sediment yield at regional scales: where do we stand? Earth-Sci. Rev. (2013), 10.1016/j.earscirev.2013.08.014
28. Déqué, 2007 M. DéquéFrequency of precipitation and temperature extremes over France in an anthropogenic scenario: model results and statistical correction according to observed values Glob. Planet.

Change, 57 (2007), pp. 16-26, 10.1016/j.gloplacha.2006.11.030

29. Duijsings, 1986 J.J.H.M. DuijsingsSeasonal variation in the sediment delivery ratio of a forested drainage basin in Luxembourg R.F. Hadley (Ed.), Drainage Basin Sediment Delivery. IAHS Publ 159, IAHS Press, Wallingford (1986), pp. 153-164
30. Evans et al., 2015 R. Evans, A.L. Collins, I.D.L. Foster, R.J. Rickson, S.G. Anthony, T. Brewer, L. Deeks, J.P.Newell-Price, I.G. Truckell, Y. ZhangExtent, frequency and rate of water erosion of arable land in Britain – benefits and challenges for modelling Soil Use Manag. (2015), 10.1111/sum.12210
31. Farkas et al., 2013 C. Farkas, S. Beldring, M. Bechmann, J. DeelstraSoil erosion and phosphorus losses under variable land use as simulated by the INCA-P model Soil Use Manage., 29 (2013), pp. 124-137, 10.1111/j.1475-2743.2012.00430.x
32. Foster et al., 2012 I.D.L. Foster, K.M. Rowntree, J. Boardman, T.M. MighallChanging sediment yield and sediment dynamics in the Karoo uplands, South Africa; post-European impacts Land Degr. Dev., 23 (2012), pp. 508-522, 10.1002/ldr.2180
33. Fowler et al., 2010 H.J. Fowler, D. Cooley, S.R. Sain, M. ThurstonDetecting change in UK extreme precipitation using results from the climateprediction.net BBC climate change experiment Extremes, 13 (2010), pp. 241-267, 10.1007/s10687-010-0101-y
34. Fowler and Ekström, 2009 H.J. Fowler, M. EkströmMulti-model ensemble estimates of climate change impacts on UK seasonal precipitation extremes Int. J. Climatol., 29 (2009), pp. 385-416, 10.1002/joc.1827
35. Francipane et al., 2015 A.Francipane, S. Fatichi, V.Y. Ivanov, L.V. NotoStochastic assessment of climate impacts on hydrology and geomorphology of semiarid headwater basins using a physically based model J. Geophys. Res. Earth Surf., 120 (2015), pp. 507-533, 10.1002/2014JF003232
36. Fronzek et al., 2011 S. Fronzek, T.R. Carter, M. LuotoEvaluating sources of uncertainty in modelling the impact of probabilistic climate change on sub-arctic palsa mires Nat. Hazards Earth Syst. Sci., 11 (2011), pp. 2981-2995, 10.5194/nhess-11-2981-2011
37. Fuller et al., 2002 R.M. Fuller, G.M. Smith, J.M. Sanderson, R.A.

Hill, A.G. ThomsonThe UK Land Cover Map 2000: construction of a parcel-based vector map from satellite images Cartogr. J., 39 (2002), pp. 15-25

38. Futter et al., 2007 M.N. Futter, D. Butterfield, B.J. Cosby, P.J. Dillon, A.J. Wade, P.G. WhiteheadModeling the mechanisms that control in-stream dissolved organic carbon dynamics in upland and forested catchments Water Resour. Res., 43 (2007), 10.1029/2006WR004960
39. Futter et al., 2014 M.N. Futter, M.A. Erlandsson, D. Butterfield, P.G. Whitehead, S.K. Oni, A.J. WadePERSiST: a flexible rainfall-runoff modelling toolkit for use with the INCA family of models Hydrol. Earth Syst. Sci., 18 (2014), pp. 855-873, 10.5194/hess-18-855-2014
40. Gallart and Llorens, 2004 F. Gallart, P. LlorensObservations on land cover changes and water resources in the headwaters of the Ebro catchment, Iberian Peninsula Phys. Chem. Earth, 29 (2004), pp. 769-773, 10.1016/j.pce.2004.05.004
41. Glenis et al., 2015 V. Glenis, V. Pinamonti, J. Hall, C. KilsbyA transient stochastic weather generator incorporating climate model uncertainty Adv. Water Resour. (2015), 10.1016/j.advwatres.2015.08.002
42. González-Hidalgo et al., 2013 J.C. González-Hidalgo, R.J. Batalla, A. CerdàCatchment size and contribution of the largest daily events to suspended sediment load on a continental scale Catena, 102 (2013), pp. 40-45, 10.1016/j.catena.2010.10.011
43. González-Hidalgo et al., 2010 J.C. González-Hidalgo, R.J. Batalla, A. Cerdà, M. de LuisContribution of the largest events to suspended sediment transport across the USA Land Degr. Dev., 21 (2010), pp. 83-91, 10.1002/ldr.897
44. Gordon et al., 2000 C. Gordon, C. Cooper, C.A. Senior, H. Banks, J.M. Gregory, T.C. Johns, J.F.B. Mitchell, R.A.WoodThe simulation of SST, sea ice extents and ocean heat transports in a version of the Hadley Centre coupled model without flux adjustments Clim. Dyn., 16 (2000), pp. 147-168, 10.1007/s003820050010
45. Haines-Young et al., 2014 R. Haines-Young, J. Paterson, M. Patschin, A. Wilson, G. KassThe UK NEA scenarios: development of storylines and analysis of outcomes S. Albon, K. Turner, R. Watson (Eds.), UK National Ecosystem Assesment, UNEP (2014),

pp. 1195-1264

46. Hay et al., 2000 L.E. Hay, R.L. Wilby, G.H. Leavesley A comparison of delta change and downscaled GCM scenarios for three mountainous basins in the United States JAWRA J. Am. Water Resour. Assoc., 36 (2000), pp. 387-397, 10.1111/j.1752-1688.2000.tb04276.x
47. Howden et al., 2013 N.J.K. Howden, T.P. Burt, F. Worrall, S.A. Mathias, M.J. WhelanFarming for water quality: balancing food security and nitrate pollution in UK river basins Ann. Assoc. Am. Geogr., 103 (2013), pp. 397-407, 10.1080/00045608.2013.754672
48. Ito, 2007 A.ItoSimulated impacts of climate and land-cover change on soil erosion and implication for the carbon cycle, 1901–2100 Geophys. Res. Lett., 34 (2007), pp. 1-5, 10.1029/2007GL029342
49. Jackson-Blake and Starrfelt, 2015 L.A. Jackson-Blake, J. StarrfeltDo higher data frequency and Bayesian auto-calibration lead to better model calibration? Insights from an application of INCA-P, a process-based river phosphorus model J. Hydrol., 527 (2015), pp. 641-655, 10.1016/j.jhydrol.2015.05.001
50. Jarritt and Lawrence, 2007 N.P. Jarritt, D.S.L. LawrenceFine sediment delivery and transfer in lowland catchments: modelling suspended sediment concentrations in response to hydrological forcing Hydrol. Process., 21 (2007), pp. 2729-2744, 10.1002/hyp.6402
51. Jarritt and Lawrence, 2006 N.P. Jarritt, D.S.L. LawrenceSimulating fine sediment delivery in lowland catchments: model development and application of INCA-Sed P.N. Owens, A.J. Collins (Eds.), Soil Erosion and Sediment Redistribution in River Catchments: Measurement Measurement: Modelling and Management, CAB International (2006), pp. 207-2016
52. Jin et al., 2012 L. Jin, P.G. Whitehead, M.N. Futter, Z. LuModelling the impacts of climate change on flow and nitrate in the River Thames: assessing potential adaptation strategies Hydrol. Res., 43 (2012), pp. 902-916, 10.2166/nh.2011.080
53. Julien, 2010 P.Y. JulienErosion and Sedimentation (2nd ed.), Cambridge University Press (2010)

54. Julien and Simons, 1985 P.Y. Julien, D.B. SimonsSediment transport capacity of overland flow Trans. ASAE (1985)
55. Kinniburgh and Barnett, 2009 J.H. Kinniburgh, M. BarnettOrthophosphate concentrations in the River Thames: reductions in the past decade Water Environ. J., 24 (2009), pp. 107-115, 10.1111/j.1747-6593.2008.00161.x
56. Kirkby et al., 2008 M.J. Kirkby, B.J. Irvine, R.J.A. Jones, G. GoversThe PESERA coarse scale erosion model for Europe. I.-Model rationale and implementation Eur. J. Soil Sci., 59 (2008), pp. 1293-1306, 10.1111/j.1365-2389.2008.01072.x
57. Lázár et al., 2010 A.N. Lázár, D. Butterfield, M.N. Futter, K. Rankinen, M. Thouvenot-Korppoo, N.P. Jarritt, D.S.L. Lawrence, A.J. Wade, P.G. WhiteheadAn assessment of the fine sediment dynamics in an upland river system: INCA-Sed modifications and implications for fisheries Sci. Total Environ., 408 (2010), pp. 2555-2566, 10.1016/j.scitotenv.2010.02.030
58. López-Tarazón et al., 2011 J.A. López-Tarazón, R.J. Batalla, D. VericatIn-channel sediment storage in a highly erodible catchment: the River Isábena (Ebro Basin, Southern Pyrenees) Z. Geomorphol., 55 (2011), pp. 365-382, 10.1127/0372-8854/2011/0045
59. Lu et al., 2016 Q. Lu, M.N. Futter, L. Nizzetto, G. Bussi, M.D. Jürgens, P. WhiteheadFate and transport of polychlorinated biphenyls (PCBs) in the River Thames catchment – insights from a coupled multimedia fate and hydrobiogeochemical transport model Sci. Total Environ. (2016), 10.1016/j.scitotenv.2016.03.029
60. Met Office, 2012 Met OfficeMet Office Integrated Data Archive System (MIDAS) Land and Marine Surface Stations Data (1853-Current) NCAS British Atmospheric Data Centre (2012)
61. Moriasi et al., 2007 D.N. Moriasi, J.G. Arnold, M.W. Van Liew, R.L. Bingner, R.D. Harme, T.L. VeithModel evaluation guidelines for systematic quantification of accuracy in watershed simulations Trans. ASAE, 50 (2007), pp. 885-900
62. Mouri, 2015 G. MouriAssessment of spatiotemporal variations in the fluvial wash-load component in the 21st century with regard to GCM climate change scenarios Sci. Total Environ., 533 (2015), pp. 238-246, 10.1016/j.scitotenv.2015.06.118

63. Mullan, 2013 D. MullanSoil erosion under the impacts of future climate change: Assessing the statistical significance of future changes and the potential on-site and off-site problems Catena, 109 (2013), pp. 234-246, 10.1016/j.catena.2013.03.007

64. Mullan et al., 2012 D. Mullan, D. Favis-Mortlock, R. FealyAddressing key limitations associated with modelling soil erosion under the impacts of future climate change Agric. For. Meteorol., 156 (2012), pp. 18-30, 10.1016/j.agrformet.2011.12.004

65. Murphy et al., 2007 J.M. Murphy, B.B.B. Booth, M. Collins, G.R. Harris, D.M.H. Sexton, M.J. WebbA methodology for probabilistic predictions of regional climate change from perturbed physics ensembles Philos. Trans. A. Math. Phys. Eng. Sci., 365 (2007), pp. 1993-2028, 10.1098/rsta.2007.2077

66. Murphy et al., 2009 J.M. Murphy, D.M.H. Sexton, G.J. Jenkins, B.B.B. Booth, C.C. Brown, R.T. Clark, M. Collins, G.R. Harris, E.J. Kendon, R.A. Betts, S.J. Brown, K.A. Humphrey, M.P. McCarthy, R.E. McDonald, A.Stephens, C. Wallace, R. Warren, R.L. Wilby, R.A. WoodUK Climate Projections Science Report Climate Change Projections, Met Office Hadley Centre, Exeter (2009)

67. Nash and Sutcliffe, 1970 J.E. Nash, J.V. SutcliffeRiver flow forecastin through conceptual models – Part 1 – A discussion of principles J. Hydrol., 10 (1970), pp. 282-290, 10.1016/0022-1694(70)90255-6

68. Neal et al., 2006 C. Neal, M. Neal, G.J.L. Leeks, G.H. Old, L.K. Hill, H.D. WickhamSuspended sediment and particulate phosphorus in surface waters of the upper Thames Basin, UK J. Hydrol., 330 (2006), pp. 142-154, 10.1016/j.jhydrol.2006.04.016

69. Nearing et al., 2005 M.A. Nearing, V. Jetten, C. Baffaut, O. Cerdan, A. Couturier, M. Hernandez, Y. Le Bissonnais, M.H. Nichols, J.P. Nunes, C.S. Renschler, V. Souchère, K. van OostModeling response of soil erosion and runoff to changes in precipitation and cover Catena, 61 (2005), pp. 131-154, 10.1016/j.catena.2005.03.007

70. Nearing et al., 2004 M.A. Nearing, F.F. Pruski, M.R. O'NeillExpected climate change impacts on soil erosion rates: a review J. Soil Water Conserv., 59 (2004), pp. 43-50

71. Nizzetto et al., 2016 L. Nizzetto, G. Bussi, M.N. Futter, D. Butterfield, P.G. WhiteheadA theoretical assessment of microplastic transport in

river catchments and their retention by soils and river sediments Environ. Sci. Process. Impacts (2016), 10.1039/C6EM00206D

72. Nunes et al., 2009 J.P. Nunes, J. Seixas, J.J. Keizer, A.J.D. FerreiraSensitivity of runoff and soil erosion to climate change in two Mediterranean watersheds. Part II: Assessing impacts from changes in storm rainfall, soil moisture and vegetation cover Hydrol. Process., 23 (2009), pp. 1212-1220, 10.1002/hyp.7250
73. Palmer and Räisänen, 2002 T.N. Palmer, J. RäisänenQuantifying the risk of extreme seasonal precipitation events in a changing climate Nature, 415 (2002), pp. 512-514, 10.1038/415512a
74. Paroissien et al., 2015 J.-B. Paroissien, F. Darboux, A. Couturier, B. Devillers, F. Mouillot, D. Raclot, Y. Le BissonnaisA method for modeling the effects of climate and land use changes on erosion and sustainability of soil in a Mediterranean watershed (Languedoc, France) J. Environ. Manage., 150 (2015), pp. 57-68, 10.1016/j.jenvman.2014.10.034
75. Peizhen et al., 2001 Z. Peizhen, P. Molnár, W.R. DownsIncreased sedimentation rates and grain sizes 2–4 Myr ago due to the influence of climate change on erosion rates Nature, 410 (2001), pp. 891-897, 10.1038/35073504
76. Piqué et al., 2014 G. Piqué, J.A. López-Tarazón, R.J. BatallaVariability of in-channel sediment storage in a river draining highly erodible areas (the Isábena, Ebro Basin) J. Soils Sediments, 14 (2014), pp. 2031-2044, 10.1007/s11368-014-0957-6
77. Poff et al., 2015 N.L. Poff, C.M. Brown, T.E. Grantham, J.H. Matthews, M.A. Palmer, C.M. Spence, R.L. Wilby, M. Haasnoot, G.F. Mendoza, K.C. Dominique, A. BaezaSustainable water management under future uncertainty with eco-engineering decision scaling Nat. Clim. Change, 1–10 (2015), 10.1038/nclimate2765
78. Prudhomme et al., 2013 C. Prudhomme, S. Crooks, A.L. Kay, N. ReynardClimate change and river flooding: Part 1 classifying the sensitivity of British catchments Clim. Change, 119 (2013), pp. 933-948, 10.1007/s10584-013-0748-x
79. Prudhomme et al., 2015 C. Prudhomme, E. Sauquet, G. WattsLow flow response surfaces for drought decision support: a case study from the UK J. Extrem. Events, 2 (2015), 10.1142/S2345737615500050

80. Prudhomme et al., 2010 C. Prudhomme, R.L. Wilby, S. Crooks, A.L. Kay, N.S. ReynardScenario-neutral approach to climate change impact studies: application to flood risk J. Hydrol., 390 (2010), pp. 198-209, 10.1016/j.jhydrol.2010.06.043
81. Pruski and Nearing, 2002 F.F. Pruski, M.A. NearingClimate-induced changes in erosion during the 21st century for eight U.S. locations Water Resour. Res., 38 (2002), 10.1029/2001WR000493 34-11
82. Rankinen et al., 2010 K. Rankinen, M. Thouvenot-Korppoo, A. Lazar, D.S.L. Lawrence, D. Butterfield, N.Veijalainen, I. Huttunen, A. LepistöApplication of catchment scale sediment delivery model INCA-Sed to four small study catchments in Finland Catena, 83 (2010), pp. 64-75, 10.1016/j.catena.2010.07.005
83. Rodríguez-Lloveras et al., 2015 X. Rodríguez-Lloveras, G. Bussi, F. Francés, E. Rodríguez-Caballero, A. Solé-Benet, M. Calle, G. BenitoPatterns of runoff and sediment production in response to land-use changes in an ungauged catchment J. Hydrol., 531 (2015), pp. 1054-1066, 10.1016/j.jhydrol.2015.11.014
84. Rodríguez-Lloveras et al., 2016 X. Rodríguez-Lloveras, W. Buytaert, G. BenitoLand use can offset climate change induced increases in erosion in Mediterranean watersheds Catena, 143 (2016), pp. 244-255, 10.1016/j.catena.2016.04.012
85. Routschek et al., 2014 A. Routschek, J. Schmidt, F. KreienkampImpact of climate change on soil erosion – a high-resolution projection on catchment scale until 2100 in Saxony/Germany Catena, 121 (2014), pp. 99-109, 10.1016/j.catena.2014.04.019
86. Serpa et al., 2015 D. Serpa, J.P. Nunes, J. Santos, E. Sampaio, R. Jacinto, S. Veiga, J.C. Lima, M. Moreira, J.Corte-Real, J.J. Keizer, N. AbrantesImpacts of climate and land use changes on the hydrological and erosion processes of two contrasting Mediterranean catchments Sci. Total Environ., 538 (2015), pp. 64-77, 10.1016/j.scitotenv.2015.08.033
87. Simonneaux et al., 2015 V. Simonneaux, A. Cheggour, C. Deschamps, F. Mouillot, O. Cerdan, Y. Le BissonnaisLand use and climate change effects on soil erosion in a semi-arid mountainous watershed (High Atlas, Morocco) J. Arid Environ., 122 (2015), pp. 64-75, 10.1016/j.jaridenv.2015.06.002

88. Singh et al., 2014 R. Singh, T. Wagener, R. Crane, M.E. Mann, L. NingA vulnerability driven approach to identify adverse climate and land use change combinations for critical hydrologic indicator thresholds: application to a watershed in Pennsylvania, USA Water Resour. Res., 50 (2014), pp. 3409-3427, 10.1002/2013WR014988
89. Smith, 2013 Smith, A., 2013. User Guide for the BGS DiGMapGB-50 data (V7). British Geological Survey Open Report, OR/13/008.
90. Smith et al., 2007 G. Smith, M. Beare, M. Boyd, T. Downs, M. Gregory, D. Morton, N. Brown, A.G. ThomsonUK land cover map production through the generalisation of OS MasterMap® Cartogr. J., 44 (2007), pp. 276-283, 10.1179/000870407X241827
91. Spear and Hornberger, 1980 R.C. Spear, G.M. HornbergerEutrophication in peel inlet—II. Identification of critical uncertainties via generalized sensitivity analysis Water Res., 14 (1980), pp. 43-49, 10.1016/0043-1354(80)90040-8
92. Turner et al., 2007 B.L. Turner, E.F. Lambin, A. ReenbergThe emergence of land change science for global environmental change and sustainability Proc. Natl. Acad. Sci. U.S.A., 104 (2007), pp. 20666-20671, 10.1073/pnas.0704119104
93. Wade et al., 2002a A.J. Wade, P. Durand, V. Beaujouan, W. Wessel, K.J. Raat, P.G. Whitehead, D. Butterfield, K. Rankinen, A. LepistoA nitrogen model for European catchments: INCA, new model structure and equations Hydrol. Earth Syst. Sci., 6 (2002), pp. 559-582, 10.5194/hess-6-559-2002
94. Wade et al., 2002b A.J. Wade, P.G. Whitehead, D. ButterfieldThe Integrated Catchments model of Phosphorus dynamics (INCA-P), a new approach for multiple source assessment in heterogeneous river systems: model structure and equations Hydrol. Earth Syst. Sci. (2002), 10.5194/hess-6-583-2002
95. Wetterhall et al., 2011 F. Wetterhall, L.P. Graham, J. Andréasson, J. Rosberg, W. YangUsing ensemble climate projections to assess probabilistic hydrological change in the Nordic region Nat. Hazards Earth Syst. Sci., 11 (2011), pp. 2295-2306, 10.5194/nhess-11-2295-2011
96. Whitehead et al., 2015a P.G. Whitehead, G. Bussi, M.J. Bowes, D.S. Read, M.G. Hutchins, J.A. Elliott, S.J.DadsonDynamic modelling

of multiple phytoplankton groups in rivers with an application to the Thames river system in the UK Environ. Model. Softw., 74 (2015), pp. 75-91, 10.1016/j.envsoft.2015.09.010

97. Whitehead et al., 2013 P.G. Whitehead, J. Crossman, B.B. Balana, M.N. Futter, S. Comber, L. Jin, D. Skuras, A.J. Wade, M.J. Bowes, D.S. ReadA cost-effectiveness analysis of water security and water quality: impacts of climate and land-use change on the River Thames system Philos. Trans. A. Math. Phys. Eng. Sci., 371 (2013), p. 20120413, 10.1098/rsta.2012.0413
98. Whitehead et al., 2010 Whitehead, P.G., Lázár, A.N., Futter, M.N., Pope, L., Wade, A.J., Willows, R., Burgess, C., 2010. Modelling sediment supply and transport in the River Lugg: strategies for controlling sediment loads, in: BHS Third International Symposium, Managing Consequences of a Changing Global Environment. Newcastle, pp. 1–6.
99. Whitehead et al., 2015b P.G. Whitehead, H. Leckie, K. Rankinen, D. Butterfield, M.N. Futter, G. BussiAn INCA model for pathogens in rivers and catchments: model structure, sensitivity analysis and application to the River Thames Catchment, UK Sci. Total Environ. (2015), 10.1016/j.scitotenv.2016.01.128
100. Whitehead et al., 1998a P.G. Whitehead, E. Wilson, D. ButterfieldA semi-distributed integrated nitrogen model for multiple source assessment in catchments (INCA): Part I — model structure and process equations Sci. Total Environ., 210–211 (1998), pp. 547-558, 10.1016/S0048-9697(98)00037-0
101. Whitehead et al., 1998b P.G. Whitehead, E. Wilson, D. Butterfield, K. SeedA semi-distributed integrated flow and nitrogen model for multiple source assessment in catchments (INCA): Part II — application to large river basins in south Wales and eastern England Sci. Total Environ., 210–211 (1998), pp. 559-583, 10.1016/S0048-9697(98)00038-2
102. Wolman and Miller, 1960 M.G. Wolman, J.P. MillerMagnitude and frequency of forces in geomorphic processes J. Geol., 668 (1960), pp. 54-74

103. Worrall et al., 2013 F. Worrall, T.P. Burt, N.J.K. HowdenThe flux of suspended sediment from the UK 1974 to 2010 J. Hydrol., 504 (2013), pp. 29-39, 10.1016/j.jhydrol.2013.09.012

104. Zhao et al., 2013 G. Zhao, X. Mu, Z. Wen, F. Wang, P. GaoSoil erosion, conservation and eco-environment changes in the Loess Plateau of China Land Degr. Dev., 24 (2013), pp. 499-510, 10.1002/ldr.2246

5 CAN WE BE CERTAIN ABOUT FUTURE LAND USE CHANGE IN EUROPE? A MULTI-SCENARIO, INTEGRATED-ASSESSMENT ANALYSIS

I.P. Holman[a], C Brown [b], V Janes [a], and D Sandars [a]

[a]School of Water, Energy and Environment, Cranfield University, Cranfield, Bedford MK43 0AL, UK

[b]School of Geosciences, University of Edinburgh, Edinburgh EH8 9XP, UK

ABSTRACT

The global land system is facing unprecedented pressures from growing human populations and climatic change. Understanding the effects these pressures may have is necessary to designing land management strategies that ensure food security, ecosystem service provision and successful climate mitigation and adaptation. However, the number of complex, interacting effects involved makes any complete understanding very difficult to achieve. Nevertheless, the recent development of integrated modelling frameworks allows for the exploration of the co-development of human and natural systems under scenarios of global change, potentially illuminating the main drivers and processes in future land system change. Here, we use one such integrated modelling framework (the CLIMSAVE Integrated Assessment Platform) to investigate the range of projected outcomes in the European land system across climatic and socio-economic scenarios for the 2050s. We find substantial consistency in locations and types of change even under the most divergent conditions, with results suggesting that climate change alone will lead to a contraction in the agricultural and forest area within Europe, particularly in southern Europe. This is partly offset by the introduction of socioeconomic changes that change both the demand for agricultural production, through changing food demand and net imports, and the efficiency of agricultural production. Simulated extensification and abandonment in the Mediterranean region is driven by future decreases in the relative profitability of the agricultural sector in southern Europe, owing to decreased productivity as a consequence of increased heat and drought stress and reduced irrigation water availability. The very low likelihood (< 33% probability) that current land use proportions in many parts of Europe will remain unchanged suggests that future policy should seek to promote and support the multifunctional role of agriculture and forests in different European regions, rather than focusing on increased productivity as a route to agricultural and forestry viability.

Keywords: Climate change, Socio-economic change, Impacts, Integrated assessment and Uncertainty

INTRODUCTION

Humans have been changing the European landscape for millennia in response to their requirements for the many benefits or ecosystem services arising from the natural environment and its constituent resources. Developments in social systems, new technologies and crops, growing populations and economies have all had dramatic effects (Antrop, 2005; Fischer-Kowalski and Haberl, 2007). Climatic changes have also had substantial impacts, both on the landscape and on human societies, driving a complex pattern of inter-related environmental changes (Messerlia et al., 2000; Büntgen et al., 2011). Now, as the pace of socio-economic and climatic change continues to quicken, their consequences for the land system are commensurately greater and more uncertain. Climate change is likely to have impacts through changes in precipitation, temperature,

CO_2 concentrations and sea level rise, affecting the suitability of land for different crops (Iglesias et al., 2012; Bindi and Olesen, 2011), tree species (Hanewinkel et al., 2013), habitats (e.g. Lehsten et al., 2015) and forms of management (e.g. irrigation - Garrote et al., 2015). Meanwhile, human activities will further modify the European landscape across scales, as populations, dietary preferences, trading patterns and management practices all change (Holman et al., 2008, 2016; Harrison et al., 2013; Rounsevell et al., 2006).

Previous research suggests that these non-climatic pressures may be more important drivers of land use change than climate change (Holman et al., 2005; Rounsevell and Reay, 2009). However, the diversity of social, economic, political and technical factors involved mean that future demands for living space and natural resources are hard to predict. This is exemplified by the breadth of potential socio-economic storylines used in impact assessments, from the global-to-regional Special Report on Emissions Scenarios (SRES; Nakicenovic and Swart, 2000) and Shared Socioeconomic Pathways (SSP; O'Neil et al., 2014) to the continental-to-national storylines developed by stakeholders (e.g. Kok et al., 2015; Metzger et al., 2010). Further uncertainty arises from differences between methods of analysis and modelling that emphasise distinct processes or sectors. These often give divergent projections even under identical climatic and socio-economic scenario conditions, suggesting that the identification and representation of major land change drivers requires significant improvement. In particular, the discrete sectoral nature of many models precludes consideration of the numerous cross-sectoral interactions that influence land use distributions (Harrison et al., 2016). Examples include changes in urban extent or coastal flood defence policy affecting agricultural land availability (Mokrech et al., 2008), and changes in population and water consumption affecting availability of water for irrigation (Henriques et al., 2008). Omitting such interactions can lead to substantial over- or under-estimation of climate impacts and direct and indirect consequences for land use (e.g. Reidsma et al., 2006; de Moel and Aerts, 2011; Di Lucia et al., 2012; Harrison et al., 2016).

One important outcome of uncertainty-focused model applications has been the identification of areas where high levels of uncertainty imply that land use is especially vulnerable to change. However, far less has been discovered about the areas where land use is robust to climatic and socio-economic pressures, and therefore the conditions which allow for the maintenance of food supplies, livelihoods, biodiversity and other ecosystem services. For this purpose, comprehensive scenarios and integrated modelling frameworks are particularly valuable because they allow for fuller, more realistic representations of the land system. Previous examples include studies focusing on future crop yields (Ewert et al., 2015; Wiebe et al., 2015), agricultural land use change (Piorra et al., 2009), broader land and/or energy usage (Verburg et al., 2008; Stürck et al., 2015; van Vuuren et al., 2016), inter-sectoral climate impacts (Fischer et al., 2005; Frieler et al., 2015), and policies for the sustainable development of land use systems (van Delden et al., 2010; Reidsma et al., 2011). Many of these studies involve integrated modelling of European land use; a particularly interesting case due to the research attention it has received in support of a coherent political system that attempts to influence land use outcomes across scales.

While Europe therefore provides a setting where advances in understanding of future land use change should be both possible and of practical value, this potential has not yet been fully realised. In particular, there has been a lack of assessment of the extent of certainty across land use categories under climatic and socio-economic changes acting directly and indirectly through representative cross-sectoral interactions. This paper addresses this gap using an integrated multi-sectoral modelling platform, the CLIMSAVE Integrated Assessment Platform (IAP), which incorporates a broader range of drivers and cross-sectoral processes than previous integrated models, and so allows for less strongly conditional projections of future land use change. We address two research questions:

1. Can highly uncertain futures lead to certain outcomes for European land use in the 2050s?
2. Where in Europe are the current distributions of land uses unlikely to change significantly in the medium term?

We use findings from the IAP to examine the conditions that generate projected stability, and their implications for political interventions intended to maintain land system functionality under global change.

METHODS

The CLIMSAVE IA Platform

The CLIMSAVE IA Platform (IAP) is an interactive, exploratory, web-based tool for simulating climate change impacts and vulnerabilities on a range of sectors (Harrison et al., 2013, 2015a). The Platform integrates a suite of models of urban development, water resources (Wimmer et al., 2015), coasts (Mokrech et al., 2015), agriculture and forests (Audsley et al., 2015), and biodiversity (Dunford et al., 2015a) to simulate the spatial effects of different climatic and socio-economic scenarios across Europe (Fig. 1). The IAP has been applied widely in climate change impact (Audsley et al., 2015; Holman et al., 2016; Wimmer et al., 2015; Harrison et al., 2015b; Mokrech et al., 2015; Harrison et al., 2016), adaptation (Dunford et al., 2015a) and vulnerability (Dunford et al., 2015b) assessments, in robust policy analysis (Jäger et al., 2015) and has been tested extensively through model sensitivity (Kebede et al., 2015) and uncertainty analyses (Dunford et al., 2014; Brown et al., 2015). The Platform operates at a spatial resolution of 10 arcmin × 10 arcmin (approximately 16 km × 16 km in Europe) grid cells, although multiple soil types are represented in each grid cell, and covers two thirty-year timeslices (2020s and 2050s).

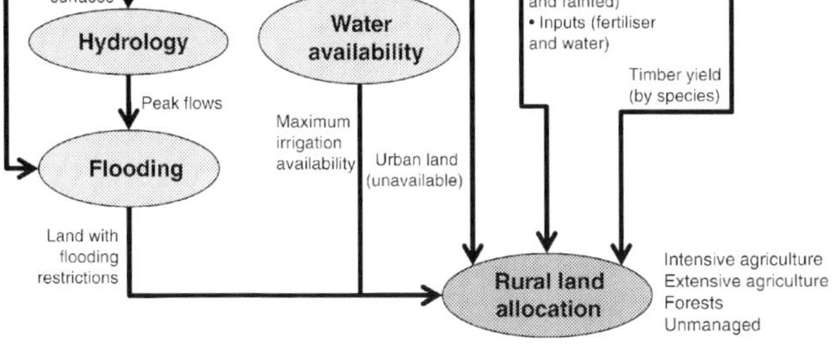

Figure 1: Schematic of the model system interlinkages determining rural land allocation within the CLIMSAVE IAP.

Climate and socio-economic scenarios

Climate scenarios

The climate change scenarios within the IA Platform are based on combinations of the IPCC emissions scenarios (A1b, A2, B1 or B2), three climate sensitivities (low, medium or high) and five global climate models (GCMs). The five GCMs (MPEH5, CSMK3, HadGEM, GFCM21 and IPCM4) were chosen from the CMIP3 database using an objective method to represent as much uncertainty as possible due to between-GCM differences (see Dubrovsky et al., 2015 for further details). Projections of Europe-wide area-average temperature change across these climate models and scenarios range from 1.1 to 4.9 °C in winter and from 1.0 to 3.6 °C in summer in the 2050s. Projections for precipitation change range from increases of between 1.1 and 12.5% in winter and decreases of between 2.0 and 29.5% in summer. The pattern of temperature and precipitation changes differs according to the GCM (see Online Resource 2 of Harrison et al., 2015b). Although we acknowledge that there are more recent scenarios available than those in the IAP, the European area-average changes across these scenarios cover at least the 25th to 75th percentile range of the European changes in summer and winter precipitation and temperature change to 2065 for the CMIP5 global models for the RCP2.6 to RCP8.5 scenarios (Christensen et al., 2013).

Socio-economic scenarios

The IAP contains four European socio-economic scenarios that were developed by stakeholders in a series of professionally-facilitated participatory workshops (see Gramberger et al., 2015). In the first and second workshops, the objectively selected stakeholder group developed and iterated qualitative socio-economic stories and dynamics according to the two drivers that they considered most important and uncertain: "[*effective* vs *ineffective*] solutions by innovation" and "[*gradual* vs *roller-coaster*] economic development". This produced four scenarios which describe the contrasting evolution of a range of social, economic, cultural, institutional and political drivers in Europe (Kok et al., 2015):

- We Are the World – effective governments change the focus from GDP to welfare, which leads to a redistribution of wealth, and thus to less inequality and more (global) cooperation;
- Should I Stay or Should I Go – a failure to address economic crises leads to an increased gap between rich and poor, political instability and conflicts;
- Icarus – short-term policy planning and a stagnating economy lead to disintegration of social fabric and the shortage of goods and services;
- Riders on the Storm – strong economic recessions hit hard, but are successfully countered with renewables and green technologies. Europe is an important player in a turbulent world.

The qualitative stories and quantitative models were linked in a transparent and reproducible way using a "fuzzy set theory methodology" (Kok et al., 2015) in the first and (refined in) the second workshops to determine scenario-specific quantitative values (Supplementary Material Table SM1), expressed relative to the value in the baseline (2010) socioeconomics. These were then used to inform expert-determined values of the remaining model inputs that were representative of the stakeholders' stories.

Models

A meta-modelling approach based on computationally efficient or reduced-form models that emulate the performance of more complex models (Harrison et al., 2013) was used to facilitate greater complexity of model linkages within the IA Platform and a relatively fast run time. Although the CLIMSAVE IAP includes a large number of interlinked models (Fig. 1), this section briefly describes those models which indirectly or directly affect spatial land allocation. For further details on the models, see Harrison et al., 2015a; Harrison et al., 2016and papers cited.

- Urban expansion: The Regional Urban Growth (RUG) metamodel consists of a look-up table of percentage of artificial surfaces per grid cell (between zero and almost 100%) derived from running the RUG model (based on Reginster and

Rounsevell, 2006) with all possible combinations of input values (population, GDP, household preference for proximity to green space versus social amenities, attractiveness of the coast (scenic value versus flood risk) and strictness of the planning regulations to limit sprawl). Development in urban and rural areas is given first priority in the allocation of land;

- Flooding: The Coastal Fluvial Flood (CFFlood) meta-model (Mokrech et al., 2015) is a simplified process-based model that identifies the area at risk of flooding based on topography, relative sea-level rise or change in peak river flow and the estimated Standard of Protection of flood defences. The probability of flood inundation constrains the allocation of land for agriculture, with land with a > 10% and > 50% annual probability of flooding being unsuitable for intensive agriculture and extensive agriculture, respectively, according to Mokrech et al. (2008).
- Water: The WaterGAP (WGMM) meta-model (Wimmer et al., 2015) uses 3D response surfaces to reproduce WaterGAP3 runs at a $5' \times 5'$ resolution for about 100 spatial units (single large river basins or clusters of smaller, neighbouring river basins with similar hydro-geographic properties). The difference between simulated water availability and projected non-agricultural water consumption determines the maximum water available for agricultural irrigation in each spatial unit;
- Forest: MetaGOTILWA + (Audsley et al., 2015) is an artificial neural network (ANN) that emulates GOTILWA + (Gracia et al., 1999). The ANN was trained on GOTILWA results for 889 grid cells across Europe, and simulates average timber yields for a range of deciduous and coniferous tree species under different management regimes and soil characteristics;
- Crops: The crop yield metamodels (Audsley et al., 2015) use ANNs to predict the average yield of a range of annual and permanent crops under rainfed and irrigated conditions. They have each been trained and validated on simulated outputs across Europe from the daily ROIMPEL model (Audsley et al., 2006) for winter and spring wheat, barley and oilseed rape,

potatoes, maize, sunflower, soya, cotton, grass and olives. The training datasets were sampled from 150,000 model data points to adequately cover the range of soil (Panagos et al., 2012) and climate predictors and the predictands.

- Rural land allocation: The SFARMOD meta-model (Audsley et al., 2015) allocates available land across Europe based on profit and other constraints (urban land use, irrigation availability; food and timber demand). It uses a series of regression equations to simulate the behaviour of the full SFARMOD-LP model, a mechanistic farm-based optimising linear programming model of long-term strategic land use. The metamodel was fitted to SFARMOD-LP outputs from 20,000 randomly selected sets of input data that fully cover the current and future parameter input space. The regression is broken into steps that estimate first the percentage of the area of each crop in each grid cell, then the costs of dairy cows (concentrates), then the fixed costs of labour and machinery, from which gross margins, net income and profit is derived. Up to 10 iterations adjust crop and livestock prices to meet the demand for food within Europe, which is a function of population, imports, food preferences and bioenergy. Where the resulting profit is above a threshold (set at €350/ha) land is deemed to be used for intensive agriculture (either arable or dairy agriculture). Otherwise the profit is re-calculated without the arable crops to represent extensive agriculture (sheep and beef) and compared with the profit from managed forests (based on the annual equivalent profit of a total Net Present Value over the life of the forest). If the resulting profit is greater than a second threshold (set at €150/ha) then this land is used for whichever of managed forest or extensive agriculture has the greatest profit. Otherwise the land is not used for productive purposes – it is assumed to be unmanaged forest if the Net Primary Productivity of unmanaged forests is positive and greater than the grass yield of extensive grass, else unmanaged land.

All metamodels were satisfactorily validated against either baseline observations or the validated outputs of the full model (see Holman and Harrison, 2011 for the validation of each model; and the sensitivity analysis of the linked models within the IA Platform in Kebede et al., 2015).

Model runs and analysis

The CLIMSAVE IA Platform was run for 300 scenarios for the 2050s timeslice to explore the effects of climate change and socio-economic change uncertainties on European land use change. The scenario combinations can be categorised into two scenario groups:

- *Climate change only* - Climate scenarios for every combination of the 4 emissions scenarios, three levels of climate sensitivity and five GCMs, combined with baseline socio-economics (60 runs);
- C*limate and socio-economic change* - Climate scenarios (60 runs above) combined with each of the four socio-economic scenarios (240 runs).

Four indicators from the IAP were analysed: 1) area of intensive agriculture; 2) area of extensive agriculture; 3) forest area; and 4) area of unmanaged land. In all analyses, it was assumed that each scenario was equally probable i.e. no prior assumptions were made regarding the likelihood of a particular magnitude of climate or socio-economic change. Grid based results for each of the 60 and 240 simulations were averaged to provide maps of the multi-scenario means for climate change only and climate and socio-economic change simulations in the 2050s, which were compared against the simulated baseline distribution to assess whether there is spatial coherency within the two scenario groups despite the diverse climate and socio-economic scenarios.

Secondly, we investigated whether there are areas in Europe in which individual land uses have future spatial certainty in the direction of change in their spatial extent. There is no common definition of what constitutes important land use change, with recent land cover change across Europe ranging from < 0.1% per annum (as a percentage of national area) in Norway, Malta, Poland and Slovenia to > 0.5% per annum in Portugal and Slovakia (European Environment Agency, 2010) and scenario studies projecting changes in the spatial extent of individual European land use classes of up to 5% by 2020 (e.g. European Centre for Nature Conservation, 2006) and 8% by 2050 (e.g. Rounsevell et al., 2005) compared to 2000. We adopted a threshold of 5%, but tested the robustness of results across a range of thresholds from 0.1% to 25%. The probability of changes in extent of each land use within each grid cell of > 5% of the simulated baseline (using 1961–90 climate and 2010 socio-economics) was calculated and classified according to the likelihood scale of Mastrandrea et al. (2010) to identify those cells in which an increase or decrease in land use extent is "likely" (66–90% probability) or "very likely" (> 90% probability). Cells in which no change in the spatial extent of each individual land use greater than ± 5% was "very likely" were also identified. Finally, the probability that none of the four land uses change in extent by > 5% was calculated for each grid square to identify areas where there is spatial certainty that all land uses will remain largely unchanged in extent. These analyses were then repeated across the range of thresholds above.

RESULTS

Multi-scenario average spatial land use allocation

Given the divergent nature of the socioeconomic scenarios and differing spatial patterns of temperature and precipitation change from the climate models, the simulated baseline distribution of each land use was compared with the multi-scenario mean of the [60] simulations with climate change only and the [240] simulations with climate and socio-economic change (Fig. 2 and Figs. SM2–4). It is apparent that there is a spatial coherency in the simulated future distribution across Europe for each land use; i.e. averaging the individual grid-level values resulting from the diverse range of input scenarios has not led to a quasi-random distribution of land use allocation. Focusing on intensive agriculture, the effect of climate change alone (with baseline economics meaning that there is no change in the food demand or imports) leads to a projected northward migration of intensive agriculture (Fig. 2), particularly into northern UK and Finland. This is also associated with reduced intensive agricultural areas in southern Europe (especially Spain and Italy) due to a combination of increased heat stress and reduced availability of irrigation water. These two factors reduce the simulated relative competitiveness of

Mediterranean agriculture in contributing towards meeting the demand for European agricultural production, so that the required demand for production can be met by a smaller agricultural area focused in a band across central and north-western Europe due to yield increases.

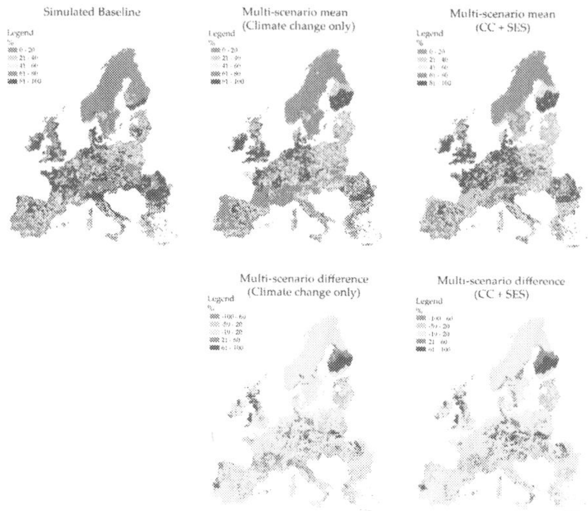

Figure 2: Comparison of the simulated baseline distribution of intensive agricultural land with the multi-scenario mean of the [60] simulations with climate change only and the [240] simulations with climate and socio-economic change for the 2050s.

The introduction of socioeconomic change affects modelled European agricultural land use requirements by changing both the demand for agricultural production, through changing food demand (due to changes in population, wealth and dietary preferences) and net imports (arising from changes in Europe's relationship with the rest of the world), and the efficiency of agricultural production (through scenario changes in mechanisation, yield development and crop breeding and irrigation efficiency). The multi-scenario mean in Fig. 2 shows the most intensive areas of agriculture (red areas) in the same locations as the climate-only multi-scenario mean (as these are the most profitable production areas) but expands the production area into regions that had lost competitiveness under climate change such as southern France and the Baltic states.

Model certainty in the direction of land use change across Europe

Fig. 3 and Fig. SM5 shows the certainty in the modelled direction of change in the spatial extent of each land use class across all of the scenario combinations, expressed as the percentage of runs in which the land use class changes in extent by > 5% (compared to the baseline simulation) in a given grid cell. The sensitivity analysis using change thresholds of between 0.1 and 25% within each grid cell (Fig. SM1 in the Supplementary material) shows that increasing the change threshold inevitably leads to an increasing spatial extent in the area with certain "no change" and decreasing extent of areas with both uncertain change and certain (increasing and decreasing) change. However, the overall percentage in each certainty class is not fundamentally changed, demonstrating the robustness of the results.

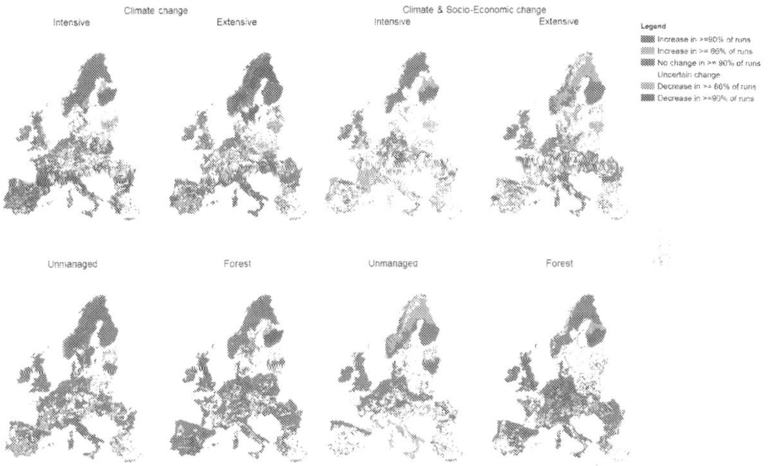

Figure 3: Multi-scenario certainty in the direction of modelled land use change (of at least 5% within a grid cell) for the 2050s for (left) climate change only and (right) climate and socioeconomic change.

With climate change alone, there are areas with significant confidence in the direction of change for all land uses, with forest area decreasing in > 90% of simulations in significant areas of all regions of Europe; unmanaged land increasing in southern Europe and Scandinavia at the expense of intensive and extensive agricultural land; extensive agricultural land increasing through a band across the centre of Europe at the expense of intensive agricultural land; and intensive agriculture increasing in parts of northern Europe at the expense of forest. There are also areas with confidence of no change – for example, intensive agriculture is unchanged within much of central England, northern France and the Benelux countries; whilst forest is little changed in large areas of Scandinavia, Spain and central Europe.

However, there is a major decrease in the confidence of the direction of change and a large expansion of the areas with uncertain change (i.e. < 66% agreement) when the socio-economic scenarios are introduced (Fig. 3 and Fig. SM5). In particular, there are no significant areas with high (> 90% of simulations) confidence in the decrease of intensive agricultural land or increase in unmanaged land. There are only areas within Scandinavia, Italy, France and Hungary in which there is at least a 66% agreement of an increase in unmanaged land. The areas with certainty of no change (change of < 5%) decrease slightly for most land covers.

Model certainty in stable land use patterns across Europe

Fig. 3 showed that there are areas across Europe in which there are high levels of certainty in the direction of change (either increases or decreases) in the extent of individual land use classes due to climate and socioeconomic change. However, it also showed extensive areas across Europe with either certainty of little change or uncertainty in the magnitude or direction of change. Fig. 4 therefore shows the percentage of simulations in which all four land use classes within a given grid cell change by less than ± 5% from the baseline proportions, classed according to the likelihood scale of Mastrandrea et al. (2010). Approximately 20% of the simulated cells across Europe are very likely (90–99% probability) or virtually certain (99–100% probability) to maintain their baseline land use proportions despite the effects of climate change on land suitability and crop and timber yields (Table 1). These stable cells include significant areas within the UK, northern France, northern Spain, Germany and Scandinavia, and account for > 50% of the areas of three countries: The Netherlands (88%), Republic of Ireland (54%) and Norway (51%).

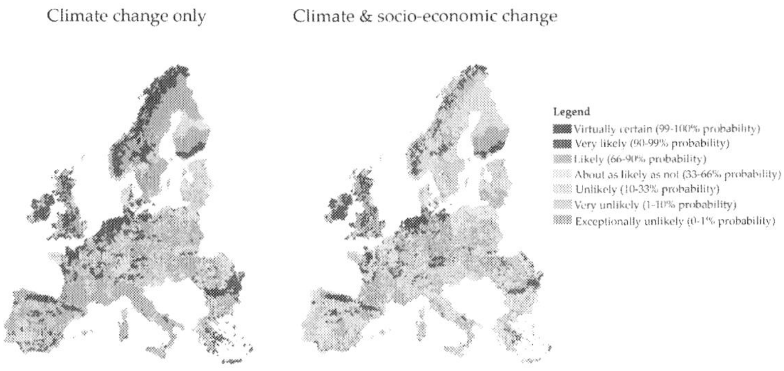

Figure 4: Certainty of simulated baseline extent of land covers remaining unchanged (± 5%) under simulations with climate change only [60] and simulations with climate and socio-economic change [240] in the 2050s.

Table 1: Likelihood of baseline land cover proportions remaining unchanged (± 5%) across modelled grids under 2050s climate and socioeconomic change

Likelihood scale	Probability	Climate change (% of grids)	Climate and socio-economic change (% of grids)
Virtually certain	99–100%	17	8
Very likely	90–99%	3	4
Likely	66–90%	8	12
About as likely as not	33–66%	11	16
Unlikely	10–33%	12	28
Very unlikely	1–10%	7	13
Exceptionally unlikely	0–1%	42	20

The introduction of the socio-economic scenarios decreases the extent of areas very likely or virtually certain to retain baseline land use proportions from 20% to about 12% of the simulated grid cells (Table 1). Only the Netherlands (54%) and Republic of Ireland (50%) now have > 50% of grid cells in this category. At a national scale, the socio-economic scenarios lead to a reduction in areas of certainty in all countries, with the exception of the Czech Republic and Austria where there is an increase of 1% and 5%, respectively. There are also smaller areas where the certainty of maintaining the current land use distribution increases; for example eastern England, northern Romania and the Po valley in Italy. However, whilst socio-economic scenarios reduce overall land use certainty in

Europe, the percentage of cells in which baseline land use proportions are exceptionally unlikely (0–1% probability) to remain unchanged also decreases, from 42% to 20%, with associated increases in cells that are very unlikely (1–10% probability) or unlikely (10–33% probability) to remain unchanged.

DISCUSSION

Changes in global land systems in the coming decades will be strongly influenced by a range of interacting climatic and socio-economic factors. This complexity makes projections of future change hard to achieve, and reliant on integrated modelling approaches that respect the dynamics that occur within and between individual sectors. As a result, assessments of future conditions, made largely in the absence of well-developed integrated approaches of this kind, have so far focused on areas of uncertainty. However, recent methodological advances allow for more confident exploration of the converse; areas that, with some level of certainty, appear robust to external drivers of change and internal complexity. Here, we applied a cross-sectoral European modelling platform spanning a wide range of climatic and socio-economic conditions, which therefore covers or exceeds the uncertainty space identified in previous studies of future land use change (e.g. Harrison et al., 2016; Brown et al., 2015; Prestele et al., 2016; Verburg et al., 2013; Stürck et al., 2015). Uncertainty in future climate conditions was addressed through sixty climate change scenarios that spanned at least the 25th to 75th percentile range of the European changes in summer and winter precipitation and temperature change to 2065 for the CMIP5 global models for the RCP2.6 to RCP8.5 scenarios, and included differing spatial patterns of change. Furthermore the four socio-economic scenarios included contrasting directions of change in key scenario inputs such as population, GDP, spatial planning policy, societal behaviour (including dietary preferences for meat and water consumption), crop breeding and agronomic improvement that influence land-take for development, food demand, irrigation water availability and agricultural productivity.

Given the extent of uncertainty space investigated here, an entirely divergent set of projected future conditions might be expected. However, we instead find substantial consistencies across results. Indeed, the strong spatial patterns within the multi-scenario means of individual land use distributions in the 2050s (Fig. 2 and Figs. SM2–4) suggest a considerable degree of predictability in future land use at the European scale. In particular, the analysis presented here suggests that there are areas of Europe with high levels of certainty in two aspects of future land use change – the direction of change for individual land uses and the probability of maintaining the current land use distribution.

These findings are consistent with Brown et al. (2015) who found significant overlap in the probability density functions of the grid-based indicators of food production per capita, land use intensity index and land use diversity index for the four contrasting CLIMSAVE scenarios. This consistency is especially notable because the analysis of Brown et al. (2015)included model uncertainty that we do not directly measure here; a potentially large source of uncertainty (e.g. Alexander et al., 2016). Furthermore, some of our broad and specific findings are consistent with those of other integrated and stand-alone models. For example, the magnitudes of change in the agricultural sector are in line with the (albeit wide) range found by the inter-model comparison of Alexander et al. (2016), while our finding that some areas of Europe appear to be robust to a wide range of scenario conditions also agrees with the studies of Verburg et al. (2010), Verburg et al. (2013), Stürck et al. (2015) and Prestele et al. (2016). These earlier studies found substantial areas of stability and/or predictability using more narrowly-based or discrete modelling approaches at European and global scales (which included the effects, for instance, of GDP and population change). These earlier findings also have some similar geographical characteristics, particularly in the locations of agricultural intensification in northern and western Europe and agricultural extensification, diversification or abandonment in parts of southern and eastern Europe (see e.g. Stürck et al., 2015, Fig. 8). These suggest that error propagation in the IAP's cross-sectoral models is not amplifying the impact of individual, sub-model uncertainty, supporting the conclusions of Kebede et al. (2015) and Dunford et al. (2014). The present study represents a valuable advance, therefore, because it encompasses a greater number of socio-economic drivers of change and of sectoral and cross-sectoral processes (also known to be large potential sources of uncertainty; Harrison et al., 2016), thereby allowing for the identification of particular characteristics that lead to certainty in land use futures.

Some of the most informative and novel findings presented here relate to the ways in which specific drivers affect, or fail to affect, land use patterns. For instance, we found that climate change was associated with a number of key simulated changes. A decrease in forest area was likely (66–100% probability) in many regions of Europe due to large increases in tree growth simulated by metaGOTILWA + under climate change. This suggests a direction of change and degree of certainty not identified by previous studies (e.g. Stürck et al., 2015), which may result from differences in underlying assumptions about climate impacts on tree growth and socio-economic impacts on wood demand (Eggers et al., 2008). In our simulations, climate change also drove agricultural changes (Fig. 2) that appear to reinforce the current trends of intensification of agriculture in northern and western Europe and extensification and abandonment in the Mediterranean region (Levers et al., 2015), as shown by the likely (66–100% probability) decreases in intensive and extensive agriculture and likely increase in unmanaged or very low intensity agriculture in many parts of southern Europe (Fig. 3). The combination of increased heat stress due to higher temperatures, increased summer drought stress in rainfed systems and reduced availability of irrigation water in the region all contribute to reducing profitability and competitiveness of Mediterranean agriculture compared to that in central and northwestern Europe. This suggests that established findings about the impacts of climate change on European agriculture (e.g. Olesen and Bindi, 2002; Bindi and Olesen, 2011) are largely robust to socio-economic change and cross-sectoral interactions, as also indicated by earlier integrated modelling studies (e.g. Verburg et al., 2013; Stürck et al., 2015).

We also find that socio-economic drivers of land use change play a crucial role, sometimes even dominating over climatic drivers. For instance, the introduction of socio-economic scenarios partly offsets the loss of agricultural land in southern Europe leading to a (small) increased likelihood of stable land systems in very unstable areas (as shown by grids shifting from "exceptionally unlikely" to remain unchanged to "unlikely or "very unlikely" – Table 1). This arises as large increases in European food demand in some scenarios (due to population growth, increased affluence and reduced net imports) allow more marginal areas to continue to contribute profitably to production. This is also true of areas such as the Po valley in Italy with its deep water-retentive soils.

Nevertheless, the total proportion of grid cells which are very likely (90–99% probability) or virtually certain (99–100% probability) to maintain baseline land use proportions decrease from 20% of grid cells under climate change alone to just 12% with both climate and socio-economic change as a consequence of the major differences in food and timber demand, agricultural productivity and food trade across the divergent socio-economic scenarios, which together determine the requirements for land . These stable areas are mostly in north-western Europe and Scandinavia, while many parts of the Mediterranean have a high likelihood of changing land use distributions (with the main exception of the Pyrenees and Cantabrian mountain chains in northern and north-eastern Spain respectively). In fact, extensification or abandonment in currently marginal areas, as agronomic conditions become less favourable through increased summer drought stress, reduced availability of irrigation water and decreasing profitability, give a high likelihood of land use change in much of Spain, southern France and Italy (Fig. 3). Elsewhere (e.g. Scandinavia), climatically-driven increases in timber yields act alongside increasing demands for agricultural land to substantially reduce the projected area of forestry, especially where timber prices also fall (Brown et al., 2015).

It is apparent from these results that some specific factors promote or impede land use change in general. Particularly influential were population growth rates, dietary preferences, yields and imports (Kebede et al., 2015; Audsley et al., 2015), with each having direct consequences for the extent of agricultural land required. Areas that are robust to variations in these factors, and therefore relatively certain of maintaining current land uses, have a number of key characteristics – (1) mountainous areas in which topography (slope) and soil depth constrain land use options, such as the Pyrenees and the Pico de Europa in Spain; (2) relatively wet and mild areas with maritime climate influences which support strong grass growth but remain unsuitable for conversion to arable agriculture, such as parts of the Republic of Ireland and central England; and (3) intensive agricultural areas with deep, water-retentive soils and favourable climates that assure continuing profitability and competitiveness, such as the Netherlands. In contrast, autonomous adaptation within much of southern Europe is likely to be insufficient to offset reduced relative profitability compared to northern and western

Europe, where competitiveness increases through increases in yield and expansion of climatically suitable areas. Therefore, further adaptation strategies (autonomous or planned) in the Mediterranean and south-eastern parts of Europe should not focus on trying to increase productivity as a route to agricultural profitability but rather strike a balance between the economic, environmental and social functions of different European regions that recognises the multifunctionality of land use (Bindi and Olesen, 2011; Brown et al., 2016a).

While our results are valid across a large area of uncertainty space due to our use of several integrated models and socio-economic scenarios, there is some residual potential for uncertainty. In particular, model uncertainty (in terms of the design and form of integration of models) is known to be an important source of uncertainty in land change projections (Alexander et al., 2016; Dormann et al., 2007; Olesen et al., 2007; Harrison et al., 2016), and one that we do not consider beyond the use of several distinct GCMs in our simulations. This may be particularly significant where the decision-making of land managers and policy-makers is concerned, given the importance of these in translating drivers into realised changes in land use (van Vliet et al., 2015; Brown et al., 2016a, 2016b). The profit-based assumptions about decision-making used in CLIMSAVE models do not allow for a variety of effects that may occur in reality, from disjointed, protectionist responses to local or national changes in livelihoods, to migration, conflict or other societal disruptions (e.g. O'Brien et al., 2006; Black et al., 2011). Therefore, while these results do not depend upon an artificially fragmented conceptualisation of land system dynamics, they remain conditional upon aspects of model design that are not explored through our simulations or through comparisons with other models.

Nevertheless, it is clear that the challenges associated with climate change will be far larger in some parts of Europe than in others. While robustly productive or unproductive areas will have relatively little need or scope for adaptation, areas that are currently marginal may face complex and difficult choices, which do not realistically include the maintenance of existing land uses. In these places, the opportunities and processes of adaptation need to be carefully considered if the essential contributions of land use to human society are to be preserved.

Policy has an important role here, in smoothing transitions to more multi-functional management that may not otherwise arise without considerable disruption to livelihoods and environments. Policies that ease preferred changes in land use, rather than attempting to prevent change, are likely to be more successful in maintaining livelihoods, food supplies, and essential ecosystem services. Indeed, consideration of the full range of land-based services will be essential if synergies and trade-offs between them are to be appropriately managed, and it will also be necessary for policy-makers to consider a wide range of possible future conditions, as well as broad European and global contexts. In Europe, the extent of coherency between national policies will be important, and a decisive factor is likely to be the extent to which the future Common Agricultural Policy encourages the provision of societal goods such as climate mitigation, biodiversity preservation and livelihood diversification over economic production. In the absence of such support, reactive adaptations to the impacts of socio-economic and climatic changes are likely to be detrimental at local and European scales.

CONCLUSIONS

The CLIMSAVE IAP has been used to simulate the spatial distribution of four broad land use classes (intensive agriculture, extensive agriculture, very low intensity/unmanaged land and forest) across Europe in the 2050s under a broad range of climatic and socioeconomic conditions. Sixty climate scenarios spanned the equivalent of at least the 25th to 75th percentile range of changes in summer and winter precipitation and temperature change to 2065 within RCP2.8 to RCP8.5. These were combined with baseline socio-economics (climate change only) and four socioeconomic scenarios (climate and socioeconomic change) to understand the certainty in the direction of change of future land use allocation and the certainty in maintaining an unchanged proportion of land use classes into the future. Results suggested that climate change alone will lead to a contraction in the agricultural and forest area within Europe, particularly in southern Europe, which is partly offset by socioeconomic changes in both the demand for agricultural production, through changing food demand and net imports, and the efficiency of agricultural production. Whilst this modelling shows significant areas in northern and western Europe in which it is likely (> 66% probability) that current land use proportions will remain largely unchanged, the results reinforce current trends and previous integrated model findings of intensification of agriculture in northern and western Europe and extensification and abandonment in the Mediterranean region. These changes are driven by decreases in the relative profitability of the agricultural sector in southern Europe, owing to decreased productivity as a consequence of increased heat and drought stress and reduced irrigation water availability. The very low likelihood (< 33% probability) that current land use proportions in many parts of Europe will remain unchanged in the future suggests that policy should seek to promote and support the multifunctional role of agriculture and forests that recognises their contribution to the economic, environmental and economic functions of different European regions rather than focusing on supporting increased productivity as a route to agricultural and forestry viability.

ACKNOWLEDGEMENTS

The research leading to these results has received funding from the Biotechnology and Biological Sciences Research Council (BBSRC) grant no BB/K010301/1 and BB/N00485X/1(as part of the MACSUR knowledge hub within FACCE Joint Programming Initiative for Agriculture, Climate Change, and Food Security) and the European Commission Seventh Framework Programme under Grant Agreement No. 244031 (The CLIMSAVE Project;Climate change integrated assessment methodology for cross-sectoral adaptation and vulnerability in Europe; www.climsave.eu). CLIMSAVE is an endorsed project of the Global Land Project of the IGBP. The authors would like to thank all CLIMSAVE partners for their contributions to the IAP. The CLIMSAVE IAP model outputs can be accessed by contacting researchdata@cranfield.ac.uk.

REFERENCES

1. Antrop, 2005 M. AntropWhy landscapes of the past are important for the future
2. Landsc. Urban Plan., 70 (1–2) (2005), pp. 21-34
3. Audsley et al., 2006 E. Audsley, K.R. Pearn, C. Simota, G. Cojocaru, E. Koutsidou, M.D.A. Rounsevell, M.Trnka, V. AlexandrovWhat can

scenario modelling tell us about future European scale agricultural land use, and what not? Environ. Sci. Pol., 9 (2006), pp. 148-162

4. Audsley et al., 2015 E. Audsley, M. Trnka, S. Sabaté, A. SanchezInteractive modelling of land profitability to estimate European agricultural and forest land use under future scenarios of climate, socio-economics and adaptation Clim. Chang., 128 (2015), pp. 215-227
5. Bindi and Olesen, 2011 M. Bindi, J.E. OlesenThe responses of agriculture in Europe to climate change Reg. Environ. Chang., 11 (2011), pp. S151-S158
6. Black et al., 2011 R. Black, S.R.G. Bennett, S.M. Thomas, J.R. BeddingtonClimate change: migration as adaptation Nature, 478 (2011), pp. 447-449
7. Brown et al., 2015 C. Brown, E. Brown, D. Murray-Rust, G. Cojocaru, C. Savin, M. RounsevellAnalysing uncertainties in climate change impact assessment across sectors and scenarios Clim. Chang., 128 (2015), pp. 293-306
8. Brown et al., 2016a C. Brown, S. Holzhauer, M. Metzger, J. Paterson, M. RounsevellLand managers' behaviours modulate pathways to visions of future land systems Reg. Environ. Chang. (2016), 10.1007/s10113-016-0999-y (in press)
9. Brown et al., 2016b C. Brown, K. Brown, M. RounsevellA philosophical case for process-based modelling of land use change Model. Earth Syst. Environ., 2 (2016), p. 50
10. Büntgen et al., 2011 U. Büntgen, W. Tegel, K. Nicolussi, M. McCormick, D. Frank, V. Trouet, J.O. Kaplan, F.Herzig, K. Heussner, H. Wanner, J. Luterbacher, J. Esper2500 years of European climate variability and human susceptibility Science, 331 (6017) (2011), pp. 578-582
11. Christensen et al., 2013 Christensen, J.H., K. Krishna Kumar, E. Aldrian, S.-I. An, I.F.A. Cavalcanti, M. de Castro, W. Dong, P. Goswami, A. Hall, J.K. Kanyanga, A. Kitoh, J. Kossin, N.-C. Lau, J. Renwick, D.B. Stephenson, S.-P. Xie and T. Zhou, 2013: Climate Phenomena and their Relevance for Future Regional Climate Change Supplementary Material. In: Climate Change 2013: The Physical Science Basis. Contribution of Working Group I to the

Fifth Assessment Report of the Intergovernmental Panel on Climate Change [Stocker, T.F., D. Qin, G.-K. Plattner, M. Tignor, S.K. Allen, J. Boschung, A. Nauels, Y. Xia, V. Bex and P.M. Midgley, eds.]. Available from www.climatechange2013.org and www.ipcc.ch.

12. de Moel and Aerts, 2011 H. de Moel, J.C.J.H. AertsEffect of uncertainty in land use, damage models and inundation depth on flood damage estimates Nat. Hazards, 58 (1) (2011), pp. 407-425
13. Di Lucia et al., 2012 L. Di Lucia, S. Ahlgren, K. EricssonThe dilemma of indirect land-use changes in EU biofuel policy – an empirical study of policy-making in the context of scientific uncertainty Environ. Sci. Pol., 16 (2012), pp. 9-19
14. Dormann et al., 2007 C.F. Dormann, O. Schweiger, P. Arens, I. Augenstein, S. Aviron, D. Bailey, J. Baudry, R.Billeter, R. Bugter, R. Bukácek, F. Burel, M. Cerny, R. De Cock, G. De Blust, R. DeFilippi, T. Diekötter, J.Dirksen, W. Durka, P.J. Edwards, M. Frenzel, R. Hamersky, F. Hendrickx, F. Herzog, S. Klotz, B.Koolstra, A. Lausch, D. Le Coeur, J. Liira, J.P. Maelfait, P. Opdam, M. Roubalova, A. Schermann-Legionnet, N. Schermann, T. Schmidt, M.J.M. Smulders, M. Speelmans, P. Simova, J. Verboom, W. van Wingerden, M. ZobelPrediction uncertainty of environmental change effects on temperate European biodiversity Ecol. Lett., 11 (3) (2007), pp. 235-244
15. Dubrovsky et al., 2015 M. Dubrovsky, M. Trnka, E. Svobodova, I.P. HolmanDeveloping a reduced-form ensemble of climate change scenarios for Europe and its application to selected impact indicators Clim. Chang., 128 (3–4) (2015), pp. 169-186
16. Dunford et al., 2014 R. Dunford, P.A. Harrison, M.D.A. RounsevellExploring scenario and model uncertainty in cross-sectoral integrated assessment approaches to climate change impacts Clim. Chang., 132 (3) (2014), pp. 417-432
17. Dunford et al., 2015a R. Dunford, A. Smith, P.A. Harrison, D. HangauEcosystem service provision in a changing Europe: adapting to the impacts of combined climate and socio-economic change Landsc. Ecol., 30 (2015), pp. 443-461
18. Dunford et al., 2015b R. Dunford, P.A. Harrison, J. Jäger, M.D.A. Rounsevell, R. TinchExploring climate change vulnerability across

sectors and scenarios using indicators of impacts and coping capacity Clim. Chang., 128 (2015), pp. 339-354

19. Eggers et al., 2008 J. Eggers, M. Lindner, S. Zudin, S. Zaehle, J. LiskiImpact of changing wood demand, climate and land use on European forest resources and carbon stocks during the 21st century Glob. Chang. Biol., 14 (10) (2008), pp. 2288-2303
20. European Centre for Nature Conservation, 2006 European Centre for Nature ConservationThe scenario study on agriculture and the rural world (Scenar 2020) European Commission Contract No. 30 – CE – 0040087/00-08 (2006) Available from http://ec.europa.eu/agriculture/agrista/2006/scenar2020/final_report/scenar2020final.pdf (accessed 19/09/16)
21. European Environment Agency, 2010 European Environment AgencyThe Eurpean Environment State and Outlook 2010: Landuse European Environment Agency, Copenhagen (2010)
22. Ewert et al., 2015 F. Ewert, R.P. Rötter, M. Bindi, H. Webber, M. Trnka, K.C. Kersebaum, J.E. Olesen, M.K. van Ittersum, S. Janssen, M. Rivington, M.A. Semenov, D. Wallach, J.R. Porter, D. Stewart, J. Verhagen, T.Gaiser, T. Palosuo, F. Tao, C. Nendel, P.P. Roggero, L. Bartošová, S. AssengCrop modelling for integrated assessment of risk to food production from climate change Environ. Model Softw., 72 (2015), pp. 287-303
23. Fischer et al., 2005 G. Fischer, M. Shah, F.N. Tubiello, H. van VelhuizenSocio-economic and climate change impacts on agriculture: an integrated assessment, 1990–2080 Philos. Trans. R. Soc. B, 360 (1463) (2005), pp. 2067-2084
24. Fischer-Kowalski and Haberl, 2007 M. Fischer-Kowalski, H. Haberl (Eds.), Socioecological Transitions and Global Change: Trajectories of Social Metabolism and Land Use, Edward Elgar Publishing (2007)
25. Frieler et al., 2015 K. Frieler, A. Levermann, J. Elliott, J. Heinke, A. Arneth, M.F.P. Bierkens, P. Ciais, D.B. Clark, D. Deryng, P. Döll, P. Falloon, B. Fekete, C. Folberth, A.D. Friend, C. Gellhorn, S.N. Gosling, I.Haddeland, N. Khabarov, M. Lomas, Y. MasakiA framework for the cross-sectoral integration of multi-model impact projections: land use decisions under climate impacts uncertainties

Earth Syst. Dynam., 6 (2) (2015), pp. 447-460

26. Garrote et al., 2015 L. Garrote, A. Iglesias, A. Granados, L. Mediero, F. Martin-CarrascoQuantitative assessment of climate change vulnerability of irrigation demands in Mediterranean Europe Water Resour. Manag., 29 (2) (2015), pp. 325-338
27. Gracia et al., 1999 C. Gracia, E. Tello, S. Sabaté, J. BellotGOTILWA: An integrated model of water dynamics and forest growth F. Rodà, J. Retana, C. Gracia, J. Bellot (Eds.), Ecology of Mediterranean Evergreen Oak Forests, Springer, Berlin (1999), pp. 163-179
28. Gramberger et al., 2015 M. Gramberger, K. Zellmer, K. Kok, M. MetzgerStakeholder Integrated Research, STIR: a new approach tested in climate change adaptation research Clim. Chang., 128 (2015), pp. 201-214
29. Hanewinkel et al., 2013 M. Hanewinkel, D.A. Cullmann, M.J. Schelhaas, G.J. Nabuurs, N.E. ZimmermannClimate change may cause severe loss in the economic value of European forest land Nat. Clim. Chang., 3 (3) (2013), pp. 203-207
30. Harrison et al., 2013 P.A. Harrison, I.P. Holman, G. Cojocaru, K. Kok, A. Kontogianni, M.J. Metzger, M.GrambergerCombining qualitative and quantitative understanding for exploring cross-sectoral climate change impacts, adaptation and vulnerability in Europe Reg. Environ. Chang., 13 (2013), pp. 761-780
31. Harrison et al., 2015a P.A. Harrison, I.P. Holman, P.M. BerryAssessing cross-sectoral climate change impacts, vulnerability and adaptation: an introduction to the CLIMSAVE project Clim. Chang., 128 (2015), pp. 153-167
32. Harrison et al., 2015b P.A. Harrison, R. Dunford, C. Savin, M.D.A. Rounsevell, I.P. Holman, A.S. Kebede, B.StuchCross-sectoral impacts of climate change and socio-economic change for multiple European land- and water-based sectors Clim. Chang., 128 (2015), pp. 279-292
33. Harrison et al., 2016 P.A. Harrison, R. Dunford, I.P. Holman, M.D.A. RounsevellClimate change impact modelling needs to include cross-sectoral interactions Nat. Clim. Chang. (2016), 10.1038/nclimate3039
34. Henriques et al., 2008 C. Henriques, I.P. Holman, E. Audsley, K.

PearnAn interactive multi-scale integrated assessment of future regional water availability for agricultural irrigation in East Anglia and North West England Clim. Chang., 90 (1–2) (2008), pp. 89-111

35. Holman and Harrison, 2011 I.P. Holman, P.A. Harrison (Eds.), Report describing the development and validation of the sectoral meta-models for integration into the IA platform, CLIMSAVE Deliverable, 2 (2011), p. 2 http://www.climsave.eu/climsave/doc/Report_on_the_Meta-models.pdf (Accessed 13 March 2015)
36. Holman et al., 2005 I.P. Holman, R.J. Nicholls, P.M. Berry, P.A. Harrison, E. Audsley, S. Shackley, M.D.A.RounsevellA regional, multi-sectoral and integrated assessment of the impacts of climate and socio-economic change in the UK: II results Clim. Chang., 71 (2005), pp. 43-73
37. Holman et al., 2008 I.P. Holman, M.D.A. Rounsevell, P.M. Berry, R.J. NichollsDevelopment and application of participatory integrated assessment software to support local/regional impact and adaptation assessment Clim. Chang., 90 (1–2) (2008), pp. 1-5
38. Holman et al., 2016 I.P. Holman, P.A. Harrison, M.J. MetzgerCross-sectoral impacts of climate and socio-economic change in Scotland - implications for adaptation policy Reg. Environ. Chang., 16 (1) (2016), pp. 97-109
39. Iglesias et al., 2012 A.Iglesias, L. Garrote, S. Quiroga, M. MoneoA regional comparison of the effects of climate change on agricultural crops in Europe Clim. Chang., 112 (1) (2012), pp. 29-46
40. Jäger et al., 2015 J. Jäger, M.D.A. Rounsevell, P.A. Harrison, I. Omann, R. Dunford, M. Kammerlander, G.PatakiAssessing policy robustness of climate change adaptation measures across sectors and scenarios Clim. Chang., 128 (2015), pp. 395-407
41. Kebede et al., 2015 A.S. Kebede, R. Dunford, E. Audsley, P.A. Harrison, I.P. Holman, M. Mokrech, R.J. Nicholls, S. Rickebusch, M.D.A. Rounsevell, S. Sabaté, F. Sallaba, A. Sanchez, C.-M. Savin, M. Trnka, F. WimmerThe sensitivity of cross-sectoral impacts to climate and socio-economic drivers for key European sectors Clim. Chang., 128 (2015), pp. 261-277
42. Kok et al., 2015 K. Kok, J. Sendzimir, I. Bärlund, M. Flörke, M. Gramberger, K. Zellmer, B. Stuch, I.P. HolmanEuropean

participatory scenario development: strengthening the link between stories and models Clim. Chang., 128 (3–4) (2015), pp. 187-200

43. Lehsten et al., 2015 V. Lehsten, M.T. Sykes, A.V. Scott, J. Tzanopoulos, A. Kallimanis, A. Mazaris, P.H.Verburg, C.J.E. Schulp, S.G. Potts, I. VogiatzakisDisentangling the effects of land-use change, climate and CO2 on projected future European habitat types Glob. Ecol. Biogeogr., 24 (6) (2015), pp. 653-663
44. Levers et al., 2015 C. Levers, D. Müller, K. Erb, H. Haberl, M.R. Jepsen, M.J. Metzger, P. Meyfroidt, T.Plieninger, C. Plutzar, J. Stürck, P.H. Verburg, P.J. Verkerk, T. KuemmerleArchetypical patterns and trajectories of land systems in Europe Reg. Environ. Chang. (2015), 10.1007/s10113-015-0907-x
45. Mastrandrea et al., 2010 M.D. Mastrandrea, C.B. Field, T.F. Stocker, O. Edenhofer, K.L. Ebi, D.J. Frame, H.Held, E. Kriegler, K.J. Mach, P.R. Matschoss, G.-K. Plattner, G.W. Yohe, F.W. ZwiersGuidance Note for Lead Authors of the IPCC Fifth Assessment Report on Consistent Treatment of Uncertainties Intergovernmental Panel on Climate Change, IPCC (2010) Available at http://www.ipcc.ch
46. Messerlia et al., 2000 B. Messerlia, M. Grosjeana, T. Hoferb, L. Núñezc, C. PfisterdFrom nature-dominated to human-dominated environmental changes Quat. Sci. Rev., 19 (1–5) (2000), pp. 459-479
47. Metzger et al., 2010 M.J. Metzger, M.D.A. Rounsevell, H.A.R.M. Heiligenberg, M. Perez-Soba, P.S. HardimanHow personal Judgment influences scenario development: an example for future rural development in Europe Ecol. Soc., 15 (2010), p. 5 [online] Available at: http://www.ecologyandsociety.org/vol15/iss2/art5/ (accesed 20th June 2016)
48. Mokrech et al., 2008 M. Mokrech, R.J. Nicholls, J.A. Richards, C. Henriques, I.P. Holman, S. ShackleyRegional impact assessment of flooding under future climate and socio-economic scenarios for East Anglia and North West England Clim. Chang., 90 (1–2) (2008), pp. 31-55
49. Mokrech et al., 2015 M. Mokrech, A.S. Kebede, R.J. Nicholls, F. Wimmer, L. FeyenAn integrated approach for assessing flood impacts due to future climate and socio-economic conditions and

the scope of adaptation in Europe Clim. Chang., 128 (2015), pp. 245-260

50. Nakicenovic and Swart, 2000 N. Nakicenovic, R. Swart (Eds.), Special Report on Emission Scenarios, Cambridge University Press, Cambridge, UK (2000)
51. O'Brien et al., 2006 K. O'Brien, S. Eriksen, L. Sygna, L.O. NaessQuestioning complacency: climate change impacts, vulnerability, and adaptation in Norway Ambio, 35 (2) (2006), pp. 50-56
52. Olesen and Bindi, 2002 J.E. Olesen, M. BindiConsequences of climate change for European agricultural productivity, land use and policy Eur. J. Agron., 16 (4) (2002), pp. 239-262
53. Olesen et al., 2007 J.E. Olesen, T.R. Carter, C.H. Díaz-Ambrona, S. Fronzek, T. Heidmann, T. Hickler, T. Holt, M.I. Minguez, P. Morales, J.P. Palutikof, M. Quemada, M. Ruiz-Ramos, G.H. Rubæk, F. Sau, B. Smith, M.T. SykesUncertainties in projected impacts of climate change on European agriculture and terrestrial ecosystems based on scenarios from regional climate models Clim. Chang., 81 (2007), pp. 123-143
54. O'Neil et al., 2014 B.C. O'Neil, E. Kriegler, K. Riahi, K.L. Ebi, S. Hallegatte, T.R. Carter, R. Mathur, D.P. van VuurenA new scenario framework for climate change research: the concept of shared socioeconomic pathways Clim. Chang., 122 (3) (2014), pp. 387-400
55. Panagos et al., 2012 P. Panagos, M. Van Liedekerke, A. Jones, L. MontanarellaEuropean Soil Data Centre, ESDAC: response to European policy support and public data requirements Land Use Policy, 29 (2) (2012), pp. 329-338
56. Piorra et al., 2009 A.Piorra, F. Ungarob, A. Ciancaglinic, K. Happed, A. Sahrbacherd, C. Sattlera, S. Uthesa, P.ZanderaIntegrated assessment of future CAP policies: land use changes, spatial patterns and targeting Environ. Sci. Pol., 12 (8) (2009), pp. 1122-1136
57. Prestele et al., 2016 R. Prestele, P. Alexander, M. Rounsevell, A. Arneth, K. Calvin, J. Doelman, D. Eitelberg, K.Engström, S. Fujimori, T. Hasegawa, P. Havlik, F. Humpenöder, A.K. Jain, T. Krisztin, P. Kyle, P.Meiyappan, A. Popp, R.D. Sands, R. Schaldach, J. Schüngel, E. Stehfest, A. Tabeau, H. van Meijl, J. van Vliet, P.H.

VerburgHotspots of uncertainty in land use and land cover change projections: a global scale model comparison Glob. Chang. Biol. (2016), 10.1111/gcb.13337

58. Reginster and Rounsevell, 2006 I.Reginster, M. RounsevellScenarios of future urban land use in Europe Environ. Plann. B. Plann. Des., 33 (2006), pp. 619-636

59. Reidsma et al., 2006 P. Reidsma, T. Tekelenburg, M. van den Berg, R. AlkemadeImpacts of land-use change on biodiversity: an assessment of agricultural biodiversity in the European Union Agric. Ecosyst. Environ., 114 (1) (2006), pp. 86-102

60. Reidsma et al., 2011 P. Reidsma, H. König, S. Feng, I. Bezlepkina, I. Nesheim, M. Bonin, M. Sghaier, S.Purushothaman, S. Sieber, M.K. van Ittersum, F. BrouwerMethods and tools for integrated assessment of land use policies on sustainable development in developing countries Land Use Policy, 28 (3) (2011), pp. 604-617

61. Rounsevell and Reay, 2009 M.D.A. Rounsevell, D.S. ReayLand use and climate change in the UK Land Use Policy, 26S (2009), pp. S160-S169

62. Rounsevell et al., 2005 M.D.A. Rounsevell, F. Ewert, I. Reginster, R. Leemans, T.R. CarterFuture scenarios of European agricultural land use II. Projecting changes in cropland and grassland. Agriculture Ecol. Environ., 107 (2005), pp. 117-135

63. Rounsevell et al., 2006 M.D.A. Rounsevell, I. Reginster, M.B. Araújo, T.R. Carter, N. Dendoncker, F. Ewert, J.I.House, S. Kankaanpää, R. Leemans, M.J. Metzger, C. Schmit, P. Smith, G. TuckA coherent set of future land use change scenarios for Europe Agric. Ecosyst. Environ., 114 (2006), pp. 57-68

64. Stürck et al., 2015 J. Stürck, C. Levers, E.H. van der Zanden, C.J.E. Schulp, P.J. Verkerk, T. Kuemmerle, J.Helming, H. Lotze-Campen, A. Tabeau, A. Popp, E. Schrammeijer, P. VerburgSimulating and delineating future land change trajectories across Europe Reg. Environ. Chang. (2015), 10.1007/s10113-015-0876-0

65. van Delden et al., 2010 H. van Delden, T. Stuczynski, P. Ciaian, M.L. Paracchini, J. Hurkens, A. Lopatka, Y. Shi, O.G. Prieto, S. Calvo, J. van Vliet, R. Vanhout Integrated assessment of agricultural

policies with dynamic land use change modelling Ecol. Model., 221 (18) (2010), pp. 2153-2166

66. van Vliet et al., 2015 J. van Vliet, H.L.F. de Groot, P. Rietveld, P.H. VerburgManifestations and underlying drivers of agricultural land use change in Europe Landsc. Urban Plan., 133 (2015), pp. 24-36
67. van Vuuren et al., 2016 D.P. van Vuuren, E. Stehfest, D.E.H.J. Gernaat, J.C. Doelman, M. van den Berg, M.Harmsen, H.S. de Boer, L.F. Bouwman, V. Daioglou, O.Y. Edelenbosch, B. Girod, T. Kram, L. Lassaletta, P.L. Lucas, H. van Meijl, C. Müller, B.J. van Ruijven, S. van der Sluis, A. TabeauEnergy, land-use and greenhouse gas emissions trajectories under a green growth paradigm Glob. Environ. Chang. (2016), 10.1016/j.gloenvcha.2016.05.008
68. Verburg et al., 2008 P.H. Verburg, B. Eickhout, H. van MeijlA multi-scale, multi-model approach for analyzing the future dynamics of European land use Ann. Reg. Sci., 42 (2008), pp. 57-77
69. Verburg et al., 2010 P.H. Verburg, D.B. van Berkel, A.M. van Doorn, M. van Eupen, H.A.R.M. van den HeiligenbergTrajectories of land use change in Europe: a model based exploration of rural futures Landsc. Ecol., 25 (2010), pp. 217-232
70. Verburg et al., 2013 P.H. Verburg, A. Tabeau, E. HatnaAssessing spatial uncertainties of land allocation using a scenario approach and sensitivity analysis: a study for land use in Europe J. Environ. Manag., 127 (2013)
71. Wiebe et al., 2015 K. Wiebe, H. Lotze-Campen, R. Sands, A. Tabeau, D. van der Mensbrugghe, A. Biewald, B.Bodirsky, S. Islam, A. Kavallari, D. Mason-D'Croz, C. Müller, A. Popp, R. Robertson, S. Robinson, H. van Meijl, D. WillenbockelClimate change impacts on agriculture in 2050 under a range of plausible socioeconomic and emissions scenarios Environ. Res. Lett., 10 (2015), p. 085010
72. Wimmer et al., 2015 F. Wimmer, E. Audsley, C.-M. Savin, M. Malsy, R. Dunford, P.A. Harrison, R. Schaldach, M.FlörkeModelling the effects of cross-sectoral water allocation schemes in Europe Clim. Chang., 128 (2015), pp. 229-244

6 GLOMALIN ACCUMULATED IN SEAGRASS SEDIMENTS REVEALS PAST ALTERATIONS IN SOIL QUALITY DUE TO LAND-USE CHANGE

Lourdes López-Merino [a], Oscar Serrano [b,c], María Fernanda Adame [d], Miguel Ángel Mateo [b,e], and Antonio Martínez Cortizas [f]

[a]Institute of Environment, Health and Societies, Brunel University London, Uxbridge UB8 3PH, UK

[b]Centre for Marine Ecosystems Research, Edith Cowan University, 270 Joondalup Drive, Joondalup, WA 6027, Australia

[c]UWA Oceans Institute, University of Western Australia, 35 Stirling Highway, Crawley 6009, Australia

[d]Australian Rivers Institute, Griffith University, 170 Kessels Road, Nathan, QLD 4111, Australia

[e]Centro de Estudios Avanzados de Blanes, Consejo Superior de Investigaciones Científicas, Acceso a la Cala S. Francesc 14, 17300 Blanes, Spain

[f]Departamento de Edafología y Química Agrícola, Facultad de Biología, Universidad de Santiago de Compostela, Campus Sur, 15782 Santiago de Compostela, Spain

ABSTRACT

Arbuscular mycorrhizal fungi (AMF), symbionts with most terrestrial plants, produce glomalin-related soil protein (GRSP), which plays a major role in soil structure and quality. Both fungi hyphae and protein production in soils are affected by perturbations related to land-use changes, implying that GRSP is a sensitive indicator of soil quality. Unfortunately, GRSP degrades within years to decades in oxic environments, preventing its use as palaeoecological proxy. However, GRSP is transported to marine, near-shore anoxic sediments, where it accumulates and remains non-degraded, enabling the assessment of its potential as a palaeoecological proxy for soil ecosystem's health. Exploiting this fact, we have obtained for the first time a long-term record (c. 1250 years) of GRSP content using a *Posidonia oceanica* seagrass mat sediment core from the Western Mediterranean (Portlligat Bay, Spain). The trends in GRSP content matched well with land-use changes related to agrarian activities reconstructed by pollen analysis. In periods of cultivation, GRSP accumulation in the mat decreased. Given the role played by GRSP, the results suggest that agrarian intensification may have resulted in perturbations to soil quality. Thus, GRSP in seagrass mat sediments can be used to assess long-term trends in continental soil quality induced by human activities. These findings open new possibilities in long-term ecology research, as other anoxic environments could be potentially valid too. Testing them would open the possibility to identify long-term patterns in soil quality and other environmental stressors that could also affect AMF and GRSP production in soils.

Keywords: Glomalin-related soil protein, *Posidonia* mat, Land-use change, Palaeoecological proxy, and Soil quality

INTRODUCTION

Unravelling the potential of new palaeoecological proxies providing information on when a system has been disturbed is challenging. Past changes in land-use, i.e., agrarian practices, can be tracked back in time using palaeoecological techniques. Alterations in forest cover, hydrology and trophic status, as well as soil erosion, are the environmental consequences associated with the establishment of crops that are typically reconstructed in palaeoecological and palaeoenvironmental research (Dearing et al., 2006). However, other transformations that can be identified in present landscapes have passed unnoticed in the palaeoecological record, even when they could be useful to detect long-term variations in soil quality due to land-use change. One of them is the alteration of hyphae production by arbuscular mycorrhizal fungi (AMF), which are symbionts with the roots of most terrestrial plants and play a significant role in soil ecosystems' functioning (Wright et al., 2006).

Glomalin-related soil protein (GRSP) is a recently discovered glycoprotein produced by AMF (Gadkar and Rillig, 2006). GRSP is a significant component of soil organic matter, playing an important role in nitrogen and carbon storage and turnover, soil aggregate stability and soil texture (Rillig and Steinberg, 2002; Rillig et al., 2001; Staddon et al., 2003; Wright and Upadhyaya, 1998). Concentrations of GRSP in soils have been linked to land-use change, indicating that its abundance responds to ecosystems' modifications. In the Delaware Wildlife Area (Ohio, US), agricultural (corn and soybean) and afforested soils had lower GRSP concentrations than native temperate deciduous forest soils (Rillig et al., 2003). The conversion of Mexican rainforest to pasture and agricultural land (corn and common bean) decreased soil GRSP concentrations (Violi et al., 2008). In soils from south Cameroon GRSP content declined from forest to fallow and field crops (groundnuts, corn cassava, plantain; Fokom et al., 2012). As land-use alters AMF communities, affecting GRSP production and its soil concentration (Wright et al., 1999), it is considered that the more GRSP in a soil the better the soil quality (Shrestha Vaidya et al., 2011), hence providing information on the health of a given soil ecosystem.

GRSP has a turnover time of years to decades in well-aerated soils (Rillig et al., 2001). Since it rapidly degrades with time in oxic environments, most terrestrial soils are not effective environmental archives for the reconstruction of long-term GRSP trends, preventing the use of GRSP as a palaeoecological proxy. However, GRSP has been detected in superficial sediments of intertidal *Posidonia australis* seagrass meadows and wetlands, and coastal coral reefs (Adame et al., 2010, 2012). Although AMF colonise most terrestrial plants' roots, the ability of AMF to colonise plants' roots is limited by salinity (Adame et al., 2012), and AMF colonisation of seagrass roots has not been observed (Nielsen et al., 1999). The GRSP produced by AMF is found outside of the fungal structures, indicating it is transferred to the environment (Driver et al., 2005). GRSP can leach or be washed from soils into streams, and it is detected downstream in floodplains and rivers (Harner et al., 2004). Rivers and groundwater transport terrestrial material containing GRSP that is deposited in accretion areas and coastal systems, evidencing a terrestrial-marine connectivity (Adame et al., 2010, 2012). The highly anoxic conditions of marine environments, such as seagrass sediments, would prevent the degradation of GRSP enabling the reconstruction of long-term records of GRSP change.

The GRSP found in the coastal marine environment is likely to reflect landscape-scale processes, as it is not produced locally and has been transported by watershed runoff towards the sea (Adame et al., 2012; Harner et al., 2004). Thus, it is worth comparing synchronous terrestrial landscape changes and variations in GRSP content in marine sediments. For this purpose, palynological studies provide landscape-scale information on when an area was transformed and the magnitude of that transformation. This comparison would allow the assessment of whether changes in GRSP content in marine sediments could be linked to terrestrial landscape changes (i.e., land-use change).

In this study we report the last c. 1250 years of changes in GRSP content, detected in a Western Mediterranean seagrass *Posidonia oceanica* mat core (Portlligat bay, Spain, Fig. 1). The GRSP record is compared with landscape vegetation changes, including agriculture indicators, obtained from a previous palynological study performed on the same mat core (López-Sáez et al., 2009) (see Supplementary Information). Fluctuations in GRSP content in the seagrass mat and in terrestrial landscape composition and structure inferred by pollen data (i.e., turnover, PCA results, richness and evenness) were characterised using change-point modelling (Gallagher et al., 2011) to statistically detect synchronous events. The goal was to evaluate whether GRSP accumulated in coastal anoxic environments could be used as a palaeoecological proxy and to assess whether long-term GRSP variations in coastal environments accurately reflect soil alterations in quality related to past changes in land-use.

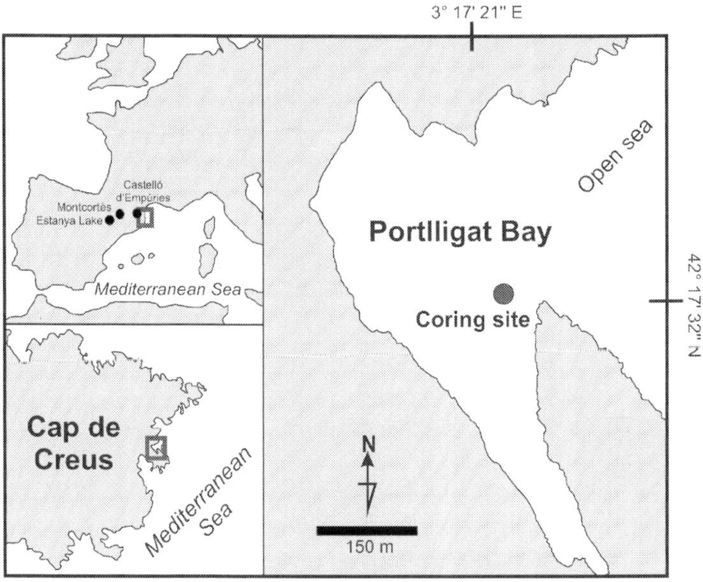

Figure 1: Location of the drilling point of core 2000 in the *P. oceanica* bed at the north-western part of the Mediterranean Sea (Portlligat Bay, Cape Creus, Spain). Other sites cited in the text are also located in the map.

METHODS

Study site and sampling

The Portlligat Bay (42°17′32″ N; 3°17′28″ E) is a small (0.14 km^2) and shallow (< 10 m deep) inlet located in NE Spain, in the Cape Creus, connected to the sea through a 213 m wide opening to the NW (Fig. 1). Living meadows of *P. oceanica* cover about 68% of the bottom of the bay. The bay receives freshwater input from a typical Mediterranean temporary stream that flows into it from its NE shore. The Cape Creus itself (eastern Pyrenees) is a geological formation originated 400 million years ago. This area presents a Mediterranean climate, with warm summers, mild winters and summer drought. The landscape has had an intense human fingerprint for a long time. Traditional activities, such as dry-land farming and fisheries, have been replaced by tourism, leaving the area occupied by abandoned terraced vineyards and pasturelands. Mediterranean scrubland dominated by *Pistacia lentiscus*, *Juniperus oxycedrus*, *Calicotome spinosa* and species of *Erica* and *Cistus* are important. Forest cover is sparse, with *Pinus halepensis* and *Quercus suber* as the most abundant tree species, while in sheltered areas *Quercus ilex* and *Quercus pubescens*, and in riparian areas *Ulmus minor*, *Fraxinus angustifolia*, *Alnus glutinosa*, *Corylus avellana* and *Salix* sp. appear (Franquesa I Codinach, 1995).

A 475 cm-long core (core 2000) was sampled in a *P. oceanica* meadow, at a water depth of 3 m, in the Portlligat bay using a floating drilling platform that combined pneumatic percussion and rotation. A detailed description of the study site, coring procedures and mat characteristics can be found in Serrano et al. (2012).

Previous palaeoenvironmental studies on the same seagrass mat

Reliable palaeoecological studies are only possible with consistent chronologies. Radiocarbon dating revealed that the 475 cm-long core mat (core 2000) was c. 4500 years old (Serrano et al., 2012). The records of trace metals matched well with phases of European mining and metalworking, confirming the accuracy of the chronology (Serrano et al., 2011, 2013). The palynological study of the uppermost section of the same mat core and spanning the last c. 1250 years (López-Sáez et al., 2009) showed that the Portlligat bay catchment area experienced farming activities. The timing of the changes agrees with other NE Iberian reconstructions (Burjachs et al., 2005; Riera et al., 2004; Rull and Vegas-Vilarrúbia, 2015; Rull et al., 2011), demonstrating the reliability of pollen studies on coastal *P. oceanica* mats in reconstructing terrestrial landscapes. We have updated the chronological model provided in López-Sáez et al. (2009) to the most accurate one developed by Serrano et al. (2012) (see Supplementary Information).

Glomalin-related soil protein extraction

Twenty three samples from the topmost 150 cm of core 2000, sampled every c. 2–10 cm and covering the last c. 1250 years, were analysed for their GRSP content. The samples were oven-dried at 60 °C before analysis, which was carried out using the procedures described by Wright and Upadhyaya (1996, 1998). Two extractions were performed to obtain two different glomalin pools: Bradford-reactive soil protein (BRSP) and immunoreactive soil protein (IRSP).

For BRSP, samples were extracted in sodium pyrophosphate, pH 9 at 120 °C by autoclaving for 60 min. Multiple extractions were performed (2–3) until the supernatant presented none of the red-brown colour typical of GRSP. For IRSP, samples were extracted in sodium citrate, pH 8.0 by autoclaving at 120 °C for 30 min. After each extraction, samples were centrifuged at 3220 *g* for 10 min. Once the process was complete, the total volume of the supernatants was recorded and sub-samples were centrifuged at 10,000 *g*. The Bradford assay was first used to determine the concentration of BRSP using bovine serum albumin as a standard. Based on BRSP concentrations, solutions containing 0.02 μg of protein were set for IRSP analysis. IRSP was analysed with an indirect enzyme-linked immunosorbent assay (ELISA) using the monoclonal antibody MAb 32B11 developed against spores of arbuscular mycorrhizal fungi (Wright and Upadhyaya, 1996). Values were compared with a standard curve calculated by Adame et al. (2012). Due to the lack of available commercial standards, IRSP values are shown as unit-less measurements, providing a comparative mean to test the relative amount of GRSP in different samples using the same standard curve. Immunoreactivity (IR%) was calculated as the IRSP/BRSP ratio and refers to the ability of the protein to attach to the antibody, informative of the decomposition state of GRSP (Lovelock et al., 2004; Wright et al., 1996).

Numerical analyses

The previously published palynological analysis performed on the same *P. oceanica* mat core (López-Sáez et al., 2009) is available at the European Pollen Database. The palynological data-set has been explored numerically to detect past landscape changes and to compare them with GRSP trends. Pollen and fern spores were considered for the calculation of the percentages and the diagram plotted using the most updated chronology developed by Serrano et al. (2012) (see Supplementary Information).

Palynological richness was estimated by rarefaction analysis, a measure of richness that standardises sample sizes to a common sum (Birks and Line, 1992), usually the lowest sum (here 213). *Evenness* was calculated using the Buzas and Gibson's index, providing information about the homogeneity in the distribution of the taxa abundances in the assemblages. *Taxonomical turnover* was assessed by detrended correspondence analysis (DCA). Percentages were squared-root transformed to stabilise variances. Detrending was by segments, a non-linear rescaling was applied and rare taxa were not down-weighted. Thus, samples scores are scaled in standard deviation (SD) units of compositional change (Birks et al., 2012). DCA gradient lengths were < 2.5 SD, indicating monotonic responses and that a linear approach is applicable (Birks et al., 2012). Thence, *principal component analysis* (PCA) was used to summarize temporal trends in the palynological dataset. Only taxa with percentages > 2% were considered, and *Juglans*, *Vitis*, *Cannabis*/*Humulus* t. and Cerealia were grouped as "Crops". Percentages were squared-root transformed, and the covariance matrix was chosen (Birks and Gordon, 1985).

Because palynological and GRSP analyses were performed at different depths/ages, *change-point modelling* was applied to both GRSP content and palynological numerical results (turnover, PC1 and PC2 factor scores, richness and evenness) to detect statistically periods of common changes in trends. The approach used here is based on Bayesian transdimensional Markov chain Monte Carlo (Gallagher et al., 2011) and it is the first time it has been used in a multi-proxy record.

Palynological richness and evenness were calculated using PAST 3.01 (Hammer et al., 2001), DCA and PCA performed with Psimpoll 4.27 (Bennett, 2009), and change-point modelling computed with the CPQtR1.0.3 statistical package (Gallagher et al., 2011).

RESULTS

Glomalin-related soil protein record in the seagrass mat

GRSP had an average content of 0.080 ± 0.005 (unit-less), with minimum and maximum values of 0.027 and 0.161 respectively. GRSP content and immunoreactivity (IR%) had a similar pattern, except for the surface sample that showed the highest IR% (Fig. 2a). This could suggest that GRSP may have lost part of its immunoreactivity after burial, although quickly reaching a post-burial stabilisation. We did not find a decreasing trend of GRSP content and IR% with depth (hence age) (Fig. 2b), which indicates that variations in GRSP might be driven by factors other than post-burial degradation kinetics. GRSP content increased from c. 1235 to 650 cal. yr BP. A sudden decrease (in about 25 years) took place at c. 580 cal. yr BP remaining low until c. 250 cal. yr BP and partially increasing afterwards.

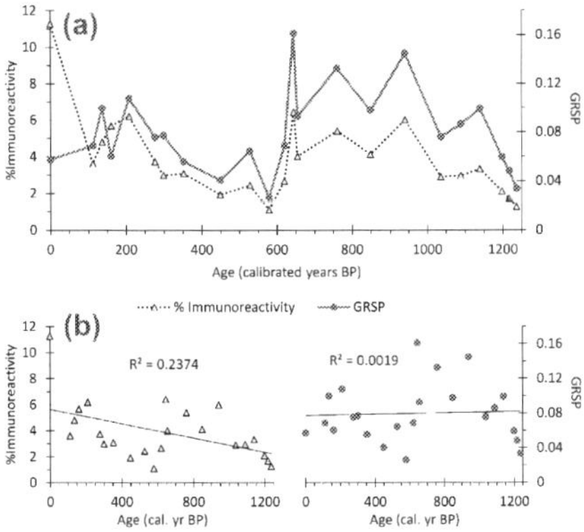

Figure 2: Glomalin-related soil protein (GRSP) results obtained for the studied *P. oceanica* mat (core 2000). (a) GRSP relative content and immunoreactivity (IR%) for the last 1200 years. (b) Age is not significantly correlated either with GRSP or IR%.

Numerical analyses of terrestrial landscape changes

Change-point modelling applied to GRSP and records of landscape change obtained from palynological data (turnover, PC1 and PC2 scores, richness and evenness) enabled the identification of six statistically significant periods of change (change-points CP-1 to 6) (Fig. 3). CP-5, the most abrupt of all of them, is contemporary with the main drop in GRSP content and important changes in the composition and structure of the terrestrial landscape vegetation. Richness and turnover, that presented similar trends ($R^2 = 0.78$), showed an abrupt increase during this event and a decrease during CP-6 (Fig. 3). Evenness showed a see-saw pattern. The principal component analysis distinguished two principal components (PC) (Fig. 4). The first principal component (PC1) accounted for 33.6% of the variance, while the second principal component (PC2) explained 13.1%. In PC1, *Q. suber*, *Erica*, *Cistus*, *Betula* and deciduous *Quercus* presented positive loadings, while evergreen *Quercus*, "Crops" and Cyperaceae had negative loadings. PC1 is interpreted as an agrarian gradient from lower (positive scores) to larger agrarian impact (negative scores). PC1 showed positive scores before CP-5, with a shift to negative scores from CP-5 to CP-6 and a return to positive scores after CP-6. In PC2, Cichorioideae and *Pinus* presented positive loadings, while *Erica* had a negative one. All samples scores for PC2 were between 0.3 and − 0.3 except for one sample. Such sample presented anomalous high percentages of Cichorioideae and low values of *Erica*, and a very positive sample score, which is probably related to a punctual perturbation. For the rest of the samples, PC2 could be indicating a larger contribution of heathlands when scores are negative and of pine woodlands when scores are positive (Fig. 4).

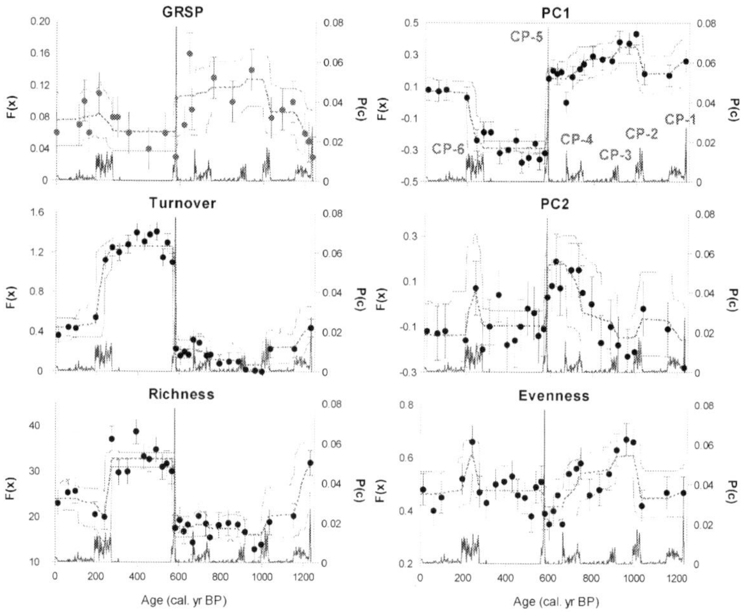

Figure 3: Changepoint functions for GRSP, turnover, PC1 and PC2 scores, richness and evenness. Dots indicate the original data, and vertical lines are the noise estimation. Thick intermittent grey curves are the average inferred functions [$F_{(x)}$] after 120,000 functions, while thinner grey lines define the 95% interval ranges. Blue curves show the probability of change-point [$P_{(c)}$], indicating that CP-5 was the most abrupt change for all the variables, while the other detected change-points were less abrupt.

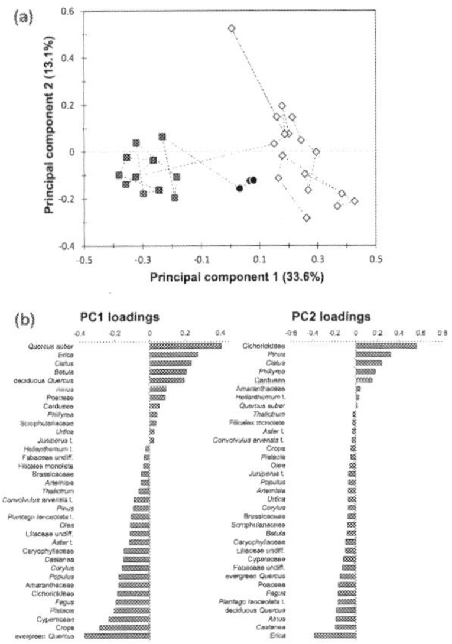

Figure 4: Principal component analysis results of core 2000 palynological data (data from López-Sáez et al. (2009) and available at the European Pollen Database). a) PCA biplot of samples scores along the first and second principal component axes. Sample points are connected in chronological order, with the base (oldest samples) of the sequence in the lower right. White squares indicate samples before CP-5, when the main compositional change occurred. Red squares indicate samples between CP-5 and CP-6, the period in which agrarian activities were more extensive, changing the landscape structure, and a drop in GRSP content is detected. Black dots indicate samples after CP-6. b) Species loadings for the two principal components. "Crops" refers to the sum of *Juglans*, *Vitis*, *Cannabis*/*Humulus* t. and Cerealia.

DISCUSSION

Land-use change affects the accumulation of GRSP in seagrass sediments

The most significant terrestrial landscape change (CP-5) occurred abruptly at c. 585–560 cal. yr BP and was related to the start of a large agrarian perturbation with increasing percentages of Cerealia, *Cannabis*/

Humulus t., *Vitis* and *Juglans*, and also a significant presence of *Castanea* and *Olea* (Fig. 5). Richness and turnover increased during this event, indicating deep transformations in landscape composition involving a fast, abrupt replacement of woodlands and scrublands by crop fields and associated weeds and other anthropogenic indicators. The end of this intensive agrarian period, indicated by CP-6 (c. 280–190 cal. yr BP), was less abrupt than the previous event and meant a partial increase in the GRSP content in the mat (Fig. 5). The impact that agrarian activities had on the structure of the landscape is also evident in the PCA biplot, as the analysis clearly differentiated samples older than CP-5, samples between CP-5 and CP-6, and samples younger than CP-6 (Fig. 4). This period of intensive farming (CP-5 to CP-6) was coeval with the phase of low GRSP content in the mat (Fig. 5), supporting the idea that agrarian activities negatively affect AMF communities and the production of GRSP in soils (Fokom et al., 2012; Rillig et al., 2003; Violi et al., 2008).

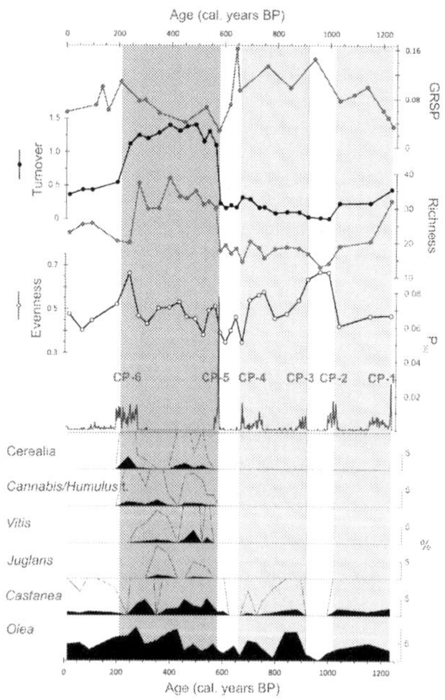

Figure 5: Synthesis of the last 1200 years of landscape change detected after the study of a *P. oceanica*mat (core 2000, Portlligat bay, NE Spain). Richness, evenness and turnover are compared with GRSP content and palynological in-

dicators of crops (data from López-Sáez et al. (2009) with the chronology of Serrano et al. (2012)). The filled silhouettes show the percentage curves of the taxa, the unfilled silhouettes show the × 10 exaggeration curves. Probability of change-point indicates significant, common events of change in GRSP contents and landscape structure. Vertical grey bands highlight phases with an importance of agrarian activities (the dark grey band for the largest period, and the two light grey bands for the two short-term periods).

Even though the period from c. 585–560 to 280–190 cal. yr BP showed the maximal expansion of cultivated land for the last c. 1250 cal. yr BP with a clear drop in GRSP content, two other short-term agricultural events were also identified. The first short-term event occurred from CP-1 (c. 1235–1150 cal. yr BP) to CP-2 (c. 1030–990 cal. yr BP), while the second one was framed between CP-3 (c. 915–880 cal. yr BP) and CP-4 (c. 750–665 cal. yr BP) (Fig. 5). These events presented increasing percentages of *Castanea* and *Olea* that, although did not completely change the structure of the landscape, were accompanied by small decreases in the content of GRSP in the mat (Fig. 5). Richness values were slightly higher during such events, although not at the level detected during the largest agrarian disturbance. Evenness, on the contrary, tended to be lower, pointing out that land-use change towards agriculture involved a more uneven landscape (Fig. 5). Our results indicate that soil quality at the Portlligat bay catchment area has varied significantly during the last twelve centuries. In particular, soil functions were affected during agrarian periods, and partially recovered when crops were no longer abundant, showing some degree of resilience.

Factors determining the GRSP terrestrial-marine transport/accumulation

After GRSP was found in surface sediments from intertidal seagrass meadows and wetlands, and coastal coral reefs, a conceptual model of the possible factors affecting the terrestrial-marine connectivity driving its transfer to the marine environment was proposed (Adame et al., 2012). Climate (mainly rainfall), runoff potential (mainly land cover and associated soil erosion) and distance of the marine environments from the shore were proposed as the main factors (Adame et al., 2012). Based on single modern sampling points distributed along different coastal settings, it was pointed out that higher content of GRSP in marine sediments should be expected when extensive forest clearance occurred, as the associated soil erosion would increase sediment runoff, enhancing the connectivity (Adame et al., 2012). However, in our long-term study, periods with lower GRSP content in the mat were coeval with declines in natural tree and shrub cover, prompted by the introduction of crops (Fig. 5). It was also proposed that the higher the rainfall, the larger the GRSP transfer to the marine environment (Adame et al., 2012). The reconstructed arid and humid phases in the nearby Estanya Lake and the precipitation patterns for NE Iberia (Morellón et al., 2011, 2012) highlight a more humid Little Ice Age (c. 650–150 cal. yr BP) than the Mediaeval Climate Anomaly (c. 1050–650 cal. yr BP) (Fig. 6). We have not detected higher GRSP content in the mat during the Little Ice Age, which partially overlaps with the period of extensive agrarian activities and lowest GRSP contents in the mat (Fig. 6). Additionally, the agrarian history detected by pollen analysis in the Portlligat bay coheres with the histories reconstructed regionally (Fig. 6), supporting that the GRSP signal recorded in the sediment core from the *Posidonia* mat has a regional character. Our findings, hence, suggest that one of the most important environmental factors driving GRSP content changes in the marine environment (i.e., *P. oceanica* mat) was the production of GRSP in the continental soils, apparently overriding the importance of soil erosion and rainfall.

Finally, although the speed of the GRSP transference and deposition from the continent to the marine environment is unknown, it appears to be fast. Firstly, due to the fast degradation of the protein, as if the GRSP transference and deposition was slow, no GRSP would have been detected in the *Posidonia* mat. Secondly, as we did not find any significant decoupling between changes in the structure of the landscape inferred by pollen analysis and changes in GRSP content in the mat (Figs. 5 and 6), confirming what ecological and soil studies state: agrarian activities negatively affect GRSP production, hence affecting soil quality (i.e., Rillig et al., 2003; Violi et al., 2008; Fokom et al., 2012).

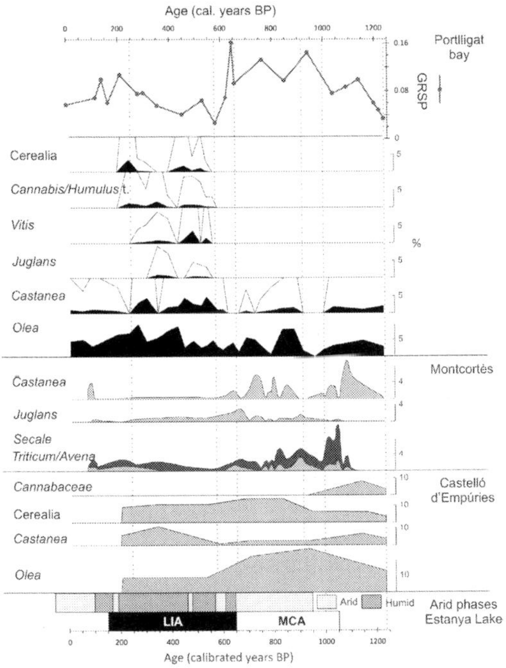

Figure 6: Comparison of the changes detected in GRSP content and palynological indicators of agrarian activities from the core 2000 (*Posidonia* mat at Portlligat bay) with some palynological indicators from other sequences located in the area: Montcortès (Rull and Vegas-Vilarrúbia, 2015) and Castelló d'Empúries (Burjachs et al., 2005). The arid and humid phases reconstructed in the Estanya Lake by Morellón et al. (2011) are also included, as well as the chronology of the Little Ice Age and the Mediaeval Climate Anomaly. Fig. 1 shows the location of the cited records.

Finally, although the speed of the GRSP transference and deposition from the continent to the marine environment is unknown, it appears to be fast. Firstly, due to the fast degradation of the protein, as if the GRSP transference and deposition was slow, no GRSP would have been detected in the *Posidonia* mat. Secondly, as we did not find any significant decoupling between changes in the structure of the landscape inferred by pollen analysis and changes in GRSP content in the mat (Figs. 5 and 6), confirming what ecological and soil studies state: agrarian activities negatively affect GRSP production, hence affecting soil quality (i.e., Rillig et al., 2003; Violi et al., 2008; Fokom et al., 2012).

A new palaeoenvironmental tool to evaluate soil quality

The negative effects that anthropogenic disturbances, especially agriculture, and habitat loss have on soils via the decrease of AMF association with roots, and the suggestion that GRSP produced by AMF can be used as indicator of soil quality, has boosted research from both farming and conservation point of views. For example, Bedini et al. (2013) showed that GRSP content was higher in organically-managed fields than in conventionally-managed ones. Moreover, Turrini and Giovannetti (2012) make an interesting evaluation of the importance of AMF preservation in protected areas worldwide, as they play a central role in the maintenance of ecosystem services, soil functions (as nutrient cycling) and plant community dynamics. These microscopic environmental changes, which can go unnoticed, are ecologically important, and provide invaluable information about the health of a given soil ecosystem.

Seagrass mat sediments have proven to be valid for unravelling past patterns in GRSP, although it is likely that other anoxic, accreting environments could be also valid. The lack of oxygen is the main feature that the palaeoenvironmental archives have to have in order to avoid the degradation of GRSP and its proper long-term accumulation for a reliable palaeoenvironmental reconstruction. However, a further understanding of the archives that are of potential use is required as changes in GRSP content could be related to other environmental stressors apart from regional land-use change. The marine coastal setting of *Posidonia* mat is ideal because there is not local signal (such as in situ plants with AMF in their roots producing GRSP). Hence, the information obtained reflects the regional landscape-scale signal that can be comparable with other landscape-scale indicators such as pollen of continental plants. However, plants growing in the fresh-water aquatic environments of peatlands or lakes do form symbiotic associations with AMF (Chaubal et al., 1982; Clayton and Bagyaraj, 1984; Bagyaraj et al., 1979; Beck-Nielsen and Madsen, 2001; Bohrer et al., 2004), so regional and local signals could be mixed.

Unravelling past patterns and processes of the AMF community via changes in GRSP could constitute a valuable tool to analyse past trends in soil quality, establish soils' ecological baselines, and assess whether changes in GRSP concentration are driven not only by changes in land-use but also by other environmental stressors. Testing more archives could open the possibility to assess past changes in soil quality at different parts of the world that faced dissimilar land-use scenarios.

CONCLUSIONS

Exploring the previously reported terrestrial-marine connectivity suggested by the accumulation of GRSP in seagrass mat sediments, this study reports the first attempt to use changes in GRSP content in a *P. oceanica* mat to reconstruct long-term changes in continental soil's quality. Unlike terrestrial soils, our findings demonstrate that the anoxic environment created in seagrass mats prevents the complete degradation of GRSP once it is buried, allowing a palaeoecological approach to be employed. Our results also suggest that the negative impact of past intensification of land-use related to agrarian activities on the AMF communities is captured in the seagrass mat via changes in GRSP content. It is likely that, apart from seagrass mat sediments, other marine sediments and continental anoxic environments such as lake sediments could be also of use. Their study could open the opportunity to reconstruct patterns and processes in past changes in soil quality in many areas, which is of interest to both conservation and the farming community.

ACKNOWLEDGEMENTS

L. López-Merino is funded by The Leverhulme Trust towards an Early Career Fellowship (ECF-2013-530, "*Posidonia* as environmental archive: long-term ecology and conservation views"). The study was also funded by the Spanish Ministry of Economy and Competitiveness, SUMILEN project (CTM2013-47728-R), and the Spanish Autonomous Organism of National Parks, PALEOPARK project (ref. 1104), and supported by the CSIRO Marine and Coastal Carbon Biogeochemistry Cluster. We are indebted to Alex Couce, Suzanne Leroy, Chris Green and Tom Hoyle for their perceptive comments on earlier drafts.

REFERENCES

1. Adame et al., 2010 M.F. Adame, D. Neil, S.F. Wright, C.E. LovelockSedimentation within and among mangrove forests along a gradient of geomorphological settings Estuar. Coast. Shelf Sci., 86 (2010), pp. 21-30
2. Adame et al., 2012 M.F. Adame, S.F. Wright, A. Grinham, K. Lobb, C.E. Reymond, C.E. LovelockTerrestrial-marine connectivity: patterns of terrestrial soil carbon deposition in coastal sediments determined by analysis of glomalin related soil protein Limnol. Oceanogr., 57 (2012), pp. 1492-1502
3. Bagyaraj et al., 1979 D.J. Bagyaraj, A. Manjunath, R.B. PatilOccurrence of vesicular–arbuscular mycorrhizas in some tropical aquatic plants Trans. Br. Mycol. Soc., 72 (1979), pp. 164-167
4. Beck-Nielsen and Madsen, 2001 D. Beck-Nielsen, T.V. MadsenOccurrence of vesicular–arbuscular mycorrhiza in aquatic macrophytes from lakes and streams Aquat. Bot., 71 (2001), pp. 141-148
5. Bedini et al., 2013 S. Bedini, L. Avio, C. Sbrana, A. Turrini, P. Migliorini, C. Vazzana, M. GiovannettiMycorrhizal activity and diversity in a long-term organic Mediterranean agroecosystem Biol. Fertil. Soils, 49 (2013), pp. 781-790
6. Bennett, 2009 K.D. BennettDocumentation for psimpoll 4.27 and pscomb 1.03. C programs for plotting pollen diagrams and analysing pollen data The 14Chrono Centre, Archaeology and Palaeoecology, Queen's University of Belfast, Belfast, UK (2009) (Available at: http://www.chrono.qub.ac.uk/psimpoll/psimpoll.html)
7. Birks and Gordon, 1985 H.J.B. Birks, A.D. GordonNumerical Methods in Quaternary Pollen Analysis Academic Press, London (1985) (317 pp.)
8. Birks and Line, 1992 H.J.B. Birks, J.M. LineThe use of rarefaction analysis for estimating palynological richness from Quaternary pollen-analytical data The Holocene, 2 (1992), pp. 1-10
9. Birks et al., 2012 H.J.B. Birks, A.F. Lotter, S. Juggings, J.P. Smol (Eds.), Tracking Environmental Change Using Lake Sediments,

vol. 5, Kluwer Academic Publishers, Dordrecht, The Netherlands (2012) (745 pp.)

10. Bohrer et al., 2004 K.E. Bohrer, C.F. Frieseb, J.P. AmonSeasonal dynamics of arbuscular mycorrhizal fungi in differing wetland habitats Mycorrhiza, 14 (2004), pp. 329-337
11. Burjachs et al., 2005 F. Burjachs, J. Bach, R. Buxó, P. Llàcer, J. McGlade, M. Picazom, R. Piqué, M.T. RosEl territori d'Emporion i les seves dades paleoambientals Empúries, 54 (2005), pp. 21-28
12. Chaubal et al., 1982 R. Chaubal, G.D. Sharma, R.R. MishraVcsicular arbuscular mycorrhiza in subtropical aquatic and marshy plant communities Proc. Indian Acad. Sci. Plant Sci., 91 (1982), pp. 69-77
13. Clayton and Bagyaraj, 1984 J.S. Clayton, D.J. BagyarajVesicular arbuscular mycorrhizas in submerged aquatic plants of New Zealand Aquat. Bot., 19 (1984), pp. 251-262
14. Dearing et al., 2006 J.A. Dearing, R.W. Battarbee, R. Dikau, I. Larocque, F. OlfieldHuman–environment interactions: learning from thc past Rcg. Environ. Chang., 6 (2006), pp. 1-16
15. Driver et al., 2005 J.D. Driver, W.E. Holben, M.C. RilligCharacterization of glomalin as a hyphal wall component of arbuscular mycorrhizal fungi Soil Biol. Biochem., 37 (2005), pp. 101-106
16. Fokom et al., 2012 R. Fokom, S. Adamou, M.C. Teugwa, A.D. Begoude Boyogueno, W.L. Nana, M.E.L.Ngonkeu, N.S. Tchameni, D. Nwaga, G. Tsala Ndzomo, P.H. Amvam ZolloGlomalin related soil protein, carbon, nitrogen and soil aggregate stability as affected by land use variation in the humid forest zone of south Cameroon Soil Tillage Res., 120 (2012), pp. 69-75
17. Franquesa i Codinach, 1995 T. Franquesa i CodinachEl paisatge vegetal de la península del Cap de Creus Arxius de la Secció de Ciències, 109, Institut d'Estudis Catalans, Barcelona (1995) (628 pp.)
18. Gallagher et al., 2011 K. Gallagher, T. Bodin, M. Sambridge, D. Weiss, M. Kylander, D. LargeInference of abrupt changes in noisy geochemical records using Bayesian transdimensional changepoint

models Earth Planet. Sci. Lett., 311 (2011), pp. 182-194

19. Gardkar and Rilling, 2006 V. Gadkar, M. RilligThe arbuscular mycorrhizal fungal protein glomalin is a putative homolog of heat shock protein 60 FEMS Microbiol. Lett., 263 (2006), pp. 93-101
20. Hammer et al., 2001 Ø. Hammer, D.A.T. Harper, P.D. RyanPAST: paleontological statistics software package for education and data analysis Palaeontol. Electron., 4 (1) (2001) (9 pp. http://palaeo-electronica.org/2001_1/past/issue1_01.htm)
21. Harner et al., 2004 M.J. Harner, P.W. Ramsey, M.C. RilligProtein accumulation and distribution in floodplain soils and river foam Ecol. Lett., 7 (2004), pp. 829-836
22. López-Sáez et al., 2009 J.A. López-Sáez, L. López-Merino, M.A. Mateo, O. Serrano, S. Pérez-Díaz, L. SerranoPalaeoecological potential of the marine organic deposits of *Posidonia oceanica*: a case study in the NE Iberian Peninsula Palaeogeogr. Palaeoclimatol. Palaeoecol., 271 (2009), pp. 215-224
23. Lovelock et al., 2004 C.E. Lovelock, S.F. Wright, D.A. Clark, R.W. RuessSoil stocks of glomalin produced in arbuscular mycorrhizal fungi across a tropical rainforest landscape J. Ecol., 92 (2004), pp. 278-287
24. Morellón et al., 2011 M. Morellón, B. Valero-Garcés, P. González-Sampériz, T. Vegas-Vilarrúbia, E. Rubio, M.Rieradevall, A. Delgado-Huertas, P. Mata, Ó. Romero, D.R. Engstrom, M. López-Vicente, A. Navas, J.SotoClimate changes and human activities recorded in the sediments of Lake Estanya (NE Spain) during the Medieval Warm Period and Little Ice Age J. Paleolimnol., 46 (2011), pp. 423-452
25. Morellón et al., 2012 M. Morellón, A. Pérez-Sanz, J.P. Corella, U. Büntgen, J. Catalán, P. González-Sampériz, J.J. González-Trueba, J.A. López-Sáez, A. Moreno, S. Pla-Rabes, M.A. Saz-Sánchez, P. Scussolini, E.Serrano, F. Steinhilber, V. Stefanova, T. Vegas-Villarrúbia, B. Valero-GarcésA multi-proxy perspective on millennium-long climate variability in the Southern Pyrenees Clim. Past, 8 (2012), pp. 683-700
26. Nielsen et al., 1999 S.L. Nielsen, I. Thingstrup, C. WigandApparent lack of vesicular-arbuscular mycorrhiza (VAM) in the seagrasses *Zostera marina* L. and *Thalassia testudinum* Banks ex König Aquat.

Bot., 63 (1999), pp. 261-266

27. Riera et al., 2004 S. Riera, G. Wansard, R. Julià2000-year environmental history of a karstic lake in the Mediterranean Pre-Pyrenees: the Estanya lakes (Spain) Catena, 55 (2004), pp. 293-324
28. Rillig and Steinberg, 2002 M.C. Rillig, P.D. SteinbergGlomalin production by an arbuscular mycorrhizal fungus: a mechanism of habitat modification Soil Biol. Biochem., 34 (2002), pp. 1371-1374
29. Rillig et al., 2001 M.C. Rillig, S.F. Wright, K.A. Nichols, W.F. Schmidt, M.S. TornLarge contribution of arbuscular mycorrhizal fungi to soil carbon pools in tropical forest soils Plant Soil, 233 (2001), pp. 167-177
30. Rillig et al., 2003 M.C. Rillig, P.W. Ramsey, S. Morris, E.A. PaulGlomalin, an arbuscular-mycorrhizal soil protein, responds to land-use change Plant Soil, 253 (2003), pp. 293-299
31. Rull and Vegas-Vilarrúbia, 2015 V. Rull, T. Vegas-VilarrúbiaCrops and weeds from the Estany de Montcortes catchment, central Pyrenees, during the last millennium: a comparison of palynological and historical records Veg. Hist. Archaeobot. (2015), 10.1007/s00334-015-0525-z
32. Rull et al., 2011 V. Rull, P. González-Sampériz, J.P. Corella, M. Morellón, S. GiraltVegetation changes in the southern Pyrenean flank during the last millennium in relation to climate and human activities: the Montcortès lacustrine record J. Paleolimnol., 46 (2011), pp. 387-404
33. Serrano et al., 2011 O. Serrano, M.A. Mateo, A. Dueñas-Bohórquez, P. Renom, J.A. López-Sáez, A. Martínez CortizasThe *Posidonia oceanica* marine sedimentary record: a Holocene archive of heavy metal pollution Sci. Total Environ., 409 (2011), pp. 4831-4840
34. Serrano et al., 2012 O. Serrano, M.A. Mateo, P. Renom, R. JuliàCharacterization of soils beneath a *Posidonia oceanica* meadow Geoderma, 185–186 (2012), pp. 26-36
35. Serrano et al., 2013 O. Serrano, A. Martínez-Cortizas, M.A. Mateo, H. Biester, R. BindlerMillennial scale impact on the marine biogeochemical cycle of mercury from early mining on the Iberian Peninsula Global Biogeochem. Cycles, 27 (2013), pp. 21-30
36. Shrestha Vaidya et al., 2011 G. Shrestha Vaidya, M.C. Rillig, H.

WallanderThe role of glomalin in soil erosion Sci. World, 9 (2011), pp. 82-85

37. Staddon et al., 2003 P.L. Staddon, C.B. Ramsey, N. Ostle, P. Ineson, A.H. FitterRapid turnover of hyphae of mycorrhizal fungi determined by AMS microanalysis of ^{14}C Science, 300 (2003), pp. 1138-1140
38. Turrini and Giovannetti, 2012 A.Turrini, M. GiovannettiArbuscular mycorrhizal fungi in national parks, nature reserves and protected areas worldwide: a strategic perspective for their in situ conservation Mycorrhiza, 22 (2012), pp. 81-97
39. Violi et al., 2008 H.A. Violi, A.F. Barrientos-Priego, S.F. Wright, E. Escamilla-Prado, J.B. Morton, J.A. Menge, C.J. LovattDisturbance changes arbuscular mycorrhizal fungal phenology and soil glomalin concentrations but not fungal spore composition in montane rainforests in Veracruz and Chiapas, Mexico Forest Ecol. Manag., 254 (2008), pp. 276-290
40. Wright and Upadhyaya, 1996 S.F. Wright, A. UpadhyayaExtraction of an abundant and unusual protein from soil and comparison with hyphal protein of arbuscular mycorrhizal fungi Soil Sci., 161 (1996), pp. 575-586
41. Wright and Upadhyaya, 1998 S.F. Wright, A. UpadhyayaA survey of soils for aggregate stability and glomalin, a glycoprotein produced by hyphae of arbuscular mycorrhizal fungi Plant Soil, 198 (1998), pp. 97-107
42. Wright et al., 1996 S.F. Wright, M. Franke-Snyder, J.B. Morton, A. UpadhyayaTime-course study and partial characterization of a protein on hyphae of arbuscular mycorrhizal fungi during active colonization of roots Plant Soil, 181 (1996), pp. 193-203
43. Wright et al., 1999 S.F. Wright, J.L. Starr, I.C. PaltineanuChanges in aggregate stability and concentration of glomalin during tillage managements transition Soil Sci. Soc. Am. J., 63 (1999), pp. 1825-1829
44. Wright et al., 2006 S.F. Wright, K.A. Nichols, W.F. SchmidtComparison of efficacy of three extractants to solubilize glomalin on hyphae and in soil Chemosphere, 64 (2006), pp. 1219-1224

7 AGRICULTURAL LAND USE CHANGES – A SCENARIO-BASED SUSTAINABILITY IMPACT ASSESSMENT FOR BRANDENBURG, GERMANY

ABSTRACT

Decisions for agricultural management are taken at farm scale. However, such decisions may well impact upon regional sustainability. Two of the likely agricultural management responses to future challenges are extended use of irrigation and increased production of energy crops. The drivers for these are high commodity prices and subsidy policies for renewable energy. However, the impacts of these responses upon regional sustainability are unknown. Thus, we conducted integrated impact assessments for agricultural intensification scenarios in the federal state of Brandenburg, Germany, for 2025. One Irrigation scenario and one Energy scenario were contrasted with the Business As Usual (BAU) scenario. We applied nine indicators to analyze the economic, social and environmental effects at the regional, in this case district scale, which is the smallest administrative unit in Brandenburg. Assessment results were discussed in a stakeholder workshop involving 16 experts from the state government.

The simulated area shares of silage maize for fodder and energy were 29%, 37% and 49% for the BAU, Irrigation, and Energy scenarios, respectively. The Energy scenario increased bio-electricity production to 41% of the demand of Brandenburg, and it resulted in CO_2savings of up to 3.5 million tons. However, it resulted in loss of biodiversity, loss of landscape scenery, increased soil erosion risk, and increased area demand for water protection requirements. The Irrigation scenario led to yield increases of 7% (rapeseed), 18% (wheat, sugar beet), and 40% (maize) compared to the BAU scenario. It also reduced the year-to-year yield variability. Water demand for irrigation was found to be in conflict with other water uses for two of the 14 districts. Spatial differentiation of scenario impacts showed that districts with medium to low yield potentials were more affected by negative impacts than districts with high yield potentials.

In this first comprehensive sustainability impact assessment of agricultural intensification scenarios at regional level, we showed that a considerable potential for agricultural intensification exists. The intensification is accompanied by adverse environmental and socio-economic impacts. The novelty lies in the multiscale integration of comprehensive, agricultural management simulations with regional level impact assessment, which was achieved with the adequate use of indicators. It provided relevant evidence for policy decision making. Stakeholders appreciated the integrative approach of the assessment, which substantiated ongoing discussions among the government bodies. The assessment approach and the Brandenburg case study may stay exemplary for other regions in the world where similar economic and policy driving forces are likely to lead to agricultural intensification.

Keywords: Scenarios, Impact assessment, Agricultural intensification, Land use change, Irrigation, Bioenergy, Social and environmental indicators

INTRODUCTION

Globally, agriculture is facing an increasing demand for food, bio-based energy and fiber products. Nonetheless, numerous studies predict that agricultural techniques will adapt to the increasing demand (Makowski et al., 2013). However, adaptation poses challenges for the integration of environmental and socio-cultural services into agricultural production.

Farmers' adaptation to the increasing demand may include changes in the choice of crops, crop rotations, utilization of crops, and intensification of production. Trends include technical solutions to remove yield limiting factors, such as water availability for crops, and increasing use of agricultural biomass as a source of renewable energy. The latter is often supported by government legislation. For example, in Germany, the introduction of the Renewable Energy Law EEG in 2000 resulted in threefold increase of the area of maize (*Zea mays*) and of rapeseed (*Brassica napus*) for bioenergy, reaching 17.5% of the German cropland area in 2012 (FNR, 2012a). Also, the use of irrigation to increase and stabilize crop yields is becoming more attractive, especially in areas with limited rainfall and with soils with limited water holding capacity. Additional reasons for these changes are related to an increasing water demand of new, more productive cultivars, and the prospect of more irregular precipitation patterns due to climatic change. For example, at the current market prices for agricultural goods, irrigation in the German federal state of Brandenburg is already on the verge of becoming economically viable (Münch et al., 2014). Also, a connection between the production of bioenergy and irrigation exists: as farmers become fuel suppliers for power plants, it becomes more important to achieve stable yields even in dry years.

Agriculture is multifunctional. This means that in addition to (private) economic production, it contributes to public goods such as the character of rural landscapes and its ecosystem services (Wiggering et al., 2006; Van Zanten et al., 2013). A simple focus of agricultural management aimed solely at maximizing economic returns can lead to depletion of groundwater resources, erosion, loss of water quality, biodiversity loss and a reduction of socio-cultural services. Although these services become evident at landscape level, which has a spatial scale larger than a farm, it is the decision-making at farm level that affects these services. Sustainable development therefore requires consideration of the balance between the economic production functions of agriculture and environmental and social services. Policies are implemented to incentivize farmers to respect this balance by remunerating for the provision of public goods.

It is important, when making policy decisions that support sustainable development, to acknowledge present and future development trends and their potential economic, social and ecological impacts. Here, we use ex-ante impact assessment (Helming et al., 2011). This integrates state-of-the-art knowledge from various disciplines in order to highlight those risks and opportunities which are inherent in expected trends. We use scenarios of agricultural management to draft different development options. These scenarios provide the opportunity to explore possible future developments through a comparative analysis of alternative driving forces and trends. In spanning a range of options they help to explore rather than predict possible developments (Milestad et al., 2014). We use impact indicators to assess economic, social and environmental effects of the agricultural management scenarios.

The objectives of the study described in this paper were as follows: (i) to develop an indicator based impact assessment method that combines expertise on agricultural management, landscape, hydrology, soil erosion, biodiversity, stakeholder interaction, sustainability to create and analyze agricultural intensification scenarios and (ii) to conduct an integrated ex-ante assessment of regional sustainability impacts induced by farm level scenarios of bioenergy production and crop irrigation. Results had to assist policy stakeholders in identifying sustainability issues that require policy steering.

The scenarios were designed to integrate currently trending assumptions of driving forces and describe their effects on crop choice and crop management at the farm level. Decisions at the farm level were translated into crop distribution patterns at the hectare scale for analysis of scenario impacts. Results were aggregated to the district (NUTS3) scale to derive policy relevance of the assessment. Also, this aggregation allowed the representation of the landscape level of scenario impacts. Scenarios were developed for the year 2025: a time frame that is sufficiently long to allow for major changes in agricultural management, while still being short enough to allow for realistic predictions of climate effects and yield trends.

We chose the state of Brandenburg, Germany, as a case study area because several characteristics may make this area an example for the application of integrated assessments of agricultural development scenarios: large, specialized farms; low soil fertility; high technological level; yield limitations by water; subsidies for agricultural energy production. We anticipate that the trend of increased use of cropland for the production of renewable energy will continue and that irrigation will become more important in the future to increase and stabilize yields. The extent and speed of these changes in agricultural management are largely unknown, as are the sustainability implications of these changes. A few studies have dealt with specific aspects of bioenergy production in Germany (e.g., Dressler et al., 2012; Hennig and Gawor, 2012) and with specific agricultural adaptation scenarios at field sites (Nendel et al., 2014). However, to our knowledge, no integrated impact assessment studies exist that analyze the effects of agricultural intensification on environmental, social and economic indicators at regional level. Such a case study will help to generate spatially explicit systems knowledge of human–environment interactions in land change processes (Magliocca et al., 2014).

MATERIALS AND METHODS

Case study Brandenburg

Brandenburg is the fifth largest German state, and it has a land surface area of 29,640 km^2, of which 45% is agricultural land, most of which (75%) is arable land. Brandenburg surrounds the German capital Berlin and comprises 14 districts. The state's mineral soils are developed on glaciofluvial, periglacial and glacial deposits, aeolian deposits and river sand. Almost two-thirds of the state's territory, mainly with sandy and sandy loamy soil, has a water holding capacity lower than 140 mm (Table 1). Peat soils are excluded from crop cultivation by decree.

Table 1: Classification of the water holding capacity (mm) of soils in Brandenburg (area: 29,485 km^2)

Cumulative area (%)	Range (mm)	Description[2]
21	≤80	Very low
34	>80–110	Low
62	>110–140	Moderate
77	>140–170	High

82	>170–200	Very high
100	>200	Extreme high

Source: Schindler et al. (2004).

Agricultural practice is dominated by large farm enterprises with an average size of 238 ha, which is four times the German average. Mechanization is high, and the labor force is only 1.7 persons per 100 ha on average. Large-scale operations, hired labor and a high mechanization rate result in highly competitive farm enterprises that are oriented towards profit maximization. A study on the regional impacts for Brandenburg of changes to the European Common Agricultural Policy (CAP) indicated that the dependency of farm enterprises on world market prices is high compared to the European average (Sieber et al., 2013).

The average rainfall in Brandenburg is 554 mm year^{-1} (1981–2010), which is low compared to the German average of 800 mm year^{-1} (German Weather Service, 2013). This limits the yield for rain-fed agriculture. Irrigation is currently used on only 2% of the cropland area (Statistical Office Brandenburg, 2012). Due to increasing world market prices for agricultural commodities and concerns about increasing water stress related to climate change, we expect that irrigation will become more important in the near future. However, there are concerns over dwindling groundwater resources related to this development. We analyzed the scenario effects for the 14 districts of Brandenburg. Table 2 lists the names, cropland areas and name codes of districts.

Table 2: Districts of Brandenburg and their cropland areas

Code	**District**	**Cropland (1000 ha)**	**Code**	**District**	**Cropland (1000 ha)**
BAR	Barnim	41.9	OPR	Ostprignitz-Ruppin	98.7
EE	Elbe-Elster	82.6	OSL	Oberspreewald-Lausitz	30.0
HVL	Havelland	67.5	PM	Potsdam-Mittel-mark	84.3
LDS	Dahme-Spreewald	67.3	PR	Prignitz	123.7

LOS	Oder-Spree	66.7	SPN	Spree-Neisse	45.7
MOL	Märkisch-Oderland	120.2	TF	Teltow-Fläming	77.1
OHV	Oberhavel	56.7	UM	Uckermark	156.9

Analytical framework

We adapted and simplified the DPSIR framework of the European Environment Agency (Gabrielsen and Bosch, 2003) to structure the ex-ante impact assessment of agricultural intensification scenarios. Three steps were involved (Fig. 1): Step 1 was the scenario definition, including the driver-pressure part of DPSIR. The driving forces in German agriculture were characterized as Drivers. Pressures represented the changes of agricultural management, in this case an increased area of silage maize to be used in biogas power plants and an increased area of irrigated cropland. Step 2 comprised the analysis of the scenario effects on several economic, social and ecological states through the use of indicators. Step 3 involved presenting the assessment results to members of the federal state ministries and local administration. This was done to inform policy decision-making and to evaluate the relevance of anticipated changes with regards to impacts on achieving policy strategies. The responsepart of DPSIR was not covered by the analysis.

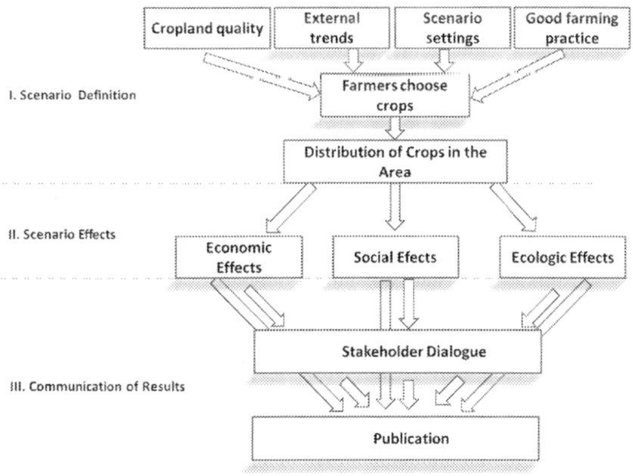

Figure 1: The analytical framework of the study, comprising three steps: (1) scenario definition by translating external driving forces into decisions of agricultural management; (2) analysis of the effects of the scenario on economic, social and environmental indicators at the district scale; (3) communication of assessment results to decision-makers.

Sources of uncertainties of such an assessment approach are found in the applied models (structural uncertainty, parameter uncertainty), in the data used to drive the models and in the basic assumptions under which the models are used. In our case, we did not attempt to quantify the uncertainty that came with the models or data as this is subject to investigation in the model developer community (Asseng et al., 2013). The use of scenarios (Step 1), however, reduces the certainty requirement for model results, since scenarios are usually compared against each other and not evaluated in absolute terms. The scenario building process involves assuming conditions that are uncertain *per se* and this uncertainty can hardly be quantified. However, in the process of result communication (Step 3) uncertainty information plays a major role and in our case, spatial variability of model results was used to increase the addressees' awareness of the uncertainty of projections and the likelihood of scenarios becoming true.

Scenarios

Scenarios of future cropping practices were chosen based on their relevance to agricultural land use and sustainable development. They were based on the driving forces: high world market prices for agricultural goods, climate change projections, and state legislation to promote energy production from agricultural sources. Three scenarios were constructed (Table 3): a Business As Usual *(*BAU*)* scenario extrapolating current trends into the year 2025, an Energy scenario in which government subsidies for biogas production lead to further increase in the cultivation area of silage maize, and an Irrigation scenario in which all areas of silage maize, winter rapeseed, winter wheat and sugar beet were irrigated. In all scenarios, we assumed that silage maize for energy purposes was only used in biogas power plants. This is the typical energetic use of silage maize in Germany. The scenarios were chosen to continue current trends (irrigation, subsidies for 1st generation biofuel) and address agricultural production decisions. The first is because the current policy frame is likely to further support the trend. It also serves as role model for other countries in search for adaption measures to energy scarcities. The second is because our intention was to assess spatially explicit sustainability impacts of agricultural land use changes. Changes happening further down in the value chain such as the use of residues for bioenergy production (2nd generation biofuel) was not the focus of the study.

Table 3: Parameters describing the three agricultural scenarios: BAU, Energy, and Irrigation for 2025. The scenarios include assumptions about changes in prices, yields and agricultural management constraints related to animal husbandry, technology, and crop rotations

Parameters	**Scenarios 2025**		
	Business As Usual (BAU)	**Energy**	**Irrigation**
Prices/cost (relative to 2011)	Net margin from agricultural goods +10%	Same as BAU and net income from silage maize for biogas +20%	Same as BAU
Irrigation	No irrigation	Same as BAU	All areas of silage maize, winter wheat, winter rapeseed and sugar beet are irrigated for optimum productivity
Fallow land	Fallow land reduced from 4% of cropland area in 2011 to 2% in 2025		
Yield development (relative to annual yields in 2011) (ranges reflect different site qualities)	(Based on expert judgement):		
	Silage maize:	+1.2 – 1.8 t ha^{-1}	
	Winter barley:	+0 t ha^{-1}	
	Winter rapeseed:	+0.55–0.60 t ha^{-1}	
	Winter rye:	+0 – 0.15 t ha^{-1}	
	Winter wheat:	+0.45–0.60 t ha^{-1}	
	Sugar beet:	+7.50 t ha^{-1}	
Animal husbandry (relative to 2011)	Constant		
Technology (relative to 2011)	Constant		
Restriction incrop rotations	50% of winter rapeseed uses winter barley as the previous crop; sugar beet area constant due to existing contracts		

Translation of scenario assumptions (Drivers) into crop shares and crop distributions (Pressures)

To translate scenario assumptions into agricultural management and cropping decisions, we used a simplified linear programming optimization model for "region farms" that consists of all farm resources of one district, taking into account established farm structures, site characteristics, markets and agricultural policy. We assumed that farmers make their cropping choices mainly by optimizing net farm income. This is an assumption that reflects long term developments and that is relevant particularly in Brandenburg since its exceptionally large agricultural enterprises with hired labor enforce economic decision making. Economic professionality of the enterprises can also be seen in the fact that, according to the agricultural census 2010, 46% of farm managers hold a university degree (Federal Statistical Office Germany, 2011).

The approach used existing crop production data for each of five soil categories representing different levels of soil fertility (LELF, 2010), and it incorporated additional expert assessments of inputs and outputs for important cropping practices, including irrigation, yield development as influenced by genetic progress, climate change and management improvements. Crops were compared for their net margin based on default machinery cost figures (KTBL, 2012). The linear programming farm model was constructed for each district. Assumptions and constraints were (i) constant livestock fodder requirements, (ii) complete use of manure in the cropping systems, (iii) constant level of contract-based cropping systems, and (iv) adherence to crop rotation restrictions. On this basis, the model maximized the total net margin of the district by allocating part of the area of each soil category to the best-performing cropping practices, taking the restrictions into account. The result was a specific distribution of crops for each district and for each of the five soil categories.

For the assessment of six of the impact indicators (see Table 4), it was necessary to disaggregate the simulated crop distribution from the district scale to the hectare scale. This was done by using a Monte Carlo approach to distribute crops on a high-resolution soil suitability map, taking into account the total area share of each crop as derived from the linear programming described above (Wenkel et al., 2013). The approach implements the idea to distribute a list of crops and its proportions of the agricultural area according to the suitability of the soil at a certain location for each crop. The result was a soil suitability-based probability map of crop distributions at a 1 ha scale for the area of Brandenburg.

Table 4: Overview of indicators used to assess scenario impacts. Nine indicators were chosen, representing two economic, one social and six environmental themes relevant to the sustainable development of agricultural landscapes. Indicators were determined at the district and/or ha level

Indicator	Unit	Sustainability dimension			Spatial reference		Model/Method used	Reference
		Econ.	Soc.	Environ.	District	ha		
Crop yields	$t\,ha^{-1}$	X				X	YIELDSTAT	Mirschel et al. (2014)
Irrigation water demand	$mm\,ha^{-1}$	X				X	ZUWABE	Mirschel et al. (2012)
Water constraints for irrigation	Irrigation water demand in % of groundwater recharge, classified: low/medium/high			X		X	Contextual data analysis	-
Electricity demand covered	%			X	X		Contextual data analysis	-
Landscape scenery	Nominal scale		X		X		Contextual data analysis	-
Greenhouse gas savings	t CO_2-equ. per district			X	X		Contextual data analysis	-
Biodiversity	Nominal scale			X		X	Habitat suitability model	Brandt and Glemnitz (2013)
Erosion risk	t per district			X		X	USLE adapted	Wishmeier and Smith (1978)
Area demand for water quality protection	km^2, % Cropland			X		X	Contextual data analysis	-

Scenario effects

For the analysis of scenario effects, we focused on issues that were considered sensitive to agricultural management and relevant to policies in the state of Brandenburg. This means that indicators had to be chosen that indicate at landscape level the impact of field scale agricultural management. Additionally, a balance between economic, social, and environmental dimensions was sought. One indicator was selected for each impact issue based on scientific credibility, sensitivity to agricultural scenario changes, applicability to regional conditions, and determinability. An overview of the chosen indicators is given in Table 4. Calculation methods for each indicator are briefly described below. The final results for all indicators were aggregated at the district level.

Indicator 1: crop yields

Crop yields were calculated by using the statistical hybrid model YIELDSTAT (YIELD estimation based on Statistics, Mirschel et al.,

2014) and a model for the spatial distribution of crops in a region. Both models were controlled via the Spatial Analysis and Modeling Tool SAMT (Spatial Analysis and Modeling Tool, Wieland et al., 2006). Spatial crop distribution on a hectare scale was calculated by using statistical information about crop production at the state level, by performing calculations of crop-specific economic benefit, by developing a soil-based suitability map for crop growth and by using a set of algorithms that combines optimization routines and stochastic (Monte Carlo) elements. The scenario simulations were run under the following assumptions: no change in crop and soil management (conventional farming, soil tillage using a plough). Future climate was provided from the STAR2 climate data set for Brandenburg (2K scenario). Annual results were calculated for the time period 2020–2030 and averaged.

Indicator 2: irrigation water demand

Water is the most limiting factor for agricultural yields in Brandenburg. For this reason, irrigation is essential to stabilize and increase yields. Irrigation water demand (IWD) was calculated by using the model ZUWABE (Zusatzwasserbedarf (irrigation water demand), Mirschel et al., 2012), which is implemented in SAMT (Wieland et al., 2006). Soil types were grouped into four categories of different water storage capacities. Crop-specific irrigation water benchmarks for the soil categories were calculated by considering crop-specific irrigation periods, climatic water balance within the irrigation period, soil water storage capacity, soil water content at the start of the irrigation period, and rooting depth. Calculations considered the anticipated climatic water balance for the period 2020–2030 and a predicted decrease in crop transpiration caused by rising atmospheric CO_2 concentrations. Input data were the medium scale agricultural site mapping for arable land (Schmidt and Diemann, 1991), simulated crop distributions at the hectare scale and the STAR2 future climate data for Brandenburg (2K scenario).

The dynamic agro-ecosystem simulation model MONICA (model for nitrogen and carbon dynamics in agro-ecosystems, Nendel et al., 2011) was used to back up the simulations of the model YIELDSTAT and ZUWABE. MONICA has been tested against crop and soil monitoring data from Brandenburg. As a process-based model, it may be more sensitive to environmental changes than YIELDSTAT and ZUWABE.

However, in a spatial application, statistical model approaches prove more robust against outliers. Therefore, MONICA results were used in an iterative process to compare and assure the results of the other two models.

Indicator 3: water constraints for irrigation

The water constraints for irrigation constitute spatially differentiated indicators for detecting those parts of the state where groundwater-borne irrigation could pose problems for other water users, including groundwater-dependent ecosystems (peatland). A critical water balance was considered in this study when IWD was above 25% of the supplied groundwater recharge. Water balance was calculated as the spatially differentiated long-term mean annual groundwater recharge over the period 1976–2005 (100%) minus IWD (see above). Calculations were performed at the spatial level of polygons of hydrogeological units based on classified properties of the upper groundwater-bearing layer (Merz et al., 2009). Polygon results were resampled and aggregated at the district level. Relative water demand was clustered into three categories: IWD < 16% ground water recharge (low), 16% ≤ IWD ≤ 25% ground water recharge (medium), and IWD > 25% groundwater recharge (high).

Indicator 4: share of electricity demand covered by the production of biogas

The production of biogas in Germany serves the purpose of increasing the share of renewable resources used in the production of electricity. The share of electricity demand covered was calculated by dividing the net electricity produced by the electricity demand projected for the state of Brandenburg in 2025.

The electricity produced was based on silage maize yields for energy generation. These were multiplied by typical values for the amount of methane produced per ton of silage maize and by the average conversion efficiency of biogas plants (FNR, 2012b). Net production of electricity was calculated by subtracting an average value of 7.9% of electricity demand for the biogas plant operations (FNR, 2010). Electricity demand for the year 2025 was based on the targets of the energy policy of the federal state of Brandenburg, which aim for a reduction of 10% between 2007 and 2030.

Indicator 5: the scenic value of landscape

The scenic value of landscape is an important factor for the quality of life in rural areas. It determines, in part, the recreational value of agricultural landscapes. In this study, scenario impacts on landscape scenery were determined by applying the principle of Ecological Risk Analysis as established in the German Environmental Impact Assessment. This analysis makes it possible to compare landscapes capacities under different human land use scenarios. Changes in landscape scenery caused by scenario differences were determined by identifying sensitive areas and demand areas. Sensitive areas have high landscape values; demand areas are designated for recreational or touristic purposes. Disposition was determined as the share of sensitive and demand cropland areas divided by the total cropland area in a district.

The effect of scenarios upon the scenic value of landscape was estimated in a three-step approach. First, the aesthetic impressions of crops were determined by using eight criteria: colors, light and shadow, exposure; odor; body and structure, border situations; dynamics; and orientation. Second, the effects of the relative area shares of crops were assessed at a three-stage ordinal scale using these criteria. As a result, the ratio of grain and rapeseed areas to maize areas (GRM) proved to be a critical indicator integrating step one and two. Third, the disposition value was divided by the GRM to form an impact indicator representing ecological risk.

Indicator 6: greenhouse gas savings

Germany intends to substitute fossil fuels and reduce total greenhouse gas emissions by increasing the production of electricity from renewable sources. Greenhouse gas savings through the substitution of fossil fuels by biogas were calculated based on the yields of silage maize for energy generation. We assumed that all biogas plants are heat and power plants. The coefficients for methane generation per ton of biomass, the efficiency of conversion and the electricity demand for biogas plant operations were based on typical values (FNR 2010, 2012b). Greenhouse gas savings per kWh produced were calculated under the assumption that only fossil fuels would be substituted.

Indicator 7: biodiversity

The biodiversity strategy of Germany (National Strategy on Biological Diversity, 2007) uses bird species as core indicators of the biodiversity of agricultural landscapes. We used the same indicators to analyze the potential conflicts between scenarios and biodiversity strategies. The indicator aggregates the population trends of ten arable bird species, of which we considered seven that either (i) breed on arable land: *Skylark* (*Alauda arvensis*), *Corn bunting* (*Miliaria calandra*), *Lapwing* (*Vanellus vanellus*) and *Winchat* (*Saxicola rubetra*), or (ii) feed regularly on arable land: red-backed shrike (*Lanius collurio*), *Yellowhammer* (*Emberiza citrinella*) and *Woodlark* (*Lullula arborea*). We applied a habitat suitability model (Brandt and Glemnitz, 2013), which calculates the 'intrinsic habitat value' according to Anderson and Fergusson (2006) over the growing season based on the following data: vegetation structure of the crop species (vegetation height and density; real field data), habitat demands of the bird species, agricultural management (sequences of agricultural activities = disturbances).

Indicator 8: erosion risk

Some crops (e.g., maize, rapeseed) are more vulnerable to soil erosion by water than others (e.g., cereals), which is why the crop distribution within the scenarios may affect soil erosion risk. Erosion risk was calculated by using a German adaptation of the universal soil loss equation model (USLE – Wischmeier and Smith, 1978). Whilst it should not be used for precise estimates of soil loss, it is suitable for appraising erosion trends resulting from changes in agricultural management (Deumlich et al., 2005). The equation comprises six factors: rainfall and runoff (R-factor), soil erodibility (K-factor), slope length (L-factor), slope steepness (S-factor), cover and management (C-factor), and support practice (P-factor). Potential long-term average soil loss is calculated as the product of the individual factors. In this study, the scenarios affect only the C-factor because every combination of crop and management is associated with a specific C-factor. The results were calculated on the scale of field blocks (as of May 2011) and aggregated at the NUTS3 level.

Indicator 9: area demand for water quality protection

An increase in maize cultivation area may enhance nutrient transport into nearby water bodies. Maize fields are considered to be hot spots of surface and groundwater pollution because of the large distance between plant rows and because maize is a preferred crop for manure application from livestock or biogas plants (Jaafar and Walling, 2010). An "end-of-pipe" method was applied to assess the water quality impairment of the scenarios, using area demand for water protection as an indicator. We calculated the area demand for riparian buffer zones against soluble and solid matter entries into the water bodies. Buffer zones of 20 m between maize fields and running waters of 1st and 2nd order (according to the German Water Law) and around standing waters are most effective with respect to both retention of solid matter transported by surface runoff and retention/transformation of dissolved substances transported by groundwater runoff (DWA, 2012). Area demand was calculated by overlaying maps of field blocks and crop distribution for the scenarios and by considering the distribution of running waters of 1st and 2nd order in the state of Brandenburg. Existing riparian buffer zones were not considered.

Involvement of stakeholders

Stakeholders from local government were engaged in the scenario study to (i) discuss scenario results with regards to their relevance, opportunities and threats, (ii) provide feedback on the scenario features, (iii) contribute to the identification of research needs, and (iv) identify future cooperation interests. Stakeholder engagement in research on regional land use can vary considerably from information sharing to interactive participation and joint decision-making (Knierim et al., 2010). In our case, the level of consultation was essential, which means that stakeholders had to be appropriately informed about and invited to discuss the results of the impact assessment study (Pretty, 1995). The selection of the stakeholders was guided by the aim to broadly represent public decision-makers, both from the investigated impact areas, and from organization generally involved in sustainable development. Among the 16 stakeholders participating in the workshop, there were 6 experts from the Ministry for Environment representing the divisions of water management, biodiversity and soil conservation, one experts from the Ministry of Agriculture and one from the Ministry of Economy. Four experts represented the two state agencies for environment and for agriculture. Four stakeholders represented non-government institutions, namely the farmers' association, the Brandenburg Sustainability Council, and a private, economic networking agency. In preparation for the stakeholder workshop, efforts were made to reduce the scenario results to seven key statements. A workshop of 2.5 h duration was designed, which included expert inputs, facilitated discussions and opinion polls. An interactive voting tool was used to determine how stakeholders perceived the degree of novelty and relevance of the results. After the meeting, minutes were disseminated. To facilitate uptake of the results, the findings of the study were compiled into a brochure for decision-makers (Gutzler and Helming, 2013).

RESULTS AND DISCUSSION

Scenario estimations of area shares of most important crops in 2025

The land use distributions for different scenarios are shown in Fig. 2. For the BAU scenario, silage maize for energy and fodder had the highest area share (29%), followed by rye (21%), wheat (13%) rapeseed (16%) and barley (8%). The Irrigation and Energy scenarios resulted in 6% and 20% increases of the maize cultivation area, respectively, compared to the BAU scenario. In both cases, the increase in maize area was a result of the increased profitability of this crop due to increased yield and yield stability in the Irrigation scenario, and due to the financial subsidies for the production of renewable energy assumed in the Energy scenario. The increase in the area share of maize was accompanied by reductions in all other crops, particularly rye and rapeseed. Rye is the least economically feasible crop, and rapeseed turned out not to be economically competitive in the Energy scenario. The share of barley was small in all three scenarios, and it was mainly cultivated for crop rotation purposes. Sugar beet was below 1% area share in all scenarios. The results imply a drastic increase in the maize share for both intensification scenarios compared to the BAU scenarios. However, even the maximum of 49% maize in the Energy scenario remained below the EU-CAP greening threshold of a maximum 70% share for one crop. This is due to the scenario settings, in which crop rotations (according to phytosanitary requirements and preceding crop effects) were a constraint for management decisions.

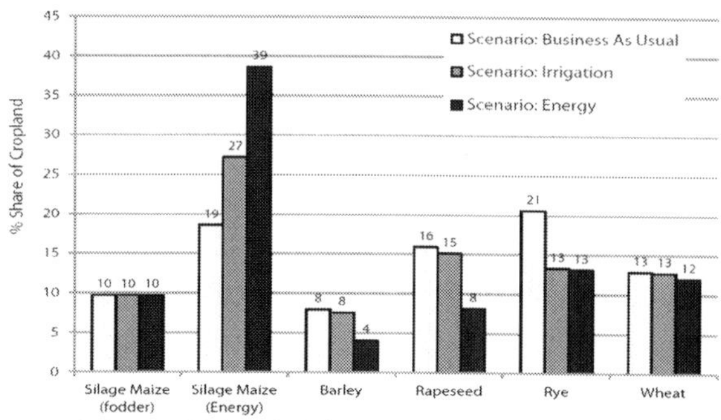

Figure 2: Simulated area shares of main crops for the three scenarios for 2025 in Brandenburg. The main scenario differences occur in the case of silage maize for energy. Increased maize share resulted in reduced shares of rye, rapeseed and barley.

Scenario effects at the district level for Brandenburg in 2025

Crop yields

Within the period 2020–2030, simulated crop yields fluctuated due to year-to-year weather variation. In the BAU scenario, the standard deviation ranged from 10 to 15% for spring crops (silage maize, sugar beet) and from 5 to 10% for winter crops. Simulated crop yields for the three scenarios and for the 14 districts are presented in Table 5. The results of the Irrigation scenario showed that irrigation was able to support significantly higher yields, to increase yield stability, and to reduce cropping risk. The yield increases were highest for silage maize (34–48%) and lowest for winter rapeseed (3–9%). For winter wheat and sugar beet, simulations resulted in yield increases of 16% and 23%, respectively. Because of the increased yields and improved yield stability, irrigation could be an effective measure to adapt to climate change. This is because climate change impacts in this study area are predicted to shift rainfall from summer to winter periods thereby aggravating water stress during the vegetation period, and to increase the rainfall variability between years, which in turn increases yield variability. In addition, irrigation is important for farms growing silage maize for biogas production, because they have to comply with long-term contracts for annual biomass delivery.

Table 5: Calculated average crop yields 2020–2030 for silage maize, sugar beet, winter rapeseed, winter rye and winter wheat in the three scenarios (BAU: Business As Usual; I: Irrigation; E: Energy) for the districts of Brandenburg. Compared to the BAU scenario, the irrigation scenario resulted in yield increases of 3–9% for rapeseed and 34–48% for silage maize in the 14 districts. No significant crop yield differences were simulated for the BAU scenario or the energy scenario

	Silage maize			**Sugar beet**			**Winter rape-seed**			**Winter rye**			**Winter wheat**		
	t ha^{-1}			**t ha^{-1}**			**t ha^{-1}**			**t ha^{-1}**			**t ha^{-1}**		
District	**BAU**	**I**	**E**	**BAU**	**I**	**E**	**BAU**	**I**	**E**	**BAU**	**I**	**E**	**BAU**	**I**	**E**
BAR	34.7	49.2	33.6	58.1	69.5	58.5	3.7	3.9	3.7	4.6	4.6	4.6	6.4	7.6	6.5
EE	37.0	49.7	35.8	60.9	70.3	60.8	3.8	4.1	3.8	4.8	4.8	4.9	6.6	7.7	6.6
HVL	35.1	49.2	33.8	59.2	69.9	59.1	3.7	3.9	3.7	4.6	4.6	4.6	6.6	7.8	6.6
LDS	31.9	46.5	30.2	60.0	70.8	60.4	3.5	3.7	3.5	4.3	4.3	4.3	6.0	7.4	6.1
LOS	31.5	45.6	30.4	60.1	69.8	60.2	3.6	3.8	3.6	4.2	4.2	4.2	6.5	7.8	6.6
MOL	35.0	49.9	34.3	58.0	69.9	57.9	3.5	3.7	3.5	4.3	4.3	4.4	6.8	7.9	6.8
OHV	33.6	48.4	32.5	60.2	71.3	60.3	3.6	3.8	3.6	4.5	4.5	4.6	6.0	7.3	6.1
OPR	33.2	49.0	30.3	61.4	71.4	61.1	3.6	3.8	3.7	4.2	4.2	4.3	6.2	7.5	6.2
OSL	36.7	50.6	34.0	59.9	69.5	59.6	3.7	4.0	3.8	4.6	4.5	4.6	6.5	7.6	6.5
PM	31.6	46.2	29.2	58.2	68.8	58.2	3.3	3.6	3.4	4.1	4.1	4.1	6.2	7.5	6.3
PR	37.5	51.0	36.8	61.6	71.2	61.6	4.0	4.2	4.0	5.0	5.0	5.0	6.8	8.0	6.9
SPN	36.6	50.4	33.3	60.4	70.2	60.2	3.6	3.8	3.6	4.4	4.4	4.4	6.6	7.8	6.7
TF	33.1	47.2	31.9	58.4	68.8	58.5	3.4	3.7	3.5	4.5	4.5	4.5	6.5	7.8	6.6
UM	35.4	49.5	34.4	58.0	69.1	58.1	3.8	3.9	3.8	4.8	4.8	4.9	7.2	8.4	7.2
Brandenburg	34.3	48.6	33.6	58.8	69.6	58.8	3.6	3.9	3.7	4.5	4.5	4.5	6.6	7.8	6.6

No significant change in crop yield was detected between the BAU and Energy scenarios. The crop yield of silage maize was reduced by 0.75 t ha^{-1} on average in the Energy scenario compared to the BAU scenario. This may be because the larger cultivation area for maize in the Energy scenario also includes sites of lower productivity.

Patterns of simulated yield levels across the 14 districts of Brandenburg showed little variation between the scenarios. The highest simulated average crop yields were in the districts PR, UM, MOL, and EE. With the highest share of arable land, PR, UM and MOL are also the districts with the highest agricultural production (Table 2). In contrast, PM, LDS and LOS had the lowest average crop yields.

Modeling results revealed that considerable yield effects can be expected with the introduction of crop irrigation.

Irrigation water demand (IWD)

The IWD calculations showed that, to achieve an unconstrained biomass and yield accumulation, the spring crops (maize, sugar beet) need much more irrigation water than the winter crops (wheat, rapeseed) (Table 6). The reason is that the time period for irrigation stretches from early June until the first decade of September for silage maize and from early July until the second decade of September for sugar beet, whereas for winter wheat it stretches only from early May until the end of June and for rapeseed it stretches from the second decade of April to the second decade of May. The between-district variation in IWD was related to the precipitation pattern within Brandenburg and to the soil quality-related yield potential. For spring crops, EE had the lowest IWD and the lowest crop yield increase, whereas OPR (silage maize) and MOL (sugar beet) had the highest IWD and crop yield increase. For winter crops, IWD and crop yield increase were highest in LDS and lowest in BAR (winter wheat) and UM (winter rapeseed). In any case, crop yield increases caused by irrigation were positively correlated with IWD.

Table 6: Irrigation scenario for 2025: calculated irrigation water demands (IWD) and respective crop yield increases for the 14 districts of Brandenburg. Spring crops need more irrigation water than winter crops. Variations of IWD between districts were related to soil quality and precipitation patterns between districts

	Silage maize		Winter wheat		Winter rapeseed		Sugar beet	
	Irrigation water demand (mm)	**Crop yield increase (t ha^{-1})**	**Irrigation water demand (mm)**	**Crop yield increase (t ha^{-1})**	**Irrigation water demand (mm)**	**Crop yield increase (t ha^{-1})**	**Irrigation water demand (mm)**	**Crop yield increase (t ha^{-1})**
BAR	124	14.9	72	1.1	19	0.2	115	10.9
EE	109	13.0	81	1.2	27	0.3	98	9.3

HVL	119	14.2	78	1.2	26	0.3	113	10.7
LDS	123	14.8	89	1.3	28	0.3	108	10.3
LOS	122	14.6	80	1.2	25	0.2	99	9.4
MOL	125	15.0	74	1.1	21	0.2	119	11.3
OHV	126	15.1	84	1.3	25	0.3	116	11.1
OPR	134	16.1	87	1.3	26	0.3	108	10.2
OSL	118	14.2	81	1.2	26	0.3	105	10.0
PM	124	14.9	83	1.2	27	0.3	113	10.7
PR	115	13.8	78	1.2	19	0.2	100	9.5
SPN	118	14.1	85	1.3	27	0.3	104	9.9
TF	119	14.3	79	1.2	27	0.3	109	10.3
UM	119	14.3	75	1.1	18	0.2	117	11.1
Bran-den-burg	121	14.5	80	1.2	24	0.2	113	10.7

Water constraints for irrigation

In Brandenburg, a total of 3350 million m^3 of groundwater is recharged per year, taken as an average of the years 1976–2005 (MUGV, 2009). The IWD in the Irrigation scenario amounted to 667 million m^3 per year, which was 20% of the annual groundwater recharge. Only in the two districts UM and MOL did the IWD amount to more than 25% of the annual groundwater recharge. In those districts, IWD was particularly high for sugar beet and maize. In seven other districts, IWD was between 16% and 25% of the groundwater recharge (Fig. 3).

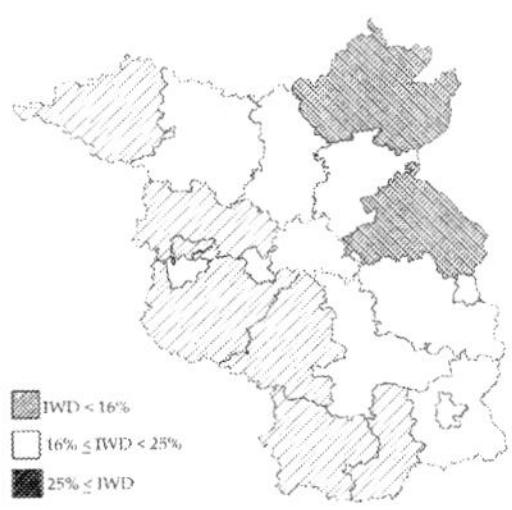

Figure 3: Water constraints for irrigation expressed as irrigation water demand (IWD) of the Irrigation scenario relative to groundwater recharge. An IWD of more than 25% of annual groundwater recharge was found to be critical with regard to competent water uses. This threshold was estimated to be reached only in two of the 14 districts in Brandenburg.

Although Brandenburg has a low annual rainfall compared to other German states, we found that there is sufficient groundwater recharge in most districts to provide a basis for crop irrigation. This was in contrast to the position taken by the federal state office for environmental protection in Brandenburg (see stakeholder workshop) that might be biased by frequent dry spells and the decline in groundwater and lake levels observed between the 1980s and approximately 2007. Water availability was defined and calculated here under the worst-case assumption of no return flow from irrigation into groundwater, meaning that all of the water pumped up would transpire or evaporate. Two factors were spatially differentiating water availability: (1) the share of arable land to be irrigated in a region, (2) the local groundwater conditions, which may account for an occasional small-scale supply shortfall. Provided that the local conditions are thoroughly taken into consideration, the results may encourage farmers to expand irrigated crop production in Brandenburg. However, local and area-specific problems are possible if todays irrigated area is increased. For example, groundwater-dependent ecosystems may locally compete with field irrigation for water use.

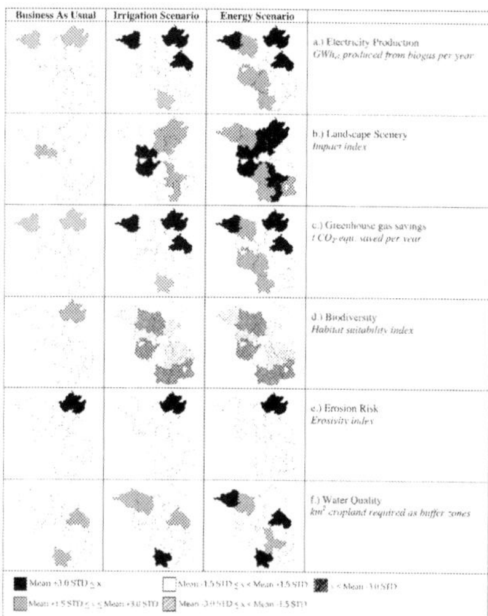

Figure 4: Impacts of three agricultural intensification scenarios: Business As Usual, Irrigation and Energy for 2025 at the district level in Brandenburg. Normalized values relative to the mean and standard deviation of the BAU scenario are displayed.

For the following indicators, the estimated scenario effects are presented at the district level in Fig. 4. To allow comparisons of the scenario effects of various indicators and to allow differentiation between the districts of Brandenburg, all indicator values were normalized. This was achieved by expressing all indicator values as values relative to the mean and standard deviation of the respective values of the BAU scenario. The results were then grouped into 5 classes of deviation from the mean BAU value (Fig. 4).

Share of electricity demand covered by the production of biogas

The share of net electricity demand of Brandenburg that can be covered by the production of biogas was estimated to rise from 21% in the BAU scenario to 34% in the Irrigation scenario and 46% in the Energy scenario. This would require between 19% (BAU) and 39% (Energy scenario) of the total cropland to be used exclusively for the production of energy. These sites would no longer be available for food, feed or fiber production.

However, this value is a rather optimistic estimate, because the net electricity demand does not account for energy losses in the distribution grid. Furthermore, the assumed demand in 2025 is based on political targets that include a 10% reduction in electricity consumption. Even in Brandenburg, where population density is low and cropland accounts for a large share of demand, 48% of the total cropland must be cultivated with maize to cover the projected electricity demand and still produce the required amount of animal fodder.

Scenic value of landscape

Assessment of the aesthetic value of different crop types showed that grain and rapeseed were perceived as beautiful, whereas the perception of maize was ambivalent, and sugar beet was considered a less attractive crop. Regarding the relative crop shares, the visual perception of the BAU scenario was best, followed by the Irrigation scenario and the Energy scenario. This was true for the federal state Brandenburg as a whole as well as for the individual districts (Fig. 4). The Irrigation scenario was particularly disadvantageous with respect to landscape scenery in the districts HVL and PM. In the Energy scenario, the landscape scenery of an additional five districts was negatively affected. Together, these districts represented 50% of all districts with negative scenery effects. However, except for UM, these districts were not those that experienced yield increases and high irrigation water demand. Obviously, sensitivity to landscape scenery deterioration was not necessarily associated with areas of high potential for yield increase.

Greenhouse gas savings

The results showed that the production of biogas can save between 8.1 t (Energy scenario) and 9.5 t (Irrigation scenario) of CO_2 equivalents per hectare of silage maize production. Higher yields led to higher greenhouse gas savings. We calculated total savings of 1.2 million tons (Irrigation scenario) and 1.7 million tons (Energy scenario) of CO_2equivalents compared to the BAU scenario. This would constitute a saving of between 4.4% and 6.2% relative to the total emissions of Brandenburg in 2011.

Using a high amount of silage maize in biogas plants to produce heat and energy could contribute to lowering the greenhouse gas emissions. However, the results must be considered as upper limits of achievable savings because a number of factors were not included in the calculations. The method used relied on official figures published by the German Ministry for the Environment. Those figures were based on the assumption that energy produced from renewable sources will replace only fossil fuels and not other renewable energy sources. Furthermore, a high percentage of cropland used for bio-energy may increase the import of agricultural goods from other countries. Such effects were not considered in the calculations. For the Irrigation scenario, we did not include any emissions caused by the irrigation machinery, which would further increase the total emissions in that scenario.

Biodiversity

All three land use scenarios were in conflict with the National Biodiversity Strategy for Germany, which seeks to halt and reverse the decline in bird populations. The habitat suitability index did not differ for the Irrigation and Energy scenarios, but it was better for the BAU scenario (Fig. 4). The simulated changes varied between bird species and between districts. Only one of the four indicator species nesting on arable land, the *Lapwing (V. vanellus),* benefited from the land use changes, whereas the *Whinchat*

(Saxiola rubreta) was hardly affected. For the *Skylark (A. arvensis)* and the *Corn bunting (M. calandra),* however, the reduction of suitable crops for breeding was dramatic. The increase of maize cultivation (Energy scenario) decreased the available habitat area by 28.2% (*Corn bunting*) and 21.3% (*Skylark*).

The variation between districts was the result of regionally varying land use change, which was a function of soil quality. The greatest decrease in the habitat suitability index was found in districts with medium to low yield potential, and the effects in the most productive arable regions were smaller. Comparing the scenario effects with the yield potential of the districts, we found that negative effects increased with decreasing yield potential, from −7.3% (breeding index change between BAU and Energy scenario) in the districts with highest yields up to −46.1% in districts with medium yields. However, for the districts with the lowest yield potential, the trend was reversed because there were only minor changes in the cropping patterns. Agricultural intensification is widely accepted as being a cause of bird population declines on farmland during the last three decades. As part of a European bird monitoring project, Guerrero et al. (2012) found that landscape factors (e.g., biotope inventory) accounted for most of the variation in ground-nesting farmland bird individual and breeding pair densities between the regions, and crop field factors (crops and their management) were found to be more important for explaining the persistence of the populations. The present approach is one of the first to integrate the habitat requirements of indicator species, regional land use structure and land management into an ex-ante analysis. Some of our findings were counterintuitive: the decline in habitat suitability was highest in the areas with medium fertility. This was due to the availability of fewer cropping options in these areas, which led to a decrease in crop species.

Erosion risk

Rainfall erosivity in Brandenburg is low compared to other European regions. However, a few heavy rainstorms produce considerable erosion at times of insufficient soil cover or in sites with heterogeneous relief. Typically, erosion is a problem in fields with steep and/or converging slopes. An increase in the area of maize cultivation or an increase in the share of maize in crop rotations increases the risk of erosion because of the late protection of the soil surface with plant cover and because of the linear structure and high distance of the maize rows. Soil erosion already exists in the BAU scenario. Any further increase in soil erosion exacerbated the situation of affected areas and neighboring areas. For the Irrigation and Energy scenarios, the largest increase in erosion risk affects the district UM, which was also identified as the most vulnerable in the BAU scenario (Fig. 4). However, individual fields can also bear particularly high erosion risk in other districts.

The higher share of maize cultivation in the Irrigation and Energy scenarios had negative effects on soil erosion. An additional effect of irrigation on soil erosion was not considered in the calculations, but such an effect may occur in cases of high irrigation on slope sites. A reduction of erosion risk requires management decisions in accordance with the site-specific conditions and with particular attention to soil protection. Even conservation tillage cannot prevent soil loss in areas with high erosion risk (high slope steepness and horizontal curvature) if crops such as maize are cultivated.

Area demand for water quality protection

The total riparian length of the running waters of the state of Brandenburg is approximately 38,000 km, of which approximately 80% is accounted for by artificial collector ditches in ameliorated lowland areas such as fen peatland. Under the scenario assumptions, the length of the required riparian zones increases from 3743 km (BAU) to 6332 km (Energy scenario). This equates to 10–16% of all water ways. At a buffer zone width of 20 m, this corresponds to 0.5–0.8% of the total arable land area in Brandenburg, or 1.9% of the arable land under maize cultivation in that zone.

The areas to be designated for water protection differed between districts in accordance with the density of their water body networks and the simulated acreages of arable land for maize cultivation. The districts EE, MOL and PR were affected most because the scenarios placed the highest portion of maize-cultivated area in those districts (1.85%, 3.16% and 1.90%, respectively). However, the density of running waters in these districts is rather low: 0.89 km km^{-2} in MOL, 1.63 km km^{-2} in EE, and 1.47 km km^{-2} in PR.

Usually, the allocation of riparian buffer zones for water protection is not appreciated by land owners because of economic losses and changes in property rights. However, when the area of maize cultivation is extended for energy production, the costs for water management will be increased anyway to implement the binding targets of EU Water Framework Directive. Our scenario calculations show that, for the state of Brandenburg, which is one of the German states with the low density of water bodies (outside of its river lowlands), the required protection area is less than 2% of the simulated maize area.

District level distributions

When the 14 districts of Brandenburg were compared across the three scenarios, an interesting pattern was found. Four districts, EE, MOL, PR, and UM, showed the highest yield potential for all three scenarios. Of those districts, MOL and UM had the highest water demand for irrigation and revealed water availability constraints in the Irrigation scenario. For UM, additional negative effects on erosion risk and landscape scenery were identified. The area demand for water quality protection was particularly high in EE, MOL and PR. In conclusion, the trade-offs of agricultural intensification with water quantity and water quality were most evident in the four districts with the highest yield potential. In contrast, the trade-offs of agricultural intensification with landscape scenery and biodiversity were most evident in districts with moderate yield potential. The identification of those regional variations in trade-offs is useful for district level decision-making because it enables the identification of those regions where agricultural intensification may cause the least negative side effects and where is requires countermeasures such as the establishment of riparian zones or habitat zones.

Stakeholder involvement in the assessment of results

All results of this comprehensive study were condensed to a few key statements to make them communicable to the invited stakeholders. The overall participation in the discussions during the stakeholder workshop was extensive and actively involved almost every participant.

Questions and comments on the study were as diverse as the group's composition and were sometimes controversial. One point of interest had to do with the social and economic impacts at the district level that will be subject to a follow-up study. Another discussion point was the energy topic–the future role of maize and other resources for biomass production. Doubts were raised with regards to the congruency of water availability estimations with older studies. Broad interest was shown in environmental indicators. Whereas 2/3 of the participants were astonished by at least part of the results (mainly water availability), the remaining third found their estimates confirmed. There were 10 participants who were interested in cooperating in further impact assessments.

The results of the stakeholder discussions are scientifically relevant for several reasons. First, adequate documentation of stakeholder involvement and perception of scientific findings is a prerequisite for a systematic appraisal of the results (Scherhaufer et al., 2013) and constitutes an essential component of a sound impact assessment. Second, the stakeholder views represent a diversity of interests that come into play when addressing changes in agricultural practices, and they underline the importance of dialogue between science, practice and policy (Knierim et al., 2010). Third, only at certain points (e.g., related to the energy maize issue) when the discussions became 'hot' did we fully understand the challenges of conveying modeling results.

There were a number of methodological gains that resulted from stakeholder involvement. First of all, it became obvious that an open learning attitude is key for all actors involved and hence, appropriate conditions have to be created by the study coordinators (Collins and Ison, 2009). In particular, scientists largely underestimated the efforts necessary to elaborate the results in a way that can be digested by stakeholders. It took several iterations of discussion and consensus finding to come up with the key statements. On the other hand, the stakeholder discussions about regional water availability showed that commonly accepted knowledge needs to be regularly tested and updated by research. Lastly, discussions about the Energy scenario showed that many stakeholders had already made up their minds before the meeting. As a majority confirmed their willingness to further cooperate in such type of studies, we conclude that the information provided through this impact assessment study is a suitable evidence base for stakeholder. But the stakeholders' involvement into assessment has to be timely so that it corresponds to both, the state of political developments and the state of scientific insights. In the case presented here, 'consultation' was appropriate, while for the future, a more procedural and interactive participation might be possible (Pretty, 1995).

CONCLUSION

In this paper, we demonstrated the application of a series of indicators in a comprehensive sustainability impact assessment of agricultural intensification at the regional level. It revealed the potential of irrigation for increasing and stabilizing yields. This information is important for both, at farm level for farmers' decision making on agricultural management, and at regional level for policy decision making on sustainability governance. Scenario assumptions were set such that good agricultural practices (crop rotations, preceding crop effects) were respected, resulting in a maximum maize share of 49% for the Energy scenario. This value is below the threshold set in the greening of the CAP (70%). In reality, without additional policy restrictions, the maize share may reach higher values. However, the simulated levels already show the adverse impacts of increasing maize share on biodiversity, landscape scenery, water quality and erosion risk. Policy-makers need to decide whether they will accept the side effects associated with an increasing maize share.

The between-district variation of the scenario results was particularly interesting. Those regions that are expected to suffer most of the adverse impacts of agricultural intensification, according to soil quality, have only medium yield expectations. High and low yield potential regions turned out to be less affected. Obviously, those medium-yield-potential regions feature areas with high landscape values that are attractive for biodiversity habitats and recreation. Additionally, they have a high share of surface water bodies, resulting in high area demands for water quality protection. In contrast, high-yield-potential regions have fewer areas of high landscape value and less connectivity between water bodies. Low-yield-potential areas were less affected by the simulated scenario changes. Therefore, the scenario impacts were also low. The landscape characteristic that determine the agricultural scenario impacts have yet to be analyzed in detail. Such analysis would facilitate the transferability of results to other regions and situations.

Our study was aimed at the science-policy interface. We sought to provide policy decision-makers with relevant state-of-the-art information about the effects of anticipated intensification trends in agriculture. This required a contextual upscaling of farm level assumptions and hectare scale yield simulations to administrative units at the regional level. Additionally, the analyzed impact issues had to be relevant at the spatial level of policy relevance. The selection of indicators was guided by this challenge. The result was an overview of regional sustainability impacts induced by farm-level decision-making. This approach proved relevant for decision-makers. Simulation results were partly new to decision makers (yield expectations), they partly confirmed their estimates (soil erosion, biodiversity), and they partly challenged common sense (water demand and water availability). In any case, important evidence was presented that formed a basis for policy decision-making. The study revealed the need for researchers at the science-policy interface to emphasize the broader sustainability impacts and focus on trade-off studies rather than analyzing single impacts without the sustainability context. With the combined use of multiple indicators this became possible.

ACKNOWLEDGEMENTS

Funding for this study was provided by the Leibniz Centre for Agricultural Landscape Research (ZALF) in the frame of the impact assessment cross-disciplinary project. The authors would like to thank the anonymous reviewers for their constructive critiques. We thank David Favis-Mortlock for his helpful comments on the writing style.

REFERENCES

1. Anderson and Fergusson, 2006 G.Q.A. Anderson, M.J. FergussonEnergy from biomass in the UK: sources, processes and biodiversity implications IBIS, 148 (2006), pp. 180-183, 10.1111/j.1474-919X.2006.00535.x
2. Asseng et al., 2013 S. Asseng, F. Ewert, C. Rosenzweig, J.W. Jones, J.L. Hatfield, A. Ruane, K.J. Boote, P.Thorburn, R.P. Rötter, D. Cammarano, N. Brisson, B. Basso, P. Martre, P.K. Aggarwal, C. Angulo, P.Bertuzzi, C. Biernath, A. Challinor, J. Doltra, S. Gayler, R. Goldberg, R. Grant, L. Heng, J. Hooker, T.Hunt, J. Ingwersen,

C. Izaurralde, K.C. Kersebaum, C. Müller, S. Naresh Kumar, C. Nendel, G. O'Leary, J.E. Olesen, T.M. Osborne, T. Palosuo, E. Priesack, D. Ripoche, M. Semenov, I. Shcherbak, P. Steduto, C. Stöckle, P. Stratonovitch, T. Streck, I. Supit, F. Tao, M. Travasso, K. Waha, D. Wallach, J. White, J.R.Williams, J. WolfQuantifying uncertainties in simulating wheat yields under climate change Nat. Clim. Change, 3 (2013), pp. 827-832

3. Brandt and Glemnitz, 2013 K. Brandt, M. GlemnitzAssessing the regional impacts of increased energy maize cultivation on farmland birds Environ. Monit. Assess. (2013), 10.1007/s10661-013-3407-9
4. Collins and Ison, 2009 K. Collins, R. IsonJumping off Arnstein's ladder: social learning as a new policy paradigm for climate change adaptation Environ. Policy Gov., 19 (2009), pp. 358-373
5. Deumlich et al., 2005 D. Deumlich, W. Mioduszewski, I. Kajewski, M. Tippl, R. DannowskiGIS-based risk assessment for identifying source areas of non-point nutrient emissions by water erosion (Odra basin and sub catchment Uecker) Arch. Agron. Soil Sci., 51 (4) (2005), pp. 447-458
6. Dressler et al., 2012 D. Dressler, A. Loewen, M. NellesLife cycle assessment of the supply and use of bioenergy: impact of regional factors on biogas production Int. J. Life Cycle Assess., 17 (9) (2012), pp. 1104-1115
7. DWA, 2012 DWA, 2012. Gewässerrandstreifen: Teil 1: Grundlagen und Funktionen. Hinweise zur Gestaltung. Deutsche Vereinigung für Wasserwirtschaft, Abwasser und Abfall e.V. Merkblatt DWA-M-612-1. (in German).
8. Federal Statistical Office Germany, 2011 Federal Statistical Office Germany, 2011. Land- und Forstwirtschaft, Fischerei: Landwirtschaftliche Berufsbildung der Betriebsleiter/ Geschäftsführer Landwirtschaftszählung/Agrarstrukturerhebung 2010. Fachserie 3, Heft 1 Online (30.07.2014) (in German).
9. FNR, 2010 FNR, 2010. Biogas-Messprogramm II. 61 Biogasanlagen im Vergleich. Fachagentur für Nachwachsende Rohstoffe, Gülzow. (in German).
10. FNR, 2012a FNR, 2012. Fachagentur für Nachwachsende Rohstoffe Anbau nachwachsender Rohstoffe 2013 auf 2,4 Millionen Hektar.

Aktuelle Nachricht 16.09.2013. Online (01.11.13.) (in German).

11. FNR, 2012b FNR, 2012. Fachagentur für Nachwachsende Rohstoffe Faustzahlen Biogas. Online (01.11.13.) (in German).
12. Gabrielsen and Bosch, 2003 P. Gabrielsen, P. Bosch Environmental Indicators: Typology and Use in Reporting. EEA Internal Working Paper European Environment Agency, Copenhagen (2003)
13. German Weather Service, 2013 German Weather Service, 2013. Time series of regional means. Online (01.11.2013).
14. Guerrero et al., 2012 I.Guerrero, M.B. Morales, J.J. Onate, F. Geiger, F. Berendse, G. de Snoo, T. TscharntkeResponse of ground-nesting farmland birds to agricultural intensification across Europe: landscape and field level management factors Biol. Conserv., 152 (2012), pp. 74-80, 10.1016/j.biocon.2012.04.001
15. Gutzler and Helming, 2013 Gutzler, C., Helming, K., 2013. Folgenabschätzung von Szenarien der landwirtschaftlichen Produktion in Brandenburg 2025: Feldberegnung und Energiemaisanbau 80S. Müncheberg (Leibniz-Zentrum für Agrarlandschaftsforschung). (in German).
16. Helming et al., 2011 K. Helming, K. Diehl, H. Bach, O. Dilly, B. König, T. Kuhlman, M. Pérez-Soba, P. Sieber s Tabbush, K. Tscherning, D. Wascher, H. WiggeringEx ante impact assessment of policies affecting land use part a: analytical framework Ecol. Soc., 27 (1) (2011)
17. Hennig and Gawor, 2012 C. Hennig, M. GaworBioenergy production and use: comparative analysis of the economic and environmental effects Energy Convers. Manage., 63 (2012), pp. 130-137
18. Jaafar and Walling, 2010 M. Jaafar, E. WallingAssessing the impact of fodder maize cultivation on soil erosion in UK P. Zdruli, M. Pagliai, A. Faz Cano (Eds.), Land Degradation and Desertification: Assessment, Mitigation and Remediation, Springer, Heidelberg (2010), pp. 581-588, 10.1007/978-90-481-8657-0
19. Knierim et al., 2010 A.Knierim, M. Pintar, S. Skerratt, H. WiggeringThe innoland approach: experimental research and action in Europe's rural regions H. Wiggering, H.P. Ende, A. Knierim, M. Pintar (Eds.), Innovations in European Rural Landscapes, Springer, Berlin (2010), pp. 11-24

20. KTBL, 2012 KTBL, 2012. Association for Technology and Structures in Agriculture: Feldarbeitsrechner. Online database: http://daten.ktbl.de/feldarbeit/entry.html (27.02.12.).
21. LELF, 2010 LELF, 2010. State Office for Rural Development, Agriculture and Rezoning. Datensammlung für die Betriebsplanung und die betriebswirtschaftliche Bewertung landwirtschaftlicher Produktionsverfahren im Land Brandenburg. Ackerbau, Grünlandwirtschaft, Tierproduktion. Schriftenreihe des Landesamtes für Ländliche Entwicklung, Landwirtschaft und Flurneuordnung, Reihe Landwirtschaft 11 (8). Landesamt für Ländliche Entwicklung, Landwirtschaft und Flurneuordnung, Frankfurt (O). (in German).
22. Makowski et al., 2013 D. Makowski, T. Nesme, F. Papy, T. DoreGlobal agronomy, a new field of research: a review Agron. Sustain. Dev. (2013), 10.1007/s13593-013-0179-0
23. Magliocca et al., 2014 N.R. Magliocca, T.K. Rudel, P.H. Verburg, W. McConnell, O. Mertz, K. Gerstner, A.Heinimann, E.C. EllisSynthesis in land change science: methodological patterns, challenges, and guidelines Reg. Environ. Change (2014), 10.1007/s10113-014-0626-8
24. Merz et al., 2009 C. Merz, J. Steidl, R. DannowskiParameterization and regionalization of redox based denitrification for GIS-embedded nitrate transport modeling in pleistocene aquifer systems Environ. Geol., 58 (2009), pp. 1587-1599, 10.1007/s00254-008-1665-6
25. Milestad et al., 2014 R. Milestad, A. Svenfelt, K.H. DreborgDeveloping integrated explorative and normative scenarios: the case of future land use in a climate-neutral Sweden Futures, 60 (2014), pp. 59-71
26. Mirschel et al., 2012 Mirschel, W., Wieland, R., Wenkel, K.O., Luzi, K., Groth, K., 2012. Regionaldifferenzierte Ausweisung fruchtartspezifischer Zusatzwasserbedarfe sowie der daraus resultierenden Mehrerträge von wichtigen Fruchtarten im Freistaat Thüringen vor dem Hintergrund der Auswirkungen des Klimawandels auf den Pflanzenbau. R&D report Leibniz Centre for Agricultural Landscape Research, Müncheberg. (in German).

27. Mirschel et al., 2014 W. Mirschel, R. Wieland, K.O. Wenkel, C. Nendel, C. GuddatYIELDSTAT – a spatial yield model for agricultural crops Eur. J. Agron., 52 (2014), pp. 33-46
28. MUGV, 2009 MUGV, 2009. Mittlere Abflussspende des Landes Brandenburg für die Zeitreihe 1976–2005. Online database: http://www.portalu.de/trefferanzeige?docuuid; Download: http://www.mugv.brandenburg.de/lua/gis/wh_abimo.zip (19.11.09.).
29. Münch et al., 2014 T. Münch, M. Berg, W. Mirschel, R. Wieland, C. NendelConsidering cost accountancy items in crop production simulations under climate change Eur. J. Argon., 52 (2014), pp. 57-68
30. National Strategy on Biological Diversity, 2007 National Strategy on Biological Diversity, 2007. Online: www.bmu.de/N41253-1/.
31. Nendel et al., 2011 C. Nendel, M. Berg, K.C. Kersebaum, W. Mirschel, X. Specka, M. Wegehenkel, K.O. Wenkel, R. WielandThe MONICA model: testing predictability for crop growth, soil moisture and nitrogen dynamics Ecol. Modell., 222 (9) (2011), pp. 1614-1625
32. Nendel et al., 2014 C. Nendel, K.C. Kersebaum, W. Mirschel, K.O. WenkelTesting farm management options as a climate change adaptation strategy using the MONICA model Eur. J. Agron., 52 (2014), pp. 47-56
33. Pretty, 1995 J.N. PrettyParticipatory learning for sustainable agriculture World Dev., 23 (8) (1995), pp. 1247-1263
34. Scherhaufer et al., 2013 P. Scherhaufer, T. Grothmann, W. LexerPartizipation in regionalen integrierten Partizipationsassessments – ein kritischer Vergleich von 14 Fallbeispielen im Bereich Klimawandel A.Knierim, S. Baasch, M. Gottschick (Eds.), Partizipation und Klimawandel. Ansprüche, Konzepte und Umsetzung, Oekom-Verlag (2013), pp. 195-214 (in German)
35. Schindler et al., 2004 Schindler, U., Thiere, J., Steidl, J., Müller, L., 2004. Bodenhydrologische Kennwerte heterogener Flächeneinheiten: Methodik der Ableitung und Anwendungsbeispiel für Nordostdeutschland. Landesumweltamt Brandenburg, MLUR, Potsdam. Heft 87-Bodenschutz 2, p. 56. (in German).

36. Schmidt and Diemann, 1991 Schmidt, R., Diemann, R., (Eds.), 1991. Erläuterungen zur Mittelmassstäbigen Landwirtschaftlichen Standortkartierung (MMK). FZB Müncheberg Bereich Bodenkunde/ Fernerkundung, Eberswalde. (in German).
37. Sieber et al., 2013 S. Sieber, T.S. Amjath-Babu, T. Jansson, K. Müller, K. Tscherning, F. Graef, D. Pohle, K.Helming, B. Rudloff, B.S. Saravia-Matus, S. Gomez y PalomaSustainability impact assessment using integrated meta-modelling: simulating the reduction of direct support under the EU common agricultural policy (CAP) Land Use Policy, 33 (2013), pp. 235-245
38. Statistical Office Brandenburg, 2012 Statistical Office Brandenburg, 2012. Bewässerung in landwirtschaftlichen Betrieben im Land Brandenburg 2009. Online (01.11.13.) (in German).
39. Van Zanten et al., 2013 B.T. Van Zanten, H. Verburg, M. Espinosa, S. Gomez-y-Paloma, G. Galimberti, J.Kantelhardt, M. Kapfer, M. Lefebvre, R. Manrique, M. Piorr a Raggi, L. Schaller, S. Targetti, I. Zasada, D.ViaggiEuropean agricultural landscapes, common agricultural policy and ecosystem services: a review Agron. Sustain. Dev. (2013), 10.1007/s13593-013-0183-4
40. Wenkel et al., 2013 K.O. Wenkel, M. Berg, W. Mirschel, R. Nendel, C. KöstnerLandCaRe DSS an interactive decision support system for climate change impact assessment and the analysis of potential agricultural land use adaptation strategies J. Environ. Manage., 127 (2013), pp. 168-183
41. Wieland et al., 2006 R. Wieland, M. Voss, X. Holtmann, W. Mirschel, I.A. AjibefunSpatial analysis and modeling tool (SAMT): 1. Structure and possibilities Ecol. Inform., 1 (2006), pp. 67-76
42. Wiggering et al., 2006 H. Wiggering, C. Dalchow, M. Glemnitz, K. Helming, K. Müller, A. Schulz, U. Stachow, P.ZanderIndicators for multifunctional land use – linking socio-economic requirements with landscape potentials Ecol Indic., 6 (2006), pp. 238-249
43. Wischmeier and Smith, 1978 W.H. Wischmeier, D.D. SmithPredicting rainfall erosion losses – a guide to conservation planning Agriculture Handbook No. 537, U.S. Gov. Printing Office, U.S. Department of Agriculture, Washington D.C(1978)

8 LOW EMISSION DEVELOPMENT STRATEGIES IN AGRICULTURE. AN AGRICULTURE, FORESTRY, AND OTHER LAND USES (AFOLU) PERSPECTIVE

AlessandroDePinto[a],ManLi[a],AkikoHaruna[a],GlennGrahamHyman[b], Mario Andrés LondoñoMartinez[c], BernardoCreamer[b], Ho-YoungKwon[a], Jhon Brayan ValenciaGarcia[b], JeimarTapasco[b], and Jesus DavidMartinez[b]

[a]International Food Policy Research Institute, USA

[b]International Center for Tropical Agriculture, Colombia

[c]Universidad de los Andes, Colombia

SUMMARY

As countries experience economic growth and choose among available development pathways, they are in a favorable position to adopt natural resource use technologies and production practices that favor efficient use of inputs, healthy soils, and ecosystems. Current emphasis on increasing resilience to climate change and reducing agricultural greenhouse gasses (GHG) emissions strengthens the support for sustainable agricultural production. In fact, reducing losses in soil fertility, reclaiming degraded lands, and promoting synergistic interaction between crop production and forests are generally seen as good climate change policies. In order for decision-makers to develop long-term policies that address these issues, they must have tools at their disposal that evaluate trade-offs, opportunities, and repercussions of the options considered. In this paper, the authors combine and reconcile the output of three models widely accessible to the public to analyze the impacts of policies that target emission reduction in the agricultural sector. We present an application to Colombia which reveals the importance of considering the full scope of interactions among the various land uses. Results indicate that investments in increasing the efficiency and productivity of the livestock sector and reducing land allocated to pasture are preferable to policies that target deforestation alone or target a reduction of emissions in crop production. Investments in livestock productivity and land-carrying capacity would reduce deforestation and provide sufficient gains in carbon stock to offset greater emissions from increased crop production while generating higher revenues.

Key words: climate change, land use change, low emission development, and AFOLU

INTRODUCTION

Resource use in many developing countries, from crop production to deforestation is responsible for the bulk of greenhouse gasses (GHG) emissions, and there are instances in which the agricultural and forestry sectors can provide low-cost climate change mitigation opportunities (Golub *et al.*, 2013; Lubowski & Rose, 2013; Smith *et al.*, 2007). From a technical point of view, reducing expected increases in GHG emissions in agriculture requires the adoption of transformative approaches in the use of resources. Emphasis has been placed on methods that increase the efficiency in the use of fertilizers, water, and fossil fuels, as well as waste reduction. A growing body of literature analyzes the effects of alternative agricultural practices (Antle & Stoorvogel, 2008; Diagana, Antle, Stoorvogel, & Gray, 2007; Gilhespy *et al.*, 2014; Schneider & Smith, 2008; Smith *et al.*, 2013; Tenningkeit, Kahrl, Wolcke, & Newcombe, 2012; Tschakert, 2007). The livestock sector has also been the target of research on mitigation opportunities (Golub *et al.*, 2013; Li *et al.*, 2012; Schils, Olesen, del Prado, & Soussana, 2007), and the mitigation potential of forests, soil, and other biomass has been amply analyzed as well (Cacho, Marshall, & Milne, 2005; Lubowski & Rose, 2013; Makundi & Sathaye, 2004; Torres, Marchant, Lovett, Smart, & Tipper, 2010). However, from a policy-making perspective, the design of low emission development strategies is an example of multi-objective decision making in which policies target the reduction of GHG emissions while other goals such as increasing agricultural productivity and food security or attaining objectives such as export goals or economic growth are preserved. It is also important to consider that all countries are part of a global economic system, and therefore it is critical that policies are devised with full recognition of the role of the international economic environment which, with its effects on commodity prices, can significantly affect the long-term viability and the budgetary implications of mitigation policies. The challenge at hand is to reconcile the limited spatial resolution of macro-level economic models that operate at a global or national level with models that function at a higher spatial resolution to properly account for changes in carbon stocks and GHG emissions. The number of analyses that confront this challenge is still small, but it is growing given the importance of the information that these studies can provide to policy makers. Schneider and Smith (2008) estimated mitigation potentials of U.S. agriculture with regionally disaggregated data and changes in welfare within the agricultural sector. Golub *et al.* (2013) examined the impact on food consumption and income of implementing mitigation policies at national and regional levels. Rutten, van Dijk, van Rooij, and Hilderink (2014)evaluated the effects of a series of climate change and economic growth scenarios on Vietnam's economy. Dace, Muizniece, Blumberga, and Kaczala (2015), used a system dynamic model to assess the effect of a group of policies on agricultural GHG emissions in Latvia. Havlík *et al.* (2014) estimated the effects of transitioning to a more efficient livestock production system on GHG mitigation and the economy, and Lubowski and Rose (2013)provided a review of a number of studies that model mitigation potentials of the Reducing Emissions from Deforestation and Forest Degradation (REDD) program along with conservation, sustainable management of forests, and enhancement of forest carbon stocks policies.

In this article we demonstrate that different models, all widely accessible to the public, can be brought together to help policymakers in their evaluation of trade-offs, opportunities, and repercussions of alternative mitigation policies in the agricultural sector. While the focus of this work is on Colombia, the analytical framework can be applied to any country interested in exploring country-wide effects and the economic viability of climate change mitigation policies in agriculture. The approach is based on the use of public and widely accessible data and we believe that the flexibility and transparency of the approach proposed in this study can increase decision-makers' trust in the results. Naturally, additional data and targeted surveys can increase the accuracy of the results and the framework does not create barriers for the inclusion of additional input. Nonetheless, it is clear from our analysis that policy-makers need substantial support in their decision-making process as the range of options they face can be very diverse and the effects of their decisions have important, and sometimes unexpected repercussions. The effects of the policies we simulated cover the entire spectrum of potential outcomes. We found win-win policies (reducing land allocated to pasture increases profits and carbon stock and reduces GHG emissions), policies with tradeoffs (limiting deforestation in the Amazon increases carbon stock, decreases emissions, but reduces profits), and policies that could generate clearly inferior results (increasing the area allocated to oil palm cultivation beyond certain amounts reduces carbon stock, increases emissions, and reduces profits). Stakeholders, from government agencies to producer and consumers' organizations to farmers will benefit from policies devised with the support of solid evidence and the effects of which can be investigated and evaluated by all the parties affected.

GREENHOUSE GAS EMISSIONS IN COLOMBIA

In 2010, Colombia presented its second National Communication on Climate Change to the United Nations Framework Convention on Climate Change. The report contains data from the last National Greenhouse Gases Inventory carried out in 2004. Colombia contributes 0.37% (180,010 Gg) of the total worldwide emissions of GHG (49 Gt). Emissions are composed of 50% carbon dioxide (CO_2), 30% methane

(CH_4), 19% nitrous oxide (N_2O), and the remaining 1% classified as chlorofluorocarbons (CFCs).

According to the last National Greenhouse Gases Inventory, agricultural activities emit 38% of total emissions, and land use, land use change, and forestry account for another 14%. Of the emissions resulting from agricultural activities, 48.5% are due to enteric fermentation, 47.5% from agricultural soil management, and 2% from emissions related to rice cultivation. Traditionally, Colombia has a large number of smallholder farmers and there is also a well-established culture of cattle ranching with both small and large livestock keepers. Urbanization and industrialization have been growing in Colombia, but agricultural and forestry activities are expected to grow and continue to claim a large share of emissions. Although the agriculture sector represents 7% of the gross national product, the sector employs 18% of the population (CIA (Central Intelligence Agency), 2014).

Colombia has developed plans and policies that address climate change mitigation identifying priority sectors with high GHG emission rates. A working group led by the Ministry of Environment and Sustainable Development (MADS) has selected target areas for low emissions development in the agriculture, forestry, and land use sectors. These include reducing emissions from deforestation and forest degradation, oil palm, livestock, forestry, and fertilizers. In December 2015, the government of Colombia presented its Intended Nationally Determined Contributions (INDCs) at the Conference of the Parties in Paris and this document includes contributions from the AFOLU sector.

According to official government statistics (IAvH *et al.*, 2007; IGAC, 2013), 52% of Colombia's 115 million hectares is covered by natural forests, mostly within the Amazon basin but also forests along the Pacific coast and in the northern part of the country. Cultivated pastures and native savanna grasslands make up 26% of the land area. These lands are characterized by cattle grazing with low stocking rates and frequent natural and anthropogenic fires. Cropland is mostly concentrated in the intermountain valleys, making up about 4% of the land surface (see Figure 1).

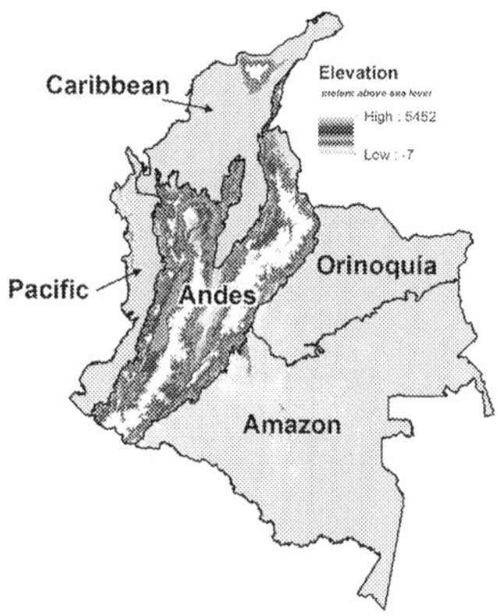

Figure 1: Main regions in Colombia.

In 2011, the Instituto Nacional de Hidrología Meteorología y Estudios Ambientales de Colombia (IDEAM) and MADS quantified national deforestation rates and trends (Table 2.1). The average annual deforestation rate over the entire period is some 238,000 hectares and the Amazon and Andes regions appear to be areas particularly at risk.

Table 2.1: Deforestation by regions

Period	Indicator	Natural region					
		Pacific	Orinoquía	Caribbean	Andes	Amazon	Total
1990–2000	Forest 1990 (Ha)	5,249,261	2,335,094	2,368,779	12,565,035	41,924,100	64,442,269
	Defor-estation (Ha)	140,426	240,580	343,019	876,597	1,198,018	2,798,640
2000–05	Forest 2000 (Ha)	5,227,673	2,182,517	2,014,227	11,716,837	40,669,967	61,811,221
	Defor-estation (Ha)	29,254	28,696	47,313	97,293	112,565	315,121
2005—2010	Forest 2005 (Ha)	5,035,400	2,123,340	1,807,073	11,151,591	40,096,203	60,213,607
	Defor-estation (Ha)	110,744	46,534	200,090	435,450	398,985	1,191,803
Annual average deforestation		22,149	9,307	40,018	87,090	79,797	238,361

Source: IDEAM (Instituto Nacional de Hidrología Meteorología y Estudios Ambientales de Colombia) 2010.

Prior to 2000, estimates of forest clearing suggested that two-thirds of this clearing was due to the pastureland encroaching into forest and one-third due to cropland expansion (Etter, McAlpine, Wilson, Phinn, & Possingham, 2006). A more recent analysis has suggested that 90% of forest clearing during 2005–10 was due to pastureland development (Nepstad, Tepper, McCann, Stickler, & McGrath, 2013).

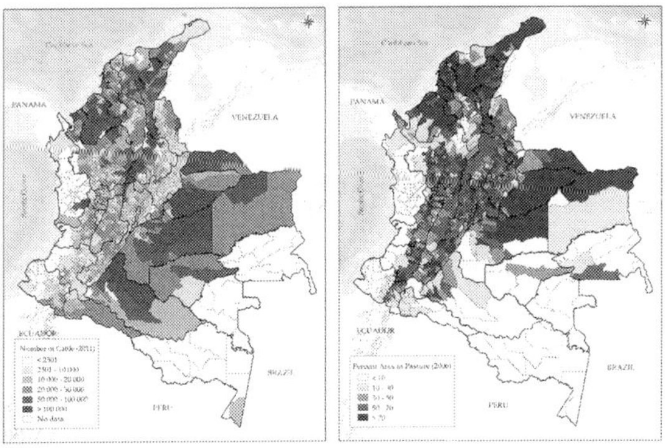

Figure 2.2: Number of cattle (left) and percent of pasture area (right) by municipality..

Source: FEDEGAN (2011) and MADR (Ministry of Agriculture & Rural Development) (2011).

Colombia is the fifth largest producer of palm oil and its production area is expected to increase. Official projections (MADR (Ministry of Agriculture & Rural Development), 2011) indicate that there will be little changes in cropland area over the coming decades, with the exception of oil palm. Oil palm is expected to increase substantially after 2016, due to its high demand for food products and biofuels. However, oil palm development mostly occurred on lands that were already cleared of their forests, a trend that according to some studies is expected to continue, at least partially (Castiblanco, Etter, & Aide, 2013). Pasturelands and livestock production may change substantially in the coming years. According to the Colombian Federation of Cattle Ranchers (Federación Colombiana de Ganaderos, FEDEGAN), the Colombian livestock inventory totals 23.5 million head of cattle and 39.2 million hectares of pasture. With less than one head of cattle per hectare, livestock occupies 26% of the total land area of Colombia. The livestock area has expanded from 14.6 to 38 million hectares in the past 50 years, mostly at the expense of tropical forest (MADS (Ministry of Environment & Sustainable Development), 2012). As seen in Figure 2.2, the majority of pasture area is located in the eastern plains and Caribbean region. However, in terms of number of animals, the northwestern part of the Amazon region also appears to hold a significant share of pasture area.

MODELING FRAMEWORK

In order to evaluate the potential for GHG emission reductions and trade-offs of alternative mitigation policies, we combined and reconciled data and outputs of economic and biophysical models. This approach included the use of the following models:

- The International Model for Policy Analysis of Agricultural Commodities and Trade Model (IMPACT; Robinson *et al.*, 2015), a global partial equilibrium agriculture model that allows for policy and agricultural productivity investment simulations;
- A spatially explicit model of land use choices to determine the possible effects of future changes in the drivers of land use choices (Li, De Pinto, Ulimwengo, You, & Robertson, 2015); and
- DeNitrification–DeComposition crop model (DNDC; Li, 2007) that estimates spatially explicit profiles of GHG emissions from cropland with varying crop genetic productivity shifters, management systems, and climate scenarios.

This suite of models produces a series of country-specific results embedded in a framework consistent with global outcomes. Using the crop model, which incorporates the most updated knowledge on GHG emissions generated by crop production, it is possible to simulate the effects on GHG emissions of current and alternative agricultural management practices, when relevant (Figure 3.1). With the goal of making the results of our analysis as relevant as possible to policy-makers, we chose to work within a relatively short time horizon, approximately 20 years (from 2008 to 2030). The start year of 2008 was dictated by data availability on land use shares at the municipal level.

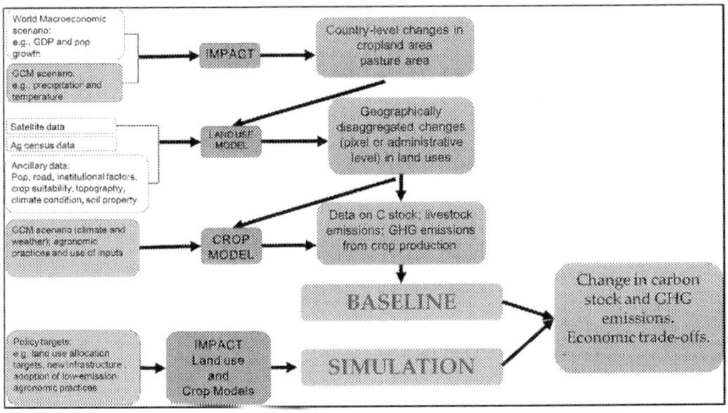

Figure 3.1: Workflow of the modeling approach.

(a). IMPACT model

IMPACT is a partial equilibrium agricultural model that uses a system of linear and nonlinear equations to approximate the underlying production and demand relationships of world agriculture (Robinson *et al.*, 2015). The model has a long record of applications and it has been employed in a wide range of analyses, from assessing the potential effects of climate change on global food production (Godfray & Robinson, 2015; Springmann *et al.*, 2016) to evaluating the global effects of biofuels production (Rosegrant, Zhu, Msangi, & Sulser, 2008). The world's food production and consumption is disaggregated into 159 countries and regional groupings with a further disaggregation in many regions to the river basin level and with the basic unit of analysis being the Food Production Unit (Figure 3.2). IMPACT models the global behavior of a competitive agricultural market for crops and livestock and is specified as a set of countries or regions, in which supply, demand, and prices for agricultural commodities are determined. Countries and regions are linked through trade so that the interactions among country-level production, consumption, and commodity prices are captured through net trade flows in global agricultural markets. Demand is a function of price, income, and population growth. Growth in crop production in each country is determined by crop prices and the rate of productivity growth from agricultural research and development, agricultural extension and education, markets, infrastructure, and irrigation. The model includes 64 commodities including all major cereals, soybeans, roots and tubers, meats, milk, eggs, oils, oilcakes and meals, vegetables, fruits, sugarcane and beets, and cotton.

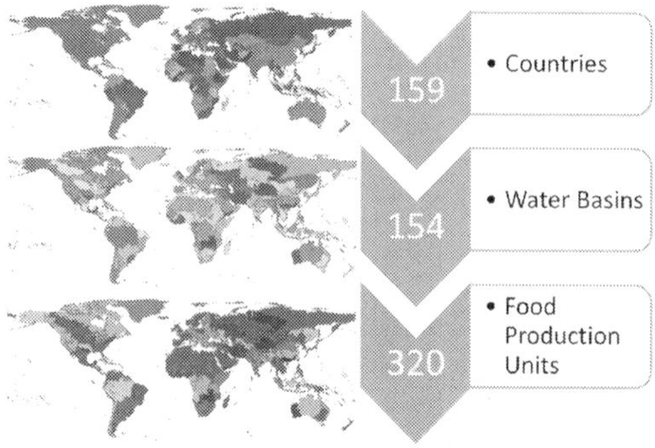

Figure 3.2: IMPACT global partitioning..

Source: Robinson et al. (2015).

IMPACT avails itself of the output from several other models. These include Global Circulation Models (GCM)[1] which simulate climate scenarios under global climate change; the Decision Support System for Agrotechnology Transfer (DSSAT) crop model suite which estimates yields with varying crop genetic productivity shifters and alternative management systems; the Spatial Production Allocation Model (SPAM) which provides an estimate of current area allocation to crop production and to management techniques such as rainfed/irrigated, and some IMPACT submodels such as the Water Simulation Model (IWSM) and the Global Hydrologic Model (IGHM).

IMPACT output is generally at a regional or national level. While this output provides information regarding changes in crop prices and in crop areas, it does not generate information with respect to the location of these changes and on the origin of new cropland. Given these limitations, it would be difficult to use the IMPACT model to quantify changes in carbon stock and GHG emissions from crop production. In order to do this, it is necessary to know the carbon content of the land uses that transitions in and out of production. Furthermore, GHG emissions from crop production are dependent on the characteristics of the land where farming takes place. It is therefore essential that the modeling output is sufficiently spatially disaggregated to account for the biophysical characteristics of each location.

(b). Land use model

In order to increase the spatial disaggregation of IMPACT output, we used an econometric model of land use choices. We followed the approach proposed by Li *et al.* (2015) based on a Maximum Likelihood method and employed a spatially explicit model which captures the main drivers of land use choice. The basic logic of the modeling approach is that the observed choices are the result of a utility-maximization process where each landholder will choose a land use that maximizes the stream of benefit deriving from the use of land. The econometric model, in this case a nested logit, attempts to explain the observed choices using a series of factors that are thought to influence the stream of benefits as explanatory variables. This class of models has been widely used in a variety of contexts (Chomitz & Gray, 1996; De Pinto & Nelson, 2009; Nelson, De Pinto, Harris, & Stone, 2004; Nelson & Hellerstein, 1997 among many) and, given the characteristics of each location, it can be used to estimate the probability for particular land use to be chosen. These probabilities are used to spatially allocate changes in cropland areas predicted by the IMPACT model.

As shown in Figure 3.3, the model structure consists of two levels of land use choices. The upper level estimates choices among large aggregation categories, such as cropland, pasture, forest, and other uses while the lower level estimates crop choices within a cropland. A detailed description of model specification and estimation method is provided in Appendix 1.

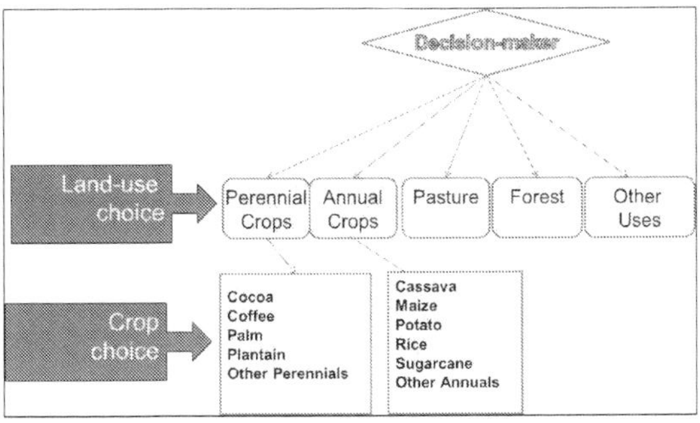

Figure 3.3: Structure of the land use model.

For the upper level model, we used a wide variety of biophysical and socioeconomic variables that influence land use choices, while for the lower level section we employed biophysical variables related to crop production as well as some economic variables such as crop prices and production costs which we considered to be key determinants of land allocation choices among crops (Table 3.1). To determine how land allocations will likely change in the year 2030, we used projections of commodity prices and the climate scenario MIROC General Circulation Model in combination with socioeconomic assumptions: AR5-SSP2 scenario (Shared Socioeconomic Pathway 2—middle of the road scenario).

Table 3.1: Summary of the explanatory variables used for the land use model

	Explanatory variables
Lower level: Choice (dependent) variable: crop shares within municipal	Biophysical suitability by crop; commodity producer price by crop; production cost by crop; elevation; terrain slope; soil pH; annual precipitation; annual mean temperature.
Upper level: Choice (dependent) variable: land use shares within municipal	Price of meat and milk; timber price; production cost for livestock (cattle); population density; travel time to major cities; elevation; coefficient of variation of elevation; terrain slope; soil pH; annual precipitation; annual mean temperature; protected areas; indigenous reserves.

Note: A table with the descriptive statistics for the explanatory variable is provided in Appendix 1.

(i). Data for upper level model

We used the Colombian government statistics to determine cropland and forest areas at the municipal level (IGAC (Instituto Geografico Agustin Codazzi), 2013) while for pastureland, we used data from the Instituto de Hidrología, Meteorología y Estudios Ambientales (IAvH (Instituto de Investigación de Recursos Biológicos Alexander Von Humboldt) *et al.*, 2007). Based on these data, we classified the land use in four categories: cropland; forest; pasture; and other land uses which include shrub and secondary vegetation.

Population density data were gathered from the Gridded Population of the World, Version 3, with a spatial resolution of 30 arc-second (~1 km) (CIESIN (Center for International Earth Science Information Network)/ Columbia University/CIAT (Centro Internacional de Agricultura Tropical), 2005). We used a lagged-population density value to mitigate endogeneity issues.

Data on market accessibility, measured by travel time to major cities (cities of 50,000 or more people in year 2000), were generated from a global map of accessibility with a resolution of 30 arc-second (Uchida and Nelson, 2010). Factors affecting the travel costs include transport network and environmental and political factors (see Nelson, 2000; Uchida and Nelson, 2010 for details).

Elevation and slope data were generated from a resampled Shuttle Radar Topography Mission (SRTM) digital elevation data with a 30 arc-second spatial resolution, produced by the U.S. National Aeronautics and Space Administration (NASA) and the U.S. National Geospatial-Intelligence Agency (NGA). The original SRTM data were available with a 90 m resolution. The original elevation data contain "no-data" observations where water or heavy shadow prevented the quantification of elevation. These observations were further processed by Jarvis, Reuter, Nelson, and Guevara (2008) to fill in the no-data voids.

Climate data, measured by mean precipitation and mean annual temperature, were generated from WorldClim (Hijmans & *et al.*, 2005). Using MIROC Global Climate Model−AR5 SSP2, we applied rates of change to extrapolate the climate condition for the year 2030. Soil pH was collected from Harmonized World Soil Database (HWSD) (FAO/IIASA/ ISRIC/ISS-CAS/JRC, 2012). Conservation practice and indigenous reserves were included as a dummy variable, extracted using Colombian government maps (SINAP (Sistema Nacional de Áreas Protegidas), 2012).

(ii). Data for lower level model

The lower level model combines local agricultural statistics with geo-referenced data. We used a dataset for cropland physical area at municipality level (in total 1,121 municipalities) for the year 2008 (MADR (Ministry of Agriculture & Rural Development), 2011). Based on these data, we identified the five major annual crops in terms of cultivated areas: cassava; maize; potato; rice; and sugarcane, and four major perennial crops: cacao; coffee; oil palm; and plantain.

Domestic commodity producer prices were collected from FAOSTAT for the year 2008. IMPACT projected rates of change for prices were utilized to extrapolate domestic commodity producer prices for the year 2030. The use of country-level commodity prices substantially reduces concerns about endogeneity between municipal crop choices and local producer prices. We assumed that domestic commodity producer prices can be observed at the major markets located in cities with populations greater than 50,000 and that farm-level producer prices move synchronously with domestic prices. With these assumptions, we estimated the spatially explicit prices, ppnj, for each location and each crop using a distance decay function:

$$pp_{nj} = PP_j \times \exp(-access_n / \max_{\{n'\}} access_{n'}), \qquad (1)$$

where PP_j represents the domestic producer price of commodity j and accessn is travel time from each location n to its nearest major market.[2]

Data on biophysical suitability were derived from the global Agro-ecological Zones (GAEZ, v1.0) assessment (Fischer, Shah, van Velthuizen, & Nachtergaele, 2001), which was developed by the International Institute for Applied Systems Analysis (IIASA) and the Food and Agriculture Organization of the United Nations (FAO). Suitability data are available at a resolution of approximately 9 km at the equator. These data were used as a proxy for the maximum attainable yield in Colombia.

(iii). Crop model

To estimate GHG emissions from crop production, we employed the biogeochemical model—DeNitrification and DeComposition (DNDC) (Li, Frolking, & Frolking, 1992)—along with coarse-resolution remote sensing data (5 arc-minute grids, approximately 10 km × 10 km) and information about local agricultural practices (Table 3.2). Carbon dioxide (CO_2), methane (CH_4), and nitrous oxide (N_2O) emitted from soil in cropland are all accounted for. Figure 3.4 illustrates key data required to run the DNDC model and Figure 3.5 shows the process followed for the computation of GHG emissions.

Table 3.2: Dataset used for crop model

Data	Spatial resolution	Source
Soil texture; soil C; pH; soil bulk density	30 arc sec grid	FAO/IIASA/ISRIC/ISS-CAS/JRC (2012)
Crop calendar	5 arc min grid	Sacks *et al.* (2010)
Inorganic N rate	Country level	FAO Fertistat (http://www.fao.org/ag/agp/fertistat/index_en.htm)
Tillage rate; residue incorporation rate; irrigation rate; rotation; potential yield (for sugarcane, cassava, potato, palm)	Country level	Agronet.gov.co, fedepapa.co, other local institutions
Precipitation and temperature	5 arc min grid	Marksim weather generator (www.ccafs-climate.org/pattern_scaling)

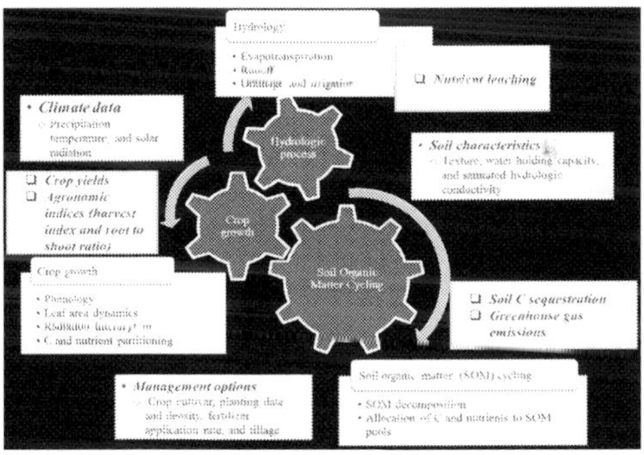

Figure 3.4: Data requirement for the use of the DNDC model.

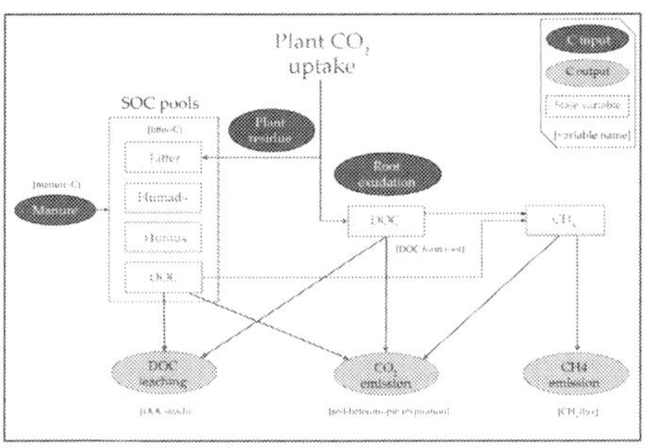

Figure 3.5: Simulation flow of GHG emissions in DNDC model.

The processes embedded in DNDC have been intensively tested in numerous conditions of soils, of climate, and against observed GHG fluxes with results generally considered sufficiently accurate (Beheydt, Boeckx, Sluetel, Li, & van Cleemput, 2007; Cai *et al.*, 2003; Deng *et al.*, 2011; Giltrap, Li, & Saggar, 2010; Grote *et al.*, 2009; Li, 2007; Pathak, Li, & Wassmann, 2005; Werner, Butterbach-Bahl, Haas, Hickler, & Kiese, 2007).

We first identified cropland and mosaic cropland land use areas categorized by National Aeronautics and Space Administration's Moderate Resolution Imaging Spectroradiometer (MODIS) and further divided them into a 5 arc-minute grid. For each cell, we applied the DNDC model to estimate crop productivity and GHG emissions. To derive the crop and location-specific data required for simulations in DNDC, information about local cropping systems and nitrogen fertilization rates were obtained through interviews and literature review. Remote sensing data were used to characterize crop calendars (that is, planting and harvesting dates) (Sacks, Deryng, Foley, & Ramankutty, 2010) and soil physical and chemical properties (such as texture, soil organic C (SOC) content, pH, and bulk density) from HWSD (FAO/IIASA/ISRIC/ISS-CAS/JRC, 2012). To derive future weather series corresponding to the MIROC climatic scenario and ensure consistency with the IMPACT scenario, we used climatological data developed by the CGIAR Research Program on Climate Change, Agriculture and Food Security and the MarkSim weather generator accompanied by the data (www.ccafs-climate.org/pattern_scaling). Note that this is an essential step given that GHG emissions and changes in SOC vary with soil types and other location-specific biophysical determinants.

Finally, DNDC simulations were run to make forward projections of GHG emissions from four annual crops (maize, rice, cassava, and potato) and two perennial crops (plantain and oil palm) for the period 2008 to 2030. We assumed current crop management practices and input levels stay constant. Global warming potential (GWP) was computed using fluxes in GHG and changes in SOC content. In order to smooth out the effects of single weather event fluctuations such as a year with abnormal precipitation or temperatures we computed the average per-hectare GWP for each crop and for each pixel. Given that total emissions were computed by multiplying per-hectare emissions by crop area estimations, changes in total emission reflect only the effect of change in crop allocation decisions.

OTHER GHG EMISSIONS AND CARBON STOCK

For the crops that could not be modeled using DNDC (coffee, cacao, and sugarcane), we used emission factors available in the literature (De Figueiredo, Panosso, Romão, & La Scala, 2010; Hergoualc'h, Skiba, Harmand, & Hénault, 2008; Palm *et al.*, 2002). In addition to cropland emissions, we considered GHG emissions from land allocated to pasture such as methane emission from enteric fermentation and manure management, as well as N_2O emissions from manure left on the ground. To generate location-specific estimations, we used national statistics for GHG emissions from livestock (FAO (Food & Agriculture Organization of the United Nations), 2014) and IMPACT values for the country herd size to generate per-head emission estimates.[3] Using total herd size and total hectares of pasture, we also estimated the stocking rate and therefore emissions per hectare of pasture. Due to lack of data on GHG emissions from standing forest (e.g., burning and decomposition) and other land uses, we disregard emissions from these sources. Above- and below-ground biomass as well as soil organic carbon (SOC) were the pools of carbon stocks estimated. Changes in stock are estimated in proportion to area changes at the municipality level for each land cover type (Figure 4.1). We assume that for uses other than cropland carbon stock does not change through time if land remains in the same land use.

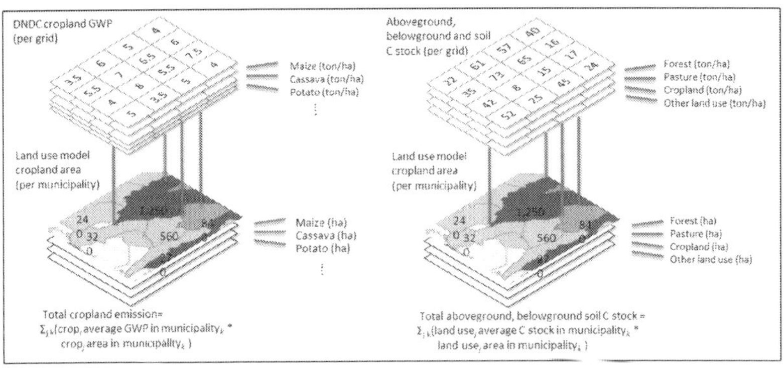

Figure 4.1: Calculation of carbon pools and GHG fluxes.

For the categories of forest; grassland; and other land uses, we used a spatial above-ground biomass dataset (Anaya, Chuvieco, & Palacios-Orueta, 2009; Saatchi *et al.*, 2011) to calculate average biomass for each land use and for each municipality. According to the country's land cover map (IAvH (Instituto de Investigación de Recursos Biológicos Alexander Von Humboldt) *et al.*, 2007), nearly 50% of "other land uses" category in Colombia consists of shrub and secondary vegetation. Biomass for this land cover category is also taken into account. For the estimation of below-ground biomass, root-shoot ratio for each land cover type is adopted from the available literature (IPCC, 2006; Saatchi *et al.*, 2011). For major perennial crops, we used biomass figures from Dossa, Fernandes, Reid, & Ezui, 2007; Henson, Ruiz, & Romero, 2012;Koskela, Nygren, Berninger, & Luukkanen, 2000. The municipal average SOC per land use type was elaborated using a global SOC dataset (Hiederer & Köchy, 2011), land cover map (IAvH (Instituto de Investigación de Recursos Biológicos Alexander Von Humboldt) *et al.*, 2007), and soil type map (Niels, 2010). Average SOC is calculated by extracting existing combinations of soil type and land cover in each municipality and weighing each combination according to the corresponding area size within the municipality. Table 4.1 provides an overview of all carbon pools and GHG fluxes accounted for in the study.

Table 4.1: Carbon pools and GHG fluxes by land use category

Land use category	**Carbon pools**			**GHG fluxes**		
	Aboveground carbon stock	**Below-ground carbon stock**	**Soil organic carbon**	**CO_2**	**N_2O**	**CH_4**
Cropland	YES[a]	YES[a]	YES	YES	YES	YES
Pasturec	YES	YES	YES		YES	YES
Forest	YES	YES	YES			
Other land uses	YES[b]	YES[b]	YES			

[a]Figures are only for perennial crops.

[b]Figures are only for areas with shrub and secondary vegetation.

[c]Emissions consider enteric fermentation and manure management.

ECONOMIC TRADE-OFFS, BASELINE, AND ALTERNATIVE SCENARIOS

Policies that aim at reducing emissions can generate trade-offs by reducing or increasing area allocated to crops, pasture, or forests or by affecting yields or production costs. Due to lack of data, we did not consider possible changes in input usage and we only account for changes in profits caused by projected changes in yields, prices, and land allocations. Forest economy was not included due to lack of data. Yields and producer prices from national statistics (FAO (Food & Agriculture Organization of the United Nations), 2014) are used for the year 2008 while growth rates for yields and prices from the IMPACT model are used for year 2030.

Figure 5.1 summarizes the methodology used to compare alternative policy scenarios with the baseline. The modeling framework we used for the analysis of low emission development strategies generated estimations and projections for the base year (2008) and the end-period (2030). Therefore, the annual rates of change had to be extrapolated from the starting and ending values. The annual rate of change is essential to compute the total amount of GHG emissions generated under a particular scenario.

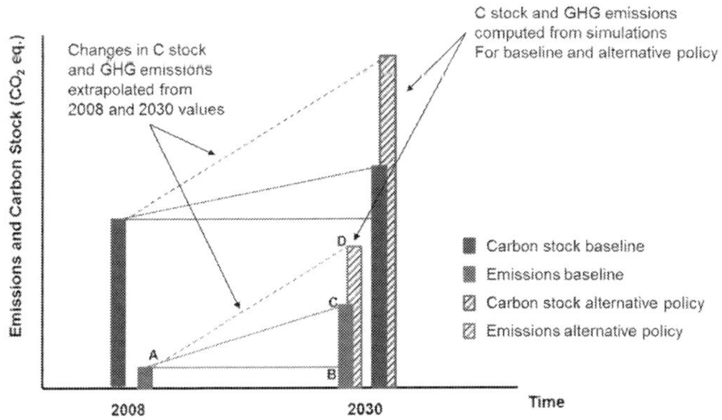

Figure 5.1: Computation of changes in GHG emissions and carbon stock for baseline and alternative policy scenarios.

The stylized example provided in Figure 5.1 is a case where both GHG emission and carbon stock increase in 2030 compared to 2008. Total change in emissions is obtained by accounting for the yearly changes throughout the time period considered (area triangle ABC). Similarly, the computation of changes in GHG emissions from implementing a particular policy is obtained by measuring the area of triangle ACD. Note that the effect of each policy is computed against the changes in emissions and carbon stock projected for the baseline.

For the baseline, we assumed that all field activities and agronomic practices stay constant, and therefore our computation of GHG emissions is based only on changes in land use due to changes in commodity prices and other underlying assumptions in IMPACT regarding changes in GDP, population growth, and climatic conditions.

Carbon stocks present in the soil and above- and below-ground biomass in each land use were also assumed constant, thus carbon stock accounting also was based only on land use change. Changes in profits were similarly computed as cumulative effects over the 21 years under consideration.

BASELINE RESULTS

IMPACT model results, reported in Table 6.1, indicate that global market forces determine significant increases in area for oil palm, plantain, and sugar cane (7%, 7%, and 25% respectively). These results are dependent on assumed population and GDP growth in Colombia (1.1% per year and 4% per year respectively during the period 2008-2030) but also on assumed global population and GDP growth (1.15% and 3.1%, respectively), technology growth, and changes in diets. For a comprehensive review of how the IMPACT model functions, including IMPACT development over time and an explanation for possible differences among published results, we refer the reader to a more detailed analysis available in Robinson *et al.*, 2015 and for comparison of IMPACT results with other global models to Nelson *et al.*, 2014.

Table 6.1: IMPACT projected changes in prices and areas for crop and livestock commodities, 2008–30

Crop	Projected change in price	Area 2008 (1,000 ha)	Area 2030 (1,000 ha)	Change 2008–30 (1,000 ha)	Change (%)
Cacao	24%	183	189	6	3%
Coffee	30%	812	824	12	1%
Oil Palm	85%	377	405	28	7%
Plantain	34%	473	505	32	7%
Other perennial	31%	181	243	62	34%
Cassava	38%	261	275	14	5%
Maize	39%	835	861	26	3%
Potato	26%	191	201	10	5%
Rice	24%	745	781	36	5%
Sugarcane	107%	346	433	87	25%
Other annual	19%	164	166	2	1%
Cow meat	23%	–	–	–	–
Cow milk	16%	–	–	–	–
Pasture	–	36,157	39,724	3,567	10%

Source: IMPACT 3.1.1.

These changes in crop areas were used as an input in the land use model. The land use model was used to spatially allocate changes in cropland projected by IMPACT and to determine how these changes affect carbon stock, GHG fluxes, and profits. The technical details on how the projected changes were used in the land use model are provided in Appendix 1. Given that we relied on the land use model for the spatial allocation of the changes in areas, it was important to evaluate how correctly the model predicts land use choices. We found that the model performed well in predicting most land uses (see Appendix 1 for a detailed assessment). As far as the upper level is concerned, the highest and lowest discrepancies were for pasture, whose share was over-predicted by 4.6% and forest which was under-predicted by 5.7%. Regarding the lower level component, the model tended to overestimate the share of perennial and annual croplands by a total of 16% and 34%, respectively. We considered this level of accuracy sufficient for the model to capture the effects of the main drivers of land use and that the parameter estimates can be used to make projections regarding future land use choices. In order to further evaluate the robustness of our econometric model, we compared different specifications (two alternatives are explored Appendix 1). We noted differences in land share predictions but very little difference in the transitions from one use to the other when policy simulations are performed. A change in these transitions could significantly alter the results, but no significant changes are noted.

Table 6.2 reports the results for the upper level of the land use model and shows how changes in agricultural product prices are projected to affect large land use categories. The most significant change is in forested area: a decrease of 4 million hectares, which is an average loss of some 190,000 hectares per year. Most of the change is caused by an increase in pastureland, 3.6 million hectares, and in lower amounts by growth in perennial and annual crop areas.

Table 6.2: Predicted land use change, 2008–30

Land use category	2008 area (Mha)	2030 area (Mha)	Change 2008–30 (Mha)	Change (%)
Perennial crops	2.0	2.2	0.2	10
Annual crops	2.5	2.7	0.2	8
Pasture	36.2	39.7	3.5	10
Forest	55.8	51.7	−4.1	−7
Other land uses	17.9	18.1	0.2	1
Total	114.4	114.4		

Note: Both 2008 and 2030 figures are model predictions.

Figure 6.1 depicts the location of deforestation and pasture expansion at the municipality level. Observing the Amazon region, it appears that pastureland expansion corresponds with the areas where deforestation occurs. A very similar pattern is apparent in the Andes region.

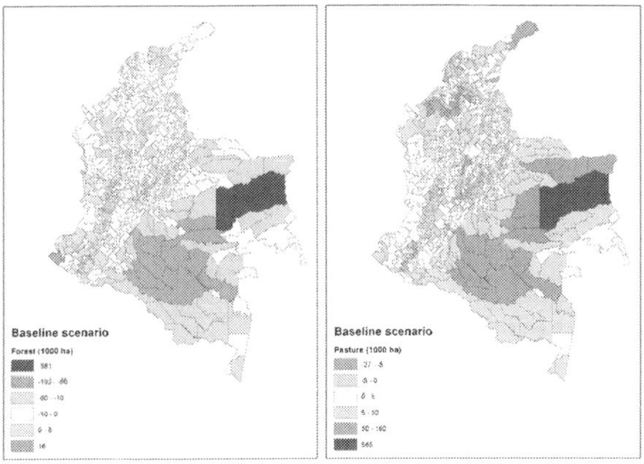

Figure 6.1: Projected changes in forest and pasture areas at municipality level, 2008–30.

Cropland (the sum of annual and perennial cropland) is projected to increase by a total of 500,000 hectares. Figure 6.2 shows projected change for two of the crops with the largest growth in area. Most of the growth for these two crops occurs outside the Amazon region and appears concentrated in the Andes and Caribbean regions. Among the annual crops considered, sugarcane shows the greatest projected change in area with a gain of 87,000 hectares. Rice and maize areas are projected to increase by some 36,000, and 26,000 hectares respectively. Area allocated to perennial crops is also projected to increase. For example, coffee, oil palm, and plantain areas increase by 12,000, 28,000, and 32,000 hectares respectively.

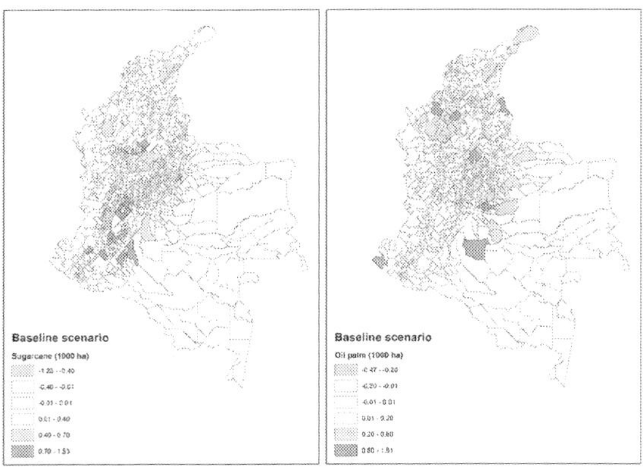

Figure 6.2: Projected changes in sugarcane and oil palm area at municipality level, 2008–30.

Projected changes in land use translate into changes in carbon stocks, and these are reported in Table 6.3. The model results show a net decrease in carbon stock, a total of 599 Tg C. This large reduction in the stock of carbon is due mostly to the decrease in forested area.

Table 6.3: Change in carbon stocks by land use, 2008–30

Land use category	Soil organic carbon 2008 (Tg C)	Above- and below-ground bio-mass 2008 (Tg C)	Total carbon stock 2008 (Tg C)	Soil organic carbon 2030 (Tg C)	Above- and below-ground biomass 2030 (Tg C)	Total carbon stock 2030 (Tg C)	Net change in carbon stock (Tg C)
Cropland	555	77	632	596	83	679	47
Pasture	4,190	327	4,517	4,604	359	4,963	446
Forest	6,012	9,369	15,381	5,576	8,690	14,266	-1,115
Other land uses	1,986	584	2,570	2,005	590	2,595	25
Total	12,743	10,358	23,101	12,780	9,721	22,501	–600

Table 6.4 reports the estimated changes in GHG emissions from cropland and pastureland. By far, the greatest change in emissions is connected to pastureland expansion and rice production. The next significant increase in GHG emissions comes from oil palm and plantain cultivation. This is projected to generate an average yearly increase in emissions equivalent to 12 Tg of CO_2 eq. Note that these changes are driven completely by changes in land use, and we assumed that land management practices, as well as livestock management practices remain unchanged for the entire period considered.

Table 6.4: GHG emissions in cropland and pasture, 2008–30

Crops	Change in area 2008–30 (1,000 ha)	Average per ha GHG emission in 2008 (Mg ha^{-1} yr^{-1})	Average per ha GHG emission in 2030 (Mg ha^{-1} yr^{-1})	Cumulative GHG emission 2008–30 (Tg CO_2 eq)
Pasture	3,567	1.1	1.1	40.2
Perennial crop				
Cacao	6	0.1[a]	0.1[a]	0.4
Coffee	12	0.7[a]	0.7[a]	12.6
Oil Palm	28	4.0	3.8	32.8
Plantain	32	3.5	3.0	35.9

Other perennial	62	–	–	–
Annual crop				
Cassava	14	2.0	1.9	11.2
Maize	26	1.7	1.8	30.3
Potato	10	3.5	3.6	14.4
Rice	36	7.2	7.2	115.3
Sugarcane	87	1.7[a]	1.7[a]	13.8
Other annual	2	-		-

[a]*Indicates values derived from existing literature.*

POLICY SIMULATIONS

Colombian public and private sectors are actively involved in finding viable strategies for reducing emissions from the agriculture sector and from land use change. A series of meetings were held with the objective of soliciting opinions and ideas on high-priority objectives from government officials, private-sector representatives, producers' organizations, and researchers. Our goal was to internalize in our simulations the interests and priorities of a wide range of stakeholders interested or involved in the design of low emission strategies. Given that some of the policy targets identified were clearly aspirational, we simulated two stages of accomplishment, (1) scenario in which the target is fully met and (2) scenario in which only half of the target is achieved. It is essential to note that the policies were evaluated with respect to the baseline which represents our best prediction of what the landscape and emissions will be in 2030.

For all scenarios, we evaluated the effects on carbon stock and GHG emissions and the impact on profits. Computing the impact on profits was particularly challenging given the lack of data on production costs. However, we believe that even limited insights into these effects can provide important information to policymakers who need to decide which policies are best to pursue. In order to accomplish this, data needed to be compiled from multiple sources. We used the change in commodity prices projected by IMPACT (Table 6.1) to compute changes in revenues. Data on production costs related to crop production (Table 7.1) are available relatively well disaggregated by region.[4] Unfortunately, key costs associated with other policy simulations are much more uncertain. Costs related to improving the quality of pasture were derived from a single study on the issue (Martínez, 2009) and from direct communications with the cattle producers' association (FEDEGAN). Costs of preventing deforestation were extracted from current studies on Nationally Appropriate Mitigation Actions and projects in the Amazon that aim at the valorization of local production compatible with standing forest.[5] The data obtained, however accurate and sound, represent very limited experiences. Therefore, results will have to be revisited as more data become available.

Table 7.1: Production costs by crop and region

Region	Cocoa[a] (\$ ha^{-1})	Coffee[b] (\$ ha^{-1})	Oil palm[c] (\$ ha^{-1})	Plantain[d] (\$ ha^{-1})	Other Perennials[e] (\$ ha^{-1})	Cassava[e] (\$ ha^{-1})	Maize[f] (\$ ha^{-1})	Potato[e] (\$ ha^{-1})	Rice[g] (\$ ha^{-1})	Sugar Cane[e] (\$ ha^{-1})	Other Annual[e] (\$ ha^{-1})
Carribe	2,101	2,106	2,063	1,103	1,843	870	415	2,373	532	1,429	606
Andena	2,577	1,906	2,158	1,772	2,103	1,402	450	2,743	1,141	1,429	998
Amazon	2,189	1,979	1,811	1,943	1,980	1,136	634	2,383	901	1,429	890
Orinoquia	2,189	1,979	1,529	2,954	2,162	1,136	813	2,383	1,029	1,429	993
Pacific	2,101	1,924	1,492	1,943	1,865	1,136	858	2,031	901	1,429	965

[a]Ministerio de Agricultura, Colombia.

[b]Corporación Colombiana de Investigación Agropecuaria.

[c]Fedepalma, Colombia. USAID.

[d]Colombia, Programa MIDAS.

[e]Departamento National de Planeation.

[f]Industria y Comercio Superintendencia.

[g]Fedearroz, Colombia.

(a). Pasture

The Colombian livestock sector association, Federación Colombiana de Ganaderos (FEDEGAN), proposed a strategy called "Plan Estratégico de la Ganadería 2019 (PEGA)", which calls for the national cattle herd to reach 48 million heads while reducing pasture area to 24 million hectares (equivalent to a reduction of more than 10-million-hectare in pastureland) by 2019 (FEDEGAN (Federación Colombiana de Ganaderos), 2006). The Ministry of Agriculture, meanwhile, proposes reducing land allocated to pasture to a total of 13 million hectares. FEDEGAN proposes to return the land to natural vegetation by promoting reforestation and silvopastoral systems (FEDEGAN (Federación Colombiana de Ganaderos), 2006). We simulated the effects of a reduction in pastureland of 10 million hectares, as well as a "middle of the road" scenario in which only 5 million hectares are removed from pasture. It is unclear how this objective would be met and what incentives would be put in place to reduce the area. It is only with a certain degree of speculation that the costs associated with achieving this goal can be assessed. The impact on profits of reaching these policy targets was evaluated by assuming that the stocking rate is increased on a portion of pastureland so that the total number of cattle in 2030 remains the same as the one projected in the baseline. On this area, stocking rate was assumed to increase to an average of 0.95 heads per hectare from the current value of 0.64 heads per hectare. This is assumed to be achieved with an improvement of pasture quality and rational use of land. The cost of this type of intervention was estimated to be between $400 and $800 per hectare (Caicedo, Escobar, & Tapasco, 2016; FEDEGAN, 2006; Martínez, 2009). We used a mid-point value of $600 per hectare in our simulations. The area on which improved pasture would have to be implemented is about 3.1 million hectare for the "middle of the road" scenario and 4.3 million hectares to achieve the full target.

(b). Forestry

Prevention of deforestation in the Amazon forest is considered a critical objective for climate change mitigation as well as biodiversity and cultural conservation. Deforestation occurs in other areas besides the Amazon forest according to historical data (Table 2.1) and to our projections (Figure 6.1); however, given its importance, the government puts an emphasis on the Amazon forest. The Colombian Low Carbon Development Strategy proposed by the government has a goal of a complete halt to deforestation in the Amazon region, while lowering the rate of deforestation in the rest of the country.

We simulated two scenarios: a complete stop to deforestation in the Amazon forest and a reduction of deforestation by half. The costs associated with achieving these targets are very uncertain. We relied on data from current studies on Nationally Appropriate Mitigation Actions and projects in the Amazon that aim at halting deforestation (Caicedo *et al.*, 2016). According to their calculations, the costs per hectare vary between $1,600 and $2,000. We used a mid-point value of $1,800 per hectare in our simulations.

(c). Oil palm cultivation

The Colombian government is planning to incentivize the cultivation of strategic crops. The list of strategic crops includes oil palm, sugarcane, and soybeans, the expansion of which has been driven by both biofuel and dietary demands. The expectation is that the majority of the area expansion would occur on pastureland which is widely considered underutilized land. We simulated two policy objectives, (1) an expansion in area allocated to oil palm cultivation by a total of 1.5 million hectares[6] and (2) a scenario in which 50% of this goal is achieved.

At the time of the consultations with stakeholders, there was no consensus on what instruments would be used to achieve the desired goals. We therefore simulated all the scenarios by altering the benefits deriving from the targeted land use in comparison to alternative uses in the land use model. In essence, we treated the policy instrument as an unobserved factor specific to the targeted land use and assumed that this policy instrument acts homogeneously across all municipalities. We added this factor to the net benefit equation of the targeted land use (Eqn. (2), Appendix 1) to reach a desired number of hectares allocated to that particular land use.

Table 7.2 shows the changes in land use induced by the implementation of these objectives. Table 7.3 shows the effects on carbon stock, GHG emissions, and profits from agricultural production.

Table 7.2: Land use change under alternative policy scenarios

Crop	Policy targets partially met						Policy targets fully met					
	Pasture reduction		Reduce deforestation		Oil Palm expansion		Pasture reduction		Zero deforestation		Oil Palm expansion	
	Area 2030 (1,000 ha)	Difference from baseline 2030 (1,000 ha)	Area 2030 (1,000 ha)	Difference from baseline 2030 (1,000 ha)	Area 2030 (1,000 ha)	Difference from baseline 2030 (1,000 ha)	Area 2030 (1,000 ha)	Difference from baseline 2030 (1,000 ha)	Area 2030 (1,000 ha)	Difference from baseline 2030 (1,000 ha)	Area 2030 (1,000 ha)	Difference from baseline 2030 (1,000 ha)
Perennial crop	*2,407*	*240*	*2,161*	*−5*	*2,238*	*72*	*2,667*	*501*	*2,156*	*−10*	*2,629*	*463*
Cacao	208	20	188	−1	149	−40	229	40	188	−1	92	−97
Coffee	910	86	823	−1	775	−50	1,001	177	822	−3	666	−158
Oil Palm	458	53	404	−1	750	345	518	113	403	−2	1,500	1,095
Plantain	559	54	503	−2	391	−114	616	111	502	−3	266	−239
Other crops	272	29	242	−1	173	−70	303	60	242	−1	104	−139
Annual crop	*3,057*	*340*	*2,707*	*−9*	*2,702*	*−14*	*3,438*	*722*	*2,699*	*−18*	*2,668*	*−49*
Cassava	310	36	273	−2	273	−2	351	76	271	−3	268	−6
Maize	969	108	858	−3	856	−5	1,090	229	854	−7	845	−16
Potato	226	25	201	0	201	0	252	51	201	0	201	0
Rice	890	109	778	−3	775	−7	1,015	234	775	−6	760	−21
Sugarcane	476	44	432	0	432	−1	522	90	432	−1	429	−4
Other crops	185	20	165	0	165	0	207	41	165	0	164	−1
Pasture	*34,724*	*−5,000*	*38,828*	*−896*	*39,599*	*−125*	*29,724*	*−10,000*	*37,933*	*−1,792*	*39,302*	*−422*

Forests	*53,996*	*2,252*	*52,947*	*1,203*	*51,761*	*17*	*56,237*	*4,493*	*54,151*	*2,407*	*51,737*	*–7*
Other land uses	*20,257*	*2,167*	*17,796*	*–294*	*18,140*	*51*	*22,374*	*4,284*	*17,502*	*–587*	*18,105*	*15*
Total	*114,440*	–	*114,440*	–	*114,440*	–	*114,440*	–	*114,440*	–	*114,440*	–

Italics indicate total values for the main land use categories.

Table 7.3: Summary of predicted total changes in carbon stock (C stock), GHG emissions, and profits, 2008–30

		Change in C stock (Tg C)	**Total change in GHG emission (Tg CO_2 eq)**	**Total change in profits (US$ billion)**
Reduce pasture by 5 million hectares	Cropland	80	16.8	2.0
	Livestock	–625	–56.7	–1.8
	Forest	621	–	–
	Other	311	–	–
	Total	387	–39.9	0.2
Reduce pasture by 10 million hectares	Cropland	168	35.0	4.0
	Livestock	–1,249	–116.0	–2.6
	Forest	1,239	–	–
	Other	614	–	–
	Total	772	–81.0	1.4

Reduce deforestation in the Amazon by 50%	Cropland	−2	−0.4	0.0
	Livestock	−112	−9.9	−0.3
	Forest	332	–	−0.5
	Other	−42	–	–
	Total	176	−10.3	−0.8
Reduce deforestation in the Amazon by 100%	Cropland	−4	−0.8	0.0
	Livestock	−224	−20.0	−0.7
	Forest	664	–	−1.0
	Other	−84	–	–
	Total	352	−20.8	−1.7
Increase area allocated to oil palm by 750,000 hectares	Cropland	10	7.5	-0.6
	Livestock	−16	−1.4	0.0
	Forest	5	–	–
	Other	7	–	–
	Total	6	6.1	−0.6
Increase area allocated to oil palm by 1.5 million hectares	Cropland	35	23.1	−1.3
	Livestock	−43	−4.7	−0.1
	Forest	−2	–	–
	Other	2	–	–
	Total	−8	18.4	−1.4

Note: Effects on profits are expressed in 2008 money and discounted using a yearly discount rate of 8% (A sensitivity analysis to the value of the discount rate was carried out and revealed that qualitatively the outcome of each policy simulation remained unchanged. One exception is the “reduction of pastureland by 5 million hectares” policy target. With a greater discount, the additional profits generated by the increased crop production would not offset the costs of improving pasture productivity, which are mostly upfront costs. A discount rate of 12% would generate a loss in profits of $0.3 billion causing a tradeoff between reduction in GHG gasses and profits. The implicit cost of abating one ton of CO_2 eq would still be very low: $0.30. Of course, increasing the discount rate high enough would cause the 10-million-hectares-reduction policy to generate tradeoffs. A discount rate of 18% would reduce the profit gains of this policy to zero.).

(d). Policy target 1: reduction in area allocated to pasture

Results of the policy simulations should be interpreted as assessing the changes in land use brought about by a policy that limits expansion and reduces pasture area by a total 10 million hectares *compared to the baseline*. Baseline results predicted a 3.6-million-hectare expansion to a total of 39.7 million hectares of pastureland, equivalent to a 10% growth during the period 2008–30. As it can be observed in Table 7.2, the effects of a 10-million-hectare reduction in pasture area would cause a growth in cropland (1.2 million hectares), forest (4.5 million hectares), and other land uses (4.2 million hectares). These changes correspond to an estimated increase in carbon stock of some 772 Tg C (Table 7.3). GHG emissions are also affected. Emissions from pastureland and the associated livestock are expected to decrease while emissions from crop production are expected to rise. The net effect would be a reduction of some 81 Tg CO_2 eq. Given our assumptions on how the reduction of pastureland was achieved, reaching this objective would lead to a reduction in profits from cattle raising compared to the baseline, which is estimated to be about $2.6 billion over the 21 years considered. However, the $4-billion increase in profits deriving from cropland expansion is expected to compensate for the loss and generate a net gain of some $1.4 billion. Of course, higher costs of implementing pasture improvements would change the outcome. A cost of $721 per hectare would reduce the profit gains to zero and costs higher than that would generate a tradeoff between economic gains and reduction of emissions. The intermediate scenario, 50% achievement of the policy goal, also returns the double benefit of a reduction of emissions and economic gains. Carbon stock is projected to increase by 387 Tg C while in total GHG emissions are reduced by a total of 39.9 Tg CO_2eq. The expected net change in profits is estimated at $0.2 billion.

(e). Policy target 2: reduction in the rate of deforestation in the Amazon forest

Results for the baseline scenario indicate a decrease in forested area of 4 million hectares during the period 2008–30, equivalent to an annual average loss of some 190,000 hectares. Results also indicate that the loss in carbon stock due to deforestation would be substantial: a total of 1,116 Tg of carbon. The implementation of a policy that stops deforestation in the Amazon would result in an additional 2.4 million hectares of forest compared to the baseline (Table 7.2). However, this would be at the expense of land that would otherwise transition into pasture and into crop production. The implementation of this policy would result in 1.8 million hectares of pastureland less than the baseline. Considered all together, the changes in land use lead to 352 Tg C of net increase in carbon stock and a total reduction in GHG emissions quantified to be about 21 Tg CO_2 eq. The avoided expansion of pasture and cropland into the forest results in a total loss of $1.7 billion in profits (Table 7.3). These losses are caused by costs incurred in protecting the forest, costs related to improving pasture quality to increase the stocking rate, and forgone crop production.[7] The scenario in which we simulate an intermediate achievement of the policy goal (deforestation reduced by half) returns similar trade-offs between mitigation and economic returns. The increase in carbon stock, which amounts to 176 Tg C, and the 10 Tg CO_2eq reduction in cropland and pasture emissions come at the expense of profits which are reduced by about $0.8 billion. Clearly, this policy generates significant tradeoffs which is not surprising given that it prevents the expansion of land uses that have a direct economic return into forests. However, the gains in carbon stock and reduction of emissions are also significant. In fact, the cost per ton of GHG abated is about $4.8 per Mg C ($1.3 Mg CO_2 eq). This is a relatively low cost considering that, despite that estimates for the social cost of carbon vary wildly and might be biased upward (Havranek, Irsova, Janda, & Zilberman, 2015), surveys and meta-analyses find values that are in the order of $43 Mg C (Yohe *et al.*, 2007). Off course, these results are dependent on the recorded costs of preventing deforestation. The greater the costs, the greater the tradeoffs, and in turn the greater the cost of a ton of GHG abated. For example, an increase in the cost of preventing deforestation by 50% would result in a cost per ton of GHG abated of $6.3 per Mg C.

(f). Policy target 3: increase the area allocated to the cultivation of oil palm

The baseline scenario projects a growth in area allocated to oil palm cultivation of about 28,000 hectares by 2030. In order to reach the policy target of 1.5 million hectares, land allocated to oil palm has to increase by more than 1 million hectares. Our simulation reveals that this policy could have significant unintended consequences if not implemented judiciously. Model results (Table 7.2) indicate that the expansion of oil palm production occurs at the expense of area allocated to perennial crops causing decreases in plantain area (239,000 hectares), coffee area (158,000 hectares), cacao area (97,000 hectares), and other perennials (139,000 hectares). Similarly, annual crops are negatively affected and total area allocated to them is reduced by a total of 49,000 hectares. Changes in land use result in a decrease in carbon stock of 8 Tg C and an increase in GHG emission of 18.4 Tg CO_2eq. It is important to note that this policy objective comes with heavy economic costs. This is because, against conventional wisdom, the expansion of oil palm cultivation only partially occurs on pastureland. The net effect of this expansion is a loss in profits equivalent to $1.4 billion for the period 2008–30.

The intermediate-goal scenario (50% of the policy goal) shows an important difference. While GHG emissions increase and profits decrease, carbon stock increases by 6 Tg C. These results indicate that there might be a role in low emission development for oil palm production but that policymakers ought to be careful to the extent of area expansion and vigilant to the land uses that might be displaced by oil palm production. A higher profit margin in palm production could potentially change results and produce a win-win outcome when area expansion is not too large. However, according to our calculations profits would have to double for this policy (additional 750,000 hectares) to achieve a "win-win" status.

DISCUSSION AND CONCLUSIONS

Results reveal the importance of considering the full scope of interactions and changes in the various land uses when planning for GHG reduction policies. As other studies have noted (Burney, Davis, & Lobell, 2010; Gockowski & Sonwa, 2011; Li *et al.*, 2015), the fate of forests matters considerably given that their carbon stock overwhelms potential changes in GHG emissions generated by crop production. Our results indicate that one additional hectare allocated to agriculture increases GHG emissions on average by some 2.5 Mg CO_2eq per year while one hectare of forest lost in the Amazon results in a loss of carbon stock (above- and below-ground biomass) equivalent to some 367 Mg CO_2 eq. Shifts in land uses determined by changes in area allocated to agricultural production can have a great effect on the existing carbon stock and these might be of greater importance than the possible changes in GHG emission from crop cultivation. To provide some perspective on the possible contribution of mitigation activities in crop production, we simulated a hypothetical reduction of GHG emissions by 50% across all crops. By the end of 2030, this would generate a total reduction of about 130 Tg CO_2 eq. Our results for the baseline scenario project a total loss of carbon stock equivalent to 2,198 Tg CO_2 eq. Therefore, even a drastic reduction in GHG emissions represents only about 6% of the change in carbon stock induced by land use transitions. This is not to say that reduction of emissions in crop production is not important. Reduction in emissions is generally obtained through an increase in efficiency and as such it represents an important goal. However, results indicate that it is paramount to consider the totality of interactions among the different land uses. Figure 7.1 provides a visual representation of the tradeoffs between profits and GHG emissions across the policies simulated. For comparison purposes, we express changes in carbon stock in CO_2 eq and add them to changes in GHG gasses. This allows us to compare overall performances for the period under consideration. Of course, one needs to keep in mind that changes in carbon stock are a one-time event while changes in emissions are yearly recurrences. “Win-win” policies are those represented in the upper left quadrant of the graph (decreased GHG emissions and increased profits). The policy that targets a reduction of land allocated to pasture performs better than all others and returns significant mitigation and economic gains. It is particularly important to note that, even though reducing pasture in our simulation is equivalent to increasing costs of production to improve carrying capacity, the increment in profits from other agricultural products offsets the loss and generates a net gain. At the same time, it is clear that halting the expansion of pastureland is key to reducing deforestation and loss of carbon stock. The policy that targets deforestation in the Amazon forest is a typical example of a policy that generates trade-offs. The significant gains in terms of carbon stock and reduction in GHG emissions generate projected losses in profits of about \$1.7 billion. Expressed in dollars per ton of GHG abated, these results indicate costs of about \$1.5 per Mg CO_2 eq. However, even if these costs are not prohibitive, this policy ranks relatively low compared to reducing pastureland: the gain in carbon stock is about one-half of what would be achieved with the pasture reduction policy and profits are reduced. This is because a policy that targets pasture prevents deforestation over the entirety of the country and it induces a more efficient use of land. Therefore, the choice between the two policies becomes an issue of political feasibility and implementation challenges.

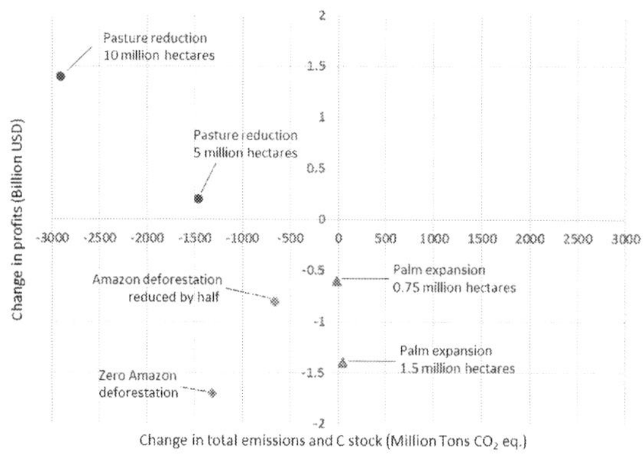

Figure 7.1: Trade-off between profits and GHG emission reduction.

Plans that promote an increase in land allocated to oil palm cultivation need to be considered with extreme caution and investigated further. Model results suggest that the past trend of oil palm cultivation taking over underproductive pastureland might not continue into the future. Area expansion beyond certain levels could lead to a detrimental reduction in land allocated to other perennial crops. This is not necessarily because oil palm cultivation replaces, for example, coffee, but because a rather complex chain of land use transitions initiated by oil palm expansion takes place. The model indicates that other important factors besides economic profit might underlie the decision of keeping land in pasture and these might impede the desired land use transition. This suggests that additional research on this issue is warranted and that particular attention should be given to the instruments and incentives that need to be put in place to achieve the desired outcome.

Some additional general considerations can be made. Given the complexity of low emission development strategies, modeling approaches, frameworks, and tools should be adaptable, open, and transparent. Modeling frameworks should be adaptable so that policy makers can explore the consequences of using different data sets and incorporate new information as it becomes available. Modeling frameworks and tools should be open to the inclusion of inputs from different models so that the robustness of the results can be assessed. In the case presented in this article, we have chosen models with which the authors were most familiar. Each one of the models chosen comes with its own set of strengths and weaknesses. For example, the use of a partial equilibrium model like IMPACT limits the scope of the analysis to the agricultural sector, and the use of a parametric model to determine land use choices cannot incorporate land uses that are not already present in the area at the time of the analysis. These limitations become more or less severe according to the specific situation and country analyzed. However, the advantage of the approach proposed is that each one of the models does have alternatives and researchers have the option to replace them or integrate them with other models. We believe that this level of transparency and openness is essential to generate trust in the results among end users.

ACKNOWLEDGEMENTS

The authors thank our colleagues in the Colombian government who kindly answered our questions and gave critical advice, without which we could not have developed this study. We are especially grateful for suggestions and advice from colleagues of the *Ministerio de Agricultura* (MADR), the *Federación Colombiana de Ganaderos (FEDEGAN),* and the *Sociedad de Agricultores de Colombia* (SAC). We are also grateful to many colleagues from the Ministry of Environment (MADS), Universidad Javieriana, Universidad Nacional, National Cereal Growers Federation (FENALCE), the National Potato Growers Association (FEDEPAPA), and the National Forestry Research and Promotion Institute (CONIF). This work was supported by a grant from the United States Agency for International Development(USAID) and received support from two Consultative Group on International Agricultural Research (CGIAR) research programs: Climate Change, Agriculture and

Food Security(CCAFS) and Forest, Trees and Agroforestry (FTA). The authors take sole responsibility for the opinions expressed within this report.

APPENDIX

Conceptual framework of the land use model

Consider how a decision-maker allocates land use in a municipal,[8] indexed by n; $n = 1, \ldots, N$. Suppose the decision-maker is a risk-neutral landowner; s/he chooses land uses to maximize the present discounted value of the stream of the expected net benefits from the land. The land grid could be allocated to K alternative major uses, indexed by k; $k = 1, \ldots, K$. Among these uses, perennial cropland and annual cropland are two nests of crop, within which the decision maker is allowed to select different crops, indexed by j; $j = 1, \ldots, J_K$. Our data for Colombia include four major perennial crops (cacao, coffee, oil palm, plantain) and five major annual crops (maize, potato, rice, sugarcane, cassava); the remaining crops are lumped into other perennial crops and other annual crops, respectively. Under these simplifying assumptions, the steady-state decision rule that emerges from the related dynamic optimization problem is to put a land parcel to the use generating the greatest present discounted value of net benefit (Lubowski, Plantinga, & Stavins 2006). That is, allocate a parcel of land to use j if

$$U_{nkj} > U_{nk'j'}, \ \forall k' \text{ and } j \neq j' \text{ if } k = k' \tag{1}$$

where U_{nkj} is the one-period expected net benefit from allocating land parcel to use j ($j \in k$). The potential value of U_{nkj} depends on attributes of the parcel, such as land quality, weather conditions, locational characteristics, and economic conditions in the surrounding area, as well as attributes associated with alternative choices, such as price and yield (see Tables 10.1 and 10.2).

Table 10.1: Summary statistics: lower level model

	Crop	Unit	*N*	Min	Max	Mean	STD
Price by crop	Cacao	US $/ kg	1120	1.4057	2.1733	2.1180	0.0861
	Coffee		1120	1.1414	1.7647	1.7198	0.0699
	Palm		1120	0.4734	0.7319	0.7133	0.0290
	Plantain		1120	0.2155	0.3331	0.3247	0.0132
	Othr_perennials		1120	0.2309	0.3570	0.3479	0.0141
	Cassava		1120	0.1378	0.2131	0.2076	0.0084
	Maize		1120	0.1984	0.3067	0.2989	0.0121
	Potato		1120	0.1214	0.1878	0.1830	0.0074
	Rice		1120	0.2001	0.3093	0.3015	0.0122
	Sugarcane		1120	0.0173	0.0268	0.0261	0.0011
	Othr_annuals		1120	0.5257	0.8128	0.7921	0.0322
Production cost by crop	Cacao	Million COP/ ha	1120	7.0031	8.5909	8.1235	0.7016
	Coffee		1120	6.3541	7.0185	6.5028	0.2587
	Palm		1120	4.9742	7.1940	6.9251	0.5828
	Plantain		1120	3.6763	9.8456	5.6645	1.2012
	Othr_perennials		1120	5.6850	6.4786	6.2684	0.3018
	Cassava		1120	2.8991	4.6735	4.2299	0.7047
	Maize		1120	1.3837	2.8610	1.5988	0.3573
	Potato		1120	6.7701	9.1446	8.7244	0.6600
	Rice		1120	1.7749	3.8029	3.3360	0.7901
	Sugarcane		1120	4.7644	4.7644	4.7644	0.0000
	Othr_annuals		1120	0.8637	7.4355	1.7311	1.6260
Biophysical suitability by crop	Cacao	0–1	1120	0.0000	0.9969	0.2284	0.2548
	Coffee		1120	0.0000	0.9738	0.3667	0.2543
	Palm		1120	0.0000	1.0000	0.1589	0.2756
	Plantain		1120	0.0000	0.9442	0.3019	0.2719
	Othr_perennials		1120	0.0000	0.8510	0.2169	0.1905
	Cassava		1120	0.0000	0.9455	0.3765	0.3123
	Maize		1120	0.0000	0.9513	0.3136	0.2224
	Potato		1120	0.0000	0.9321	0.1599	0.2048
	Rice		1120	0.0000	1.0000	0.3798	0.3264
	Sugarcane		1120	0.0000	0.9900	0.3106	0.2604

	Othr_annuals		1120	0.0000	1.7468	0.6057	0.3720
Other explanatory variables	Elev	1000 m	1120	0.0032	3.7336	1.3248	0.9953
	Slope_l	Proportion	1120	0.0084	0.8046	0.2166	0.1862
	Slope_m	Proportion	1120	0.0311	0.9793	0.5356	0.2046
	Slope_h	Proportion	1120	0.0311	0.9999	0.8338	0.2366
	Soilph		1120	0.9380	7.6794	3.8496	1.5338
	Rain2010	1000 mm	1120	0.3751	8.0898	2.0661	1.0404
	Temp2010	1 degree	1120	5.8861	28.5597	20.7905	5.5052

Table 10.2: Summary statistics: upper level model

	Crop		*N*	**Min**	**Max**	**Mean**	**STD**
Lagged land use share (explanatory variable)	Perennials	Proportion	1120	0.0000	0.0949	0.0244	0.0210
	Annuals		1120	0.0000	0.2974	0.0371	0.0372
	Pasture		1120	0.0000	0.9733	0.2497	0.2455
	Forest		1120	0.0000	0.9971	0.2540	0.2232
	Others		1120	0.0029	0.9479	0.4348	0.1756
Other explanatory variables	Production cost for pastureland	Million COP/tonne	1120	1.367	2.0910	1.7021	0.2034
	Pop. Density	1000 people/ha	1120	0	7.3498	0.1381	0.4761
	Access	1000 minutes	1120	0.0024	6.0203	0.3720	0.6302
	Elevation	1000 m	1120	0.0032	3.7336	1.3248	0.9953
	Elevation_cv		1120	0.0018	2.2420	0.3850	0.3425
	Slope	Proportion	1120	0.0084	0.8046	0.2166	0.1862
	Soil ph		1120	0.9380	7.6794	3.8496	1.5338
	Rain	1000 mm	1120	0.3751	8.0898	2.0661	1.0404
	Temp	1 degree	1120	5.8861	28.5597	20.7905	5.5052
	Park	Proportion	1120	0.0000	0.8635	0.0304	0.1002
	Afrodes	Proportion	1120	0.000	0.9920	0.0281	0.1397

	Meat	US $/kg	1120	1.8269	2.8246	2.7528	0.1118
	Timber	1000 USD/ ha	1120	0.0040	0.8513	0.2239	0.2788

A standard practice in the land use modeling literature is to decompose U_{nkj} into a deterministic component $\overline{U}_{nkj}$ and a random error term ε_{nkj}. $\overline{U}_{nkj}$ represents the expected average net benefit from allocating land use in a grid; ε_{nkj} represents the deviation from the average net benefit and is often assumed to follow a normal, logit, or type-I extreme value distribution. We further decompose $\overline{U}_{nkj}$ into two deterministic components:

$$\overline{U}_{nkj} = X_{nk} + Y_{nkj}. \tag{2}$$

X_{nk} depends on variables that describe the nest *k*; it differs over nests but not over alternatives within each nest. Y_{nkj} depends on variables that describe nested use *j* (j∈k); it varies over choices within the nest.

Under the assumption analogous to the nested logit model about ε_{nkj}, for example, that ε_{nkj} is correlated within nest *k* but uncorrelated across nests, the probability of grid *n* allocated to alternative *j* (j∈k) can be derived as a product of two multinomial logit probabilities (McFadden 1977; Train 2003):

$$P_{nk} = \frac{\exp(X_{nk} + \lambda_k I_{nk})}{\sum_{k'=1}^{K} \exp(X_{nk'} + \lambda_{k'} I_{nk'})}, \tag{3}$$

and

$$P_{nkj|k} = \frac{\exp(Y_{nkj}/\lambda_k)}{\sum_{j' \in k} \exp(Y_{nkj'}/\lambda_k)}, \tag{4}$$

where the parameter λ_k is a measure of the degree of independence in ε_{nkj} among the alternatives in each nest and $I_{nk} = \ln(\sum_{j' \in k} \exp(Y_{nkj'}/\lambda_k))$, often called inclusive value of nest *k*. A higher value of λ_k implies less correlation and λ_k=1 indicates complete independence.

Eqn. (3) defines marginal probability of choosing any alternative in nest *k* and Eqn. (4)defines conditional probability of choosing alternative *j* given that any alternative in nest *k* is chosen. We refer to the marginal probability as an upper level model and to the conditional probability as a lower level model, reflecting their relative positions in the hierarchy structure. In Eqn. (4), λ_k is treated as a scale parameter that scales coefficient parameters implicitly defined in Y_{nkj}. For those nests where there are no alternative choices inside, the conditional probability given in (4) equals 1 and the inclusive value is reduced to $I_{nk}=Y_{nkj}/\lambda_k$. Insert I_{nk} into (3) and λ_k will be canceled. Probabilities (3) and (4) are fundamental equations in the nested logit model.

Land use model specification

The average expected net benefits from allocating every grid *n* to nest *k*, X_{nk}, is specified as $X_{nk}=x_n\beta_k$, where x_n is a vector of location-specific variables describing population density, market accessibility, conservation practice, topography, soil pH level, and weather conditions; β_k is vector of coefficients on x_n. Within any nest *k*, the average expected net benefits from allocating each location *n* to alternative *j*, Y_{nkj}, is specified as $Y_{nkj}=y_{nj}\gamma_{kj}$, where y_{nj} is a vector of crop-specific variables, including the crop prices, production costs, land suitability, topography, soil pH level, and weather conditions, and an inertia variable (a lagged crop share) that captures land use conversion costs; γ_{kj} is a vector of coefficient parameters specific to crop.

Note that if the choice-specific variables perfectly captured the average expected net revenues for each crop at the level of the individual grid, then γ_{kj} should simply reflect the marginal net benefit and would not be expected to differ over crops. We allow this parameter varying across crops in our specification because both crop-specific variables are originally measured at relatively coarse resolutions and hence cannot perfectly capture the average expected net benefits for each crop at each location. To eliminate λ_k from the lower level model, we introduce $\tilde{\gamma}kj$ by dividing γkj by λk such that $\tilde{\gamma}kj=\gamma kj/\lambda k$. This scaling facilitates the estimation.

We estimate the model in a "bottom-up" sequential fashion, starting from the estimation of the lower level model and using the estimated coefficients to calculate the inclusive values, which enter the upper level model as explanatory variables. This approach exploits the fact that the choice probabilities can be decomposed into marginal and conditional probabilities that are logit functions.

Crop allocation based on IMPACT baseline projection

Assumption

(1) The growth rate of crop area between the base year and the projected year is consistent with that generated from IMPACT baseline projection.

(2) The specification of the nested land use model is correct; the parameter estimates of the land use model are consistent.

Allocating projected crop area based on the nested land use model

Let *n* be municipal index, *j* be crop index, and *t* be time period index. At the base year $t = 0$, share of crop *j* can be expressed as

$$H_{nj0} \equiv (1 + \varepsilon_{nj0})P_{nj0}, \qquad (5)$$

where P_{nj0} is the estimated crop probability from land use model and ε_{nj0} is the error between the agricultural census (*H*) and the land use model estimation (*P*). Rearranging Eqn. (9), we have

$$\varepsilon_{nj0} \equiv \frac{H_{nj0}}{P_{nj0}} - 1. \qquad (6)$$

Hence, for any *n*, the adjusted share of crop *j* is

$$\tilde{P}_{ij0} \equiv H_{nj0} \qquad (7)$$

Let's now turn to the projected year $t = 1$. Let g_j be the growth rate of crop area between time 0 and 1, derived from IMPACT baseline projection. Then the area of crop *j* at time 1 can be extrapolated as

$$H_{nj1} \equiv (1 + g_j)H_{nj0}, \qquad (8)$$

and

$$H_{nj1} \equiv (1 + \varepsilon_{nj1})P_{nj1}. \qquad (9)$$

Rearranging Eqn. (13), we have

$$\varepsilon_{nj1} \equiv \frac{H_{nj1}}{P_{nj1}} - 1 = \frac{(1+g_j)H_{nj0}}{P_{nj1}} - 1. \tag{10}$$

For any municipal, the adjusted share of crop *j* at time 1 is

$$\tilde{P}_{nj1} \equiv (1+\varepsilon_{nj1})P_{nj1} = \frac{(1+g_j)H_{nj0}}{P_{nj1}} \times P_{nj1} = (1+g_j)H_{nj0} \tag{11}$$

Therefore, changes in crop share at municipal can be derived as

$$\tilde{P}_{ij1} - \tilde{P}_{ij0} = (1+g_j)H_{nj0} - H_{nj0} = g_j H_{nj0} \tag{12}$$

Land use model performance assessment

Tables 10.3 and 10.4 present an assessment of the predictive power of the upper and lower levels of the land use model, respectively. In each table, columns 1 and 2 correspond to the observed land use derived from the data sample; columns 3 and 4 correspond to the predicted land use estimated from the land use model; columns 5 and 6 present the discrepancies by subtracting entries in columns 1 and 2 from entries in columns 3 and 4. We find that the model performs well in predicting most land uses. As far as the upper level is concerned, the model provides generally accurate in-sample predictions. The highest and lowest discrepancies, expressed in percentage terms, are −0.2% for perennial cropland and −2.3% for forests. Regarding the lower level component, the model tends to overestimate perennial and annual croplands by a total of 0.2% and 0.4%, respectively. These discrepancies arise mainly from overestimating areas allocated to oil palm, cacao, maize, and rice. Even so, the predicted errors of these four crops in terms of percentage are in a reasonable range, from 2.3% to 3.2%. The model performs well in the prediction of the remaining crops in terms of both magnitude and percentage.

Table 10.3: Assessment of the predictive power of the upper level model

Land	**Observed**		**Predicted**		**Deviation from the observed area**		**Deviation from the observed shares**	
	Mha	**Share**	**Mha**	**Share**	**Mha**	**Percent**	**Mean**	**Median**
Perennial cropland	1.7	5%	2.0	5%	0.28	16.1%	0%	2%
Annual cropland	1.9	6%	2.5	6%	0.61	31.4%	0%	1%
Pasture	34.6	43%	36.2	43%	1.59	4.6%	0%	0%
Forests	59.1	21%	55.8	21%	-3.34	-5.7%	0%	2%
Other land uses	17.1	26%	17.9	26%	0.87	5.1%	0%	0%

Source: Authors.

Table 10.4: Assessment of the predictive power of the lower level model

Crop	**Observed**		**Predicted**		**Deviation from the observed area**		**Deviation from the observed shares**	
	1000 Ha	**Percent**	**1000 Ha**	**Percent**	**1000 Ha**	**Percent**	**Mean**	**Median**
Perennial crop								
Cacao	115	6%	183	7%	68.4	59.7%	1%	2%
Coffee	770	41%	812	44%	42.1	5.5%	3%	1%
Palm	262	6%	377	16%	114.5	43.6%	9%	1%
Plantain	437	26%	473	23%	35.9	8.2%	−4%	5%
Other perennial crops	160	20%	181	10%	21.3	13.3%	−10%	2%
Annual crop								
Cassava	183	12%	261	8%	77.4	42.3%	−4%	1%
Maize	592	34%	835	30%	243.0	41.0%	−4%	1%
Potato	150	11%	191	13%	40.7	27.1%	1%	0%
Rice	505	10%	745	16%	239.8	47.5%	7%	3%
Sugarcane	371	20%	346	23%	−25.7	−6.9%	4%	3%
Other annual crops	137	14%	164	10%	27.1	19.8%	−4%	1%

Source: Authors.

Robustness check of the specification of spatially explicit price in the land use model

The major assumption made in the land use model is the functional form of spatially explicit prices with exponential decay assumed in Eqn. (1) on page 11. As a robustness check, we considered two alternative specifications—original country-level "flat prices" and spatial prices with linear decay. With the specification of flat prices, all prices are equal across the country; with the specification of linear decay, prices decline more sharply than the specification of exponential decay as the distance from major markets located in cities increases.

We compared the prediction accuracy of the three models, presenting the results in Table 10.5, Appendix. A comparison between the flat-price and the spatial-price specifications indicates (1) both specifications tend to under-predict forested area and over-predict the area of annual cropland, pastureland, and other lands; (2) the models with spatial-price specifications generate smaller prediction errors in both pasture and forestland, two dominant land types in Colombia; (3) the models with flat-price and the spatial-price specifications generate prediction errors in perennial cropland in opposite directions, but the prediction errors generated by the spatial price models are smaller. The balance of evidence, and log-likelihood ratio, suggests the spatial-price specifications perform better than the flat-price specification.

Table 10.5: Robustness check of land use prediction in thousand hectares for the year 2008

	Observed area	**Spatial price: exponential decay**		**Flat price**		**Spatial price: linear decay**	
		Area	**Deviation from (1)**	**Area**	**Deviation from (1)**	**Area**	**Deviation from (1)**
	(1)	**(2)**	**(3)**	**(4)**	**(5)**	**(6)**	**(7)**
Perennial crop	*1744*	*2026*	*16%*	*1355*	*−22%*	*2024*	*16%*
Cacao	115	183	60%	126	10%	158	38%
Coffee	770	812	5%	537	−30%	859	12%
Palm	262	377	44%	248	−5%	375	43%
Plantain	437	473	8%	321	−27%	453	4%
Other crops	160	181	13%	123	−23%	179	12%

Annual crop	*1939*	*2541*	*31%*	*2325*	*20%*	*2544*	*31%*
Cassava	183	261	42%	238	30%	257	40%
Maize	592	835	41%	765	29%	837	41%
Potato	150	191	27%	171	14%	191	27%
Rice	505	745	47%	681	35%	751	49%
Sugarcane	371	346	−7%	319	−14%	346	−7%
Other crops	137	164	20%	152	11%	161	18%
Pasture	*34568*	*36157*	*5%*	*37560*	*9%*	*36160*	*5%*
Forests	*59134*	*55790*	*−6%*	*55273*	*−7%*	*55790*	*−6%*
Other lands	*17054*	*17925*	*5%*	*17926*	*5%*	*17923*	*5%*
Total	*114440*	*114440*	–	*114440*	–	*114440*	–

Source: Authors.

We also compared the projected areas of land use among the three models for the end-period year (2030), presenting the results in Table 10.6. The comparison indicates that the two spatial-price models yield similar results. Given the fact that the exponential distance decay function allows crop prices to decay at a moderate speed and ensures a positive value at every location, this paper uses the functional form of exponential decay for the analysis.

Table 10.6: Robustness check of land use projection in thousand hectares for the year 2030

	Spatial price: exponential decay	**Flat price**		**Spatial price: linear decay**	
	Area	**Area**	**Deviation from (1)**	**Area**	**Deviation from (1)**
	(1)	**(2)**	**(3)**	**(4)**	**(5)**
Perennial crop	*2166*	*1449*	*−33%*	*2161*	*−0.3%*

Cacao	189	130	−31%	163	−13%
Coffee	824	545	−34%	872	5.7%
Palm	405	266	−34%	403	−0.5%
Plantain	505	343	−32%	483	−4.3%
Other crops	243	165	−32%	239	−1.5%
Annual crop	*2716*	*2486*	*−8%*	*2719*	*0.1%*
Cassava	275	251	−9%	270	−1.5%
Maize	861	789	−8%	863	0.3%
Potato	201	180	−11%	202	0.2%
Rice	781	714	−9%	788	0.8%
Sugarcane	433	400	−8%	433	−0.02%
Other crops	166	153	−8%	163	−1.5%
Pasture	*39724*	*41252*	*3.8%*	*39715*	*−0.02%*
Forests	*51744*	*51150*	*−1.1%*	*51759*	*0.03%*
Other lands	*18090*	*18103*	*0.07%*	*18087*	*−0.01%*
Total	*114440*	*114440*	–	*114440*	–

Source: Authors.

NOTE

[1]IMPACT generates scenarios based on four different Global Circulation Models (GCMs): Hadley, IP SL, MIROC, and GFDL. We work with IMPACT results that use MIROC GCM. While the choice of MIROC is arbitrary, the average difference among IMPACT projections for crop areas and prices for the year 2030 for Colombia using the four GCMs is rather low, about 0.5% (with some larger differences for oil palm and cocoa areas using GFDL: 3% and 2% respectively). MIROC seems to provide a "middle of the road" scenario for Colombia and for the year 2030. It should be noted that while the use of alternative GCMs returns different rates of growth for the prices of the considered agricultural commodities, changes in areas for Colombia are not significantly different across the different GCMs.

[2]The specification of distance decay function is arbitrary, but the authors tested other specifications in previous work (Li *et al.*, 2015; De Pinto & Nelson, 2009) and found this particular one to have some desirable properties. Specifically, in Eqn. (1) producer price decays at a moderate speed (the spatially explicit price is between $0.368PP_j$ and PP_j), and the price is never zero which corresponds to a subsistence value at all locations.

[3]The herd size is computed using IMPACT model's output on number of slaughtered animals and by assuming a constant ratio of slaughtered animal and herd size.

[4]We make the simplifying assumption that the ratio revenue-cost remains constant through time.

[5]The Redd Early Mover (REM) Programme active in the Colombian Amazon uses a cost per ton of CO_2 eq reduced by about $0.8. The average mitigation per hectare is considered to be about 378 tons CO_2 eq. This generates an approximate costs of about $302 per hectare.

[6]The government plans indicate a goal of over three million hectares of oil palm plantations by 2020. However, as other authors have indicated (Castiblanco *et al.*, 2013) it is highly unlikely that this goal will be met. Our IMPACT simulations also indicated that if such an increase in production were to be achieved, world prices would be depressed by about five percent, which makes the goal difficult to attain.

[7]It is assumed that primary forest does not generate a revenue stream. Although this is not likely true in all cases, lack of data prevents us from accounting for the contribution of forest products to the overall economy.

[8]We choose municipal as the unit of analysis in order to keep the units of upper and lower level models consistent. We apply the coefficient of variation of elevation to control potential heterogeneity within a municipal.

REFERENCES

1. Anaya, J. A., Chuvieco, E., & Palacios-Orueta, A. (2009). Aboveground biomass assessment in Colombia: A remote sensing approach. Forest Ecology and Management, 257(4), 1237–1246.

2. Antle, J. M., & Stoorvogel, J. J. (2008). Agricultural carbon sequestration, poverty, and sustainability. Environment and Development Economics, 13(03), 327–352.
3. Beheydt, D., Boeckx, P., Sluetel, S., Li, C., & van Cleemput, O. (2007). Validation of DNDC for 22 long-term nitrous oxide field emission measurements. Atmospheric Environment, 41, 6196–6211.
4. Burney, J., Davis, S., & Lobell, D. (2010). Greenhouse gas mitigation by agricultural intensification. Proceedings of the National Academy of Sciences of the United States of America, 107(26), 12052–12057.
5. Cacho, O. J., Marshall, G. R., & Milne, M. (2005). Transaction and abatement costs of carbon-sink projects in developing countries. Environment and Development Economics, 10(05), 597–614.
6. Caicedo, P., Escobar, D., & Tapasco, J. (2016). Ana´lisis econo´mico y de prefactibilidad. Capı´tulo 3 En CIAT-MADR, 2016. In Ana´lisis Integral de Sistemas Productivas en Colombia para la Adaptacio´n al Cambio Clima´tico, Convenio MADR-CIAT 20120382. Bogota´, Colombia.
7. Cai, Z., Sawamoto, T., Li, C., Kang, G., Boonjawat, J., Mosier, A., ... Tsuruta, H. (2003). Field validation of the DNDC model for greenhouse gas emissions in East Asian cropping systems. Global Biogeochemical Cycles, 17(4), 1107. http://dx.doi.org/10.1029/2003GB002046.
8. Castiblanco, C., Etter, A., & Aide, T. M. (2013). Oil palm plantations in Colombia: A model of future expansion. Environmental Science and Policy, 27, 172–183.
9. CIA (Central Intelligence Agency) (2014). CIA Factbook. Accessed November 14, 2014.
10. CIESIN (Center for International Earth Science Information Network)/ Columbia University/CIAT (Centro Internacional de Agricultura Tropical) (2005). Gridded population of the world, Version 3 (GPWv3): population density grid, future estimates. Palisades, NY: NASA Socioeconomic Data and Applications Center (SEDAC), Accessed March 16, 2012.
11. Chomitz, K., & Gray, D. (1996). Roads, land use, and deforestation: a spatial model applied to Belize. World Bank Economic Review, 10, 487–512.

12. Dace, E., Muizniece, I., Blumberga, A., & Kaczala, F. (2015). Searching for solutions to mitigate greenhouse gas emissions by agricultural policy decisions—Application of system dynamics modeling for the case of Latvia. Science of the Total Environment, 527, 80–90.
13. De Figueiredo, E. B., Panosso, A. R., Roma˜o, R., & La Scala, N. (2010). Greenhouse gas emission associated with sugar production in Southern Brazil. Carbon Balance and Management, 5(3), 1–7.
14. Deng, J., Zhou, Z., Zhu, B., Zheng, X., Li, C., Wang, X., et al. (2011). Modelling nitrogen loading in a small watershed in southwest China using a DNDC model with hydrological enhancements. Biogeosciences, 8, 2999–3009.
15. De Pinto, A., & Nelson, G. (2009). Land use change with spatially explicit data: a dynamic approach. Environmental and Resource Economics, 43, 209–229.
16. Diagana, B., Antle, J., Stoorvogel, J., & Gray, K. (2007). Economic potential for soil carbon sequestration in the Nioro region of Senegal's Peanut Basin. Agricultural Systems, 94(1), 26–37.
17. Dossa, E. L., Fernandes, E. C. M., Reid, W. S., & Ezui, K. (2007). Aboveand below-ground biomass, nutrient and carbon stocks contrasting an open-grown and a shaded coffee plantation. Agroforestry Systems, 72, 103–115.
18. Etter, A., McAlpine, C., Wilson, K., Phinn, S., & Possingham, H. (2006). Regional patterns of agricultural land use in deforestation in Colombia. Agriculture, Ecosystems and Environment, 114(2–4), 369–386.
19. FAO (Food and Agriculture Organization of the United Nations) (2014). FAOSTAT. accessed June 1, 2014.
20. FAO/IIASA/ISRIC/ISS-CAS/JRC (2012). Harmonized world soil database version 1.2. Rome and Laxenburg, Austria.
21. FEDEGAN (Federacio´n Colombiana de Ganaderos) (2006). Plan Estrate´gico de la Ganaderı´a Colombiana 2019. Bogota´.
22. Fischer, G., Shah, M., van Velthuizen, H., & Nachtergaele, F. (2001). Global agro-ecological assessment for agriculture in the 21st century IIASA research report 02-02. Laxenburg, Austria: International Institute for Applied Systems Analysis.

23. Gilhespy, S. L., Anthony, S., Cardenas, L., Chadwick, D., del Prado, A., Li, C., et al. (2014). First 20 years of DNDC (DeNitrification DeComposition): Model evolution. Ecological Modelling, 292, 51–62.

24. Giltrap, D. L., Li, C. S., & Saggar, S. (2010). DNDC: A process based model of greenhouse gas fluxes from agricultural soils. Agriculture, Ecosystems & Environment, 136, 292–300.

25. Gockowski, J., & Sonwa, D. (2011). Cocoa intensification scenarios and their predicted impact on CO_2 emissions, biodiversity conservation, and rural livelihoods in the Guinea rain forest of West Africa. Environ Management, 48, 307–321.

26. Godfray, H. C. J., & Robinson, S. (2015). Contrasting approaches to projecting long-run global food security. Oxford Review of Economic Policy, 31(1), 26–44.

27. Golub, A. A., Henderson, B. B., Hertel, T. W., Gerber, P. J., Rose, S. K., & Sohngen, B. (2013). Global climate policy impacts on livestock, land use, livelihoods, and food security. Proceedings of the National Academy of Sciences of the United States of America, 110(52), 20894–20899. http://dx.doi.org/10.1073/pnas.1108772109.

28. Grote, R., Lehmanna, E., Bru¨mmer, C., Bru¨ggemanna, N., Szarzynski, J., Kunstmanna, H., et al. (2009). Modelling and observation of biosphere–atmosphere interactions in natural savannah in Burkina Faso, West Africa. Physics and Chemistry of the Earth, 34, 251–260.

29. Havranek, T., Irsova, Z., Janda, K., & Zilberman, D. (2015). Selective reporting and the social cost of carbon. Energy Economics, 51, 394–406.

30. Havlı´k, P., Valin, H., Herrero, M., Obersteiner, M., Schmid, E., Rufino, M. C., et al. (2014). Climate change mitigation through livestock system transitions. Proceedings of the National Academy of Sciences of the United States of America, 111(10), 3709–3714. http://dx.doi.org/ 10.1073/pnas.1308044111.

31. Henson, I. E., Ruiz, R., & Romero, H. M. (2012). The greenhouse gas balance of the oil palm industry in Colombia: A preliminary analysis. I. Carbon sequestration and carbon offsets. Agronomia Colombiana, 30 (3), 359–369.

32. Hergoualc'h, K., Skiba, U., Harmand, J. M., & He´nault, C. (2008). Fluxes of greenhouse gases from andosols under coffee in monoculture or shaded by Inga Densiflora in Costa Rica. Biogeochemistry, 89(3), 329–345.
33. Hiederer, R., & Ko¨chy, M. (2011). Global soil organic carbon estimates and the harmonized world soil database. Italy: European Union.
34. Hijmans, Robert J. et al. (2005). Very high resolution interpolated climate surfaces for global land areas. International journal of climatology, 25 (15), 1965–1978.
35. IAvH (Instituto de Investigacio´n de Recursos Biolo´gicos Alexander Von Humboldt) IDEAM (Instituto de Hidrologı´a, Meteorologı´a y Estudios Ambientales) IIAP (Instituto de Investigaciones Ambientales del Pacı´fico Jhon Von Newman) INVEMAR (Instituto de Investigaciones Marinas y Costeras) & SINCHI (Instituto Amazo´nico de Investigaciones Cientı´ficas) (2007). Informe del Estado del Medio Ambiente y de los Recursos Naturales Renovables. Bogota´: Instituto de Hidrologı´a, Meteorologı´a y Estudios Ambientales.
36. IDEAM (Instituto Nacional de Hidrologı´a Meteorologı´a y Estudios Ambientales de Colombia) (2010). 2nd Comunicacio´ n Nacional ante la Convencio´ n Marco de las Naciones Unidas sobre Cambio Clima´tico. Bogota´.
37. IGAC (Instituto Geografico Agustin Codazzi) (2013). Sistema de Informacion Geogrfica para la Planeacion y el Ordenamiento Territorial. Accessed November 18, 2013.
38. IPCC (Intergovernmental Panel on Climate Change) (2006). 2006 IPCC guidelines for national greenhouse gas inventories. In H. S. Eggleston, L. Buendia, K. Miwa, T. Ngara, & K. Tanabe (Eds.), Prepared by the national greenhouse gas inventories programme. Hayama, Japan: Institute for Global Environmental Strategies.
39. Jarvis, A., Reuter, H. I., Nelson, A., & Guevara, E. (2008). Hole-filled SRTM for the globe version 4. Accessed August 5, 2011.
40. Koskela, J., Nygren, P., Berninger, F., & Luukkanen, O. (2000). Implications of the Kyoto Protocol for Tropical forest management

and land use: prospects and pitfalls. Tropical Forestry Reports 22. Helsinki: University of Helsinki.

41. Li, C., Frolking, S., & Frolking, T. A. (1992). A model of nitrous oxide evolution from soil driven by rainfall events: 1. Model structure and sensitivity. Journal of Geophysical Research, 97, 9759–9776.
42. Li, C. (2007). Quantifying greenhouse gas emissions from soils: Scientific basis and modeling approach. Soil Science and Plant Nutrition, 53, 344–352.
43. Li, C., Salas, W., Zhang, R., Krauter, C., Rotz, A., & Mitloehner, F. (2012). Manure-DNDC: A biogeochemical process model for quantifying greenhouse gas and ammonia emissions from livestock manure systems. Nutrient Cycling in Agroecosystems, 93(2), 163–200. http://dx. doi.org/10.1007/s10705-012-9507-z.
44. Li, M., De Pinto, A., Ulimwengo, J., You, L., & Robertson, R. (2015). Modeling land-use allocation with mixed-level data: An econometric analysis for the Democratic Republic of the Congo. Environment and Resource Economics, 60, 443–469.
45. Lubowski, R. N., Plantinga, A. J., & Stavins, R. N. (2006). Land-use change and carbon sinks: Econometric estimation of the carbon sequestration supply function. Journal of Environmental Economics and Management, 51(2), 135–152.
46. Lubowski, R. N., & Rose, S. K. (2013). The potential for REDD+: Key economic modeling insights and issues. Review of Environmental Economics and Policy, 7(1), 67–90. http://dx.doi. org/10.1093/reep/res024.
47. MADR (Ministry of Agriculture and Rural Development) (2011). Statistical data of agriculture and rural development 2001–2010. Hanoi: Agricultural Publishing House.
48. MADS (Ministry of Environment and Sustainable Development) (2012). Estrategia Colombiana de Desarrollo Bajo en Carbono. Bogota´.
49. Makundi, W. R., & Sathaye, J. A. (2004). GHG mitigation potential and cost in tropical forestry – Relative role for agroforestry. Environment, Development and Sustainability, 6, 235–260.
50. McFadden, D. L. (1977). Modeling the choice of residential location.

New Haven, CT: Yale University Cowles Foundation for Research in Economics.

51. Martı´nez, D. F. L. (2009). Estudio De La Viabilidad Econo´mica De La Implantacio´n Del Pastoreo Racional Voisin (Prv) En La Llanura Inundable Del Departamento De Casanare, Colombia Master thesis. Univesidad Politecnica de Valencia.
52. Nelson, A. (2000). Accessibility, transport and travel time information CIAT hillsides project report. Cali, Colombia: International Center for Tropical Agriculture.
53. Nelson, G., & Hellerstein, D. (1997). Do roads cause deforestation? Using satellite images in econometric analysis of land use. American Journal of Agricultural Economics, 79, 80–88.
54. Nelson, G. C., De Pinto, A., Harris, V., & Stone, S. (2004). Land use and road improvements: A spatial perspective. International Regional Science Review, 27, 297–325.
55. Nelson, G. C., Valin, H., Sands, R. D., Havlik, P., Ahammad, H., Deryng, D., ... Willenbockel, D. (2014). Climate change effects on agriculture: Economic responses to biophysical shocks. Proceedings of the National Academy of Sciences of the United States of America, 111(9), 3274–3279.
56. Nepstad, D. T. Bezerra, Tepper, D., McCann, K., Stickler, C., & McGrath, D. G. (2013). Addressing agricultural drivers of deforestation in Colombia: Increasing land-based production while reducing deforestation, forest degradation, greenhouse gas emissions and rural poverty Report to the United Kingdom, Foreign and Commonwealth Office and Department of Energy Climate Change, Forests and Climate Change Programme. Earth Innovation Institute.
57. Niels, H. B. (2010). IPCC default soil classes derived from the harmonized world soil database. Version 1.1. Retrieved from: Accessed May 27, 2014.
58. Palm, C. A., Alegre, J. C., Arevalo, L., Mutuo, P. K., Mosier, A. R., & Coe, R. (2002). Nitrous oxide and methane fluxes in six different land use systems in the Peruvian Amazon. Global Biogeochemical Cycles, 16 (4), 21-1–21-13.
59. Pathak, H., Li, C., & Wassmann, R. (2005). Greenhouse gas

emissions from Indian rice fields: Calibration and upscaling the DNDC model. Biogeosciences, 2, 113–123.

60. Robinson, Sherman, D'Croz, Daniel Mason, Islam, Shahnila, Cennachi, Nicola, Creamer, Bernardo, Gueneau, Arthur, Hareau, Guy, Kleinwechter, Ulrich, Mottaleb, Khondoker, Nedumaran, Swamikannu, Robertson, Richard D., Rosegrant, Mark W., Sika, Gbegbelegbe, Sulser, Timothy B., & Wiebe, Keith D. (2015). Climate change adaptation in agriculture: Ex ante analysis of promising and alternative crop technologies using DSSAT and IMPACT IFPRI discussion paper. International Food Policy Research Institute.
61. Rosegrant, M. W., Zhu, T., Msangi, S., & Sulser, T. B. (2008). The impact of biofuels on world cereal prices. Background brief in support of testimony to United States Congressional briefing on the world food situation. Washington, DC: IFPRI.
62. Rutten, M., van Dijk, M., van Rooij, W., & Hilderink, H. (2014). Land use dynamics, climate change, and food security in Vietnam: A globalto-local modeling approach. World Development, 59, 29–46.
63. Saatchi, S. S., Harris, N. L., Brown, S., Lefsky, M., Mitchard, E. T. A., Salas, W., et al. (2011). Benchmark map of forest carbon stocks in tropical regions across three continents. Proceedings of the National Academy of Sciences of the United States of America, 108(24), 9899–9904.
64. Sacks, W. J., Deryng, D., Foley, J. A., & Ramankutty, N. (2010). Crop planting dates: An analysis of global patterns. Global Ecology and Biogeography, 19, 607–620.
65. Schils, R. L. M., Olesen, J. E., del Prado, a., & Soussana, J. F. (2007). A review of farm level modelling approaches for mitigating greenhouse gas emissions from ruminant livestock systems. Livestock Science, 112 (3), 240–251.
66. Schneider, U. A., & Smith, P. (2008). Energy intensities and greenhouse gas emission mitigation in global agriculture. Energy Efficiency, 2(2), 195–206. http://dx.doi.org/10.1007/s12053-008-9035-5.
67. SINAP (Sistema Nacional de A´ reas Protegidas). 2012. GISDATA: Parques Nacionales Naturales de Colombia, Reservas Naturales de la Sociedad Civil, Areas protegidas de las categorı´as regionales, y

Reservas forestales protectoras nacionales. Versio´n 2. Data provided by Grupo de Sistemas de Informacio´n y Radiocomunicaciones de Parques Nacionales Naturales de Colombia.

68. Smith, P., Martino, D., Cai, Z., Gwary, D., Janzen, H., Kumar, P., et al. (2007). Policy and technological constraints to implementation of greenhouse gas mitigation options in agriculture. Agriculture, Ecosystems & Environment, 118(1–4), 6–28.
69. Smith, P., Haberl, H., Popp, A., Erb, K.-H., Lauk, C., Harper, R., et al. (2013). How much land-based greenhouse gas mitigation can be achieved without compromising food security and environmental goals? Global Change Biology, 19(8), 2285–2302. http://dx.doi.org/10.1111/gcb.12160.
70. Springmann, M., Mason-D'Croz, D., Robinson, S., Garnett, T., Godfray, H. C. J., Gollin, D., ... Scarborough, P. (2016). Global and regional health effects of future food production under climate change: A modelling study. The Lancet. http://dx.doi.org/10.1016/S0140-6736 (15), 01156-3.
71. Torres, A. B., Marchant, R., Lovett, J. C., Smart, J. C. R., & Tipper, R. (2010). Analysis of the carbon sequestration costs of afforestation and reforestation agroforestry practices and the use of cost curves to evaluate their potential for implementation of climate change mitigation. Ecological Economics, 69(3), 469–477.
72. Tschakert, P. (2007). Environmental services and poverty reduction: Options for smallholders in the Sahel. Agricultural Systems, 94(1), 75–86.
73. Tenningkeit, T., Kahrl, F., Wolcke, J., & Newcombe, K. (2012). Economics of agricultural carbon sequestration in Sub-Saharan Africa. In Ev. Wollenberg, A. Nihart, M.-L. Tapio-Bistrom, & M. Grieg-Gran (Eds.), Climate change mitigation and agriculture (pp. 144–158). New York and London: Earthscan.
74. Train, K. (2003). Discrete choice methods with simulation. New York: Cambridge University Press.
75. Uchida, H., & Nelson, A. (2010). Agglomeration index: Towards a new measure of urban concentration WIDER working paper 2010/29. UNU-WIDER.

76. Werner, C., Butterbach-Bahl, K., Haas, E., Hickler, T., & Kiese, R. (2007). A global inventory of N2O emissions from tropical rainforest soils using a detailed biogeochemical model. Global Biogeochemical Cycles, 21(3), GB3010.

77. Yohe, G. W., Lasco, R. D., Ahmad, Q. K., Arnell, N. W., Cohen, S. J., Hope, C., Janetos, A. C., & Perez, R. T. (2007). Perspectives on climate change and sustainability. In M. L. Parry, O. F. Canziani, J. P. Palutikof, P. J. van der Linden, & C. E. Hanson (Eds.), Climate change 2007: Impacts, adaptation and vulnerability. Contribution of Working Group II to the fourth assessment report of the Intergovernmental Panel on Climate Change (pp. 811–841). Cambridge: Cambridge University Press.

9 APPLYING OCCAM'S RAZOR TO GLOBAL AGRICULTURAL LAND USE CHANGE

Kerstin Engstrom [a], Mark D.A. Rounsevell [b], Dave Murray-Rust[b], Catherine Hardacre [c], Peter Alexander [b], Xufeng Cui [b], Paul I. Palmer [c], and Almut Arneth [d]

[a]Department of Physical Geography and Ecosystem Science, Lund University, Solvegatan 12, 22362 Lund, Sweden

[b]School of GeoSciences, University of Edinburgh, Geography Building, Drummond Street, EH89XP Edinburgh, UK

[c]School of GeoSciences, University of Edinburgh, Crew Building, The King's Buildings, West Mains Road, EH93JN Edinburgh, UK

[d]Karlsruhe Institute of Technology, Institute of Meteorology and Climate Research, Atmospheric Environmental Research (IMK-IFU), Kreuzeckbahnstr. 19, 82467 Garmisch-Partenkirchen, Germany

ABSTRACT

We present a parsimonious agricultural land-use model that is designed to replicate global land-use change while allowing the exploration of uncertainties in input parameters. At the global scale, the modelled uncertainty range of agricultural land-use change covers observed land-use change. Spatial patterns of cropland change at the country level are simulated less satisfactorily, but temporal trends of cropland change in large agricultural nations were replicated by the model. A variance-based global sensitivity analysis showed that uncertainties in the input parameters representing to consumption preferences are important for changes in global agricultural areas. However, uncertainties in technological change had the largest effect on cereal yields and changes in global agricultural area. Uncertainties related to technological change in developing countries were most important for modelling the extent of cropland. The performance of the model suggests that highly generalised representations of socio-economic processes can be used to replicate global land-use change.

Keywords: Global agricultural land-use, Parsimonious models, Parameter uncertainty, Sensitivity analysis and Cereal yields

INTRODUCTION

Land-use is a crucial link between human activities and the natural environment (Foley et al., 2005). It is central to the sustainable development debate (Turner et al., 2007), especially concerning food security (Brown and Funk, 2008; Fresco, 2009; Lobell et al., 2008), but also climate change (Brovkin et al., 2006), soil quality (Guo and Gifford, 2002; Lal, 2008), water resources (Rockström et al., 2009; Scanlon et al., 2005), biogeochemical cycles (Barth et al., 2005; Ramankutty et al., 2007), biodiversity (de Chazal and Rounsevell, 2009; Scharlemann and Laurance, 2008), human health (Xu et al., 2008) and fire activity (Cochrane and Barber, 2009). More than a third of the global land area is used for agriculture as cropland or pasture (FAOSTAT, 2012), with the remaining land consisting mainly of deserts, mountains, tundra, cities, and ecological reserves (Foley et al., 2011), that include forests and unmanaged grasslands. Between 1961 and 2009, the global population doubled, but agricultural area only increased by 10% (FAOSTAT, 2012). Increasing food demand was met by the intensification of agricultural production arising from the increased use of fertilizers and irrigation (Foley et al., 2011). Land-use intensification has caused detrimental environmental impacts such as salinization, biodiversity loss, and groundwater pollution and depletion (Foley et al., 2011). Expanding agricultural areas also encroach on natural ecosystems resulting in deforestation, conversion of natural grasslands and wetland drainage. These changes are predicted to have far-reaching consequences, impacting on many aspects of the Earth system (Foley et al., 2011). Knowledge about land-use change (LUC) is important, therefore, in understanding how to feed a growing global population whilst simultaneously avoiding environmental damage.

Land-use models are designed to systematically analyse these complex structures, interactions and feedbacks (Deffuant et al., 2012; Heistermann et al., 2006; Rounsevell et al., 2012) and have proven to be useful in both conceptualizing and testing our understanding of the role of different drivers and processes in LUC (Alcamo et al., 2011; Verburg et al., 2009). Because land systems are complex, a variety of land-use models have been developed that apply different methodological approaches (Schaldach and Priess, 2008) across global, continental, or regional scales (Heistermann et al., 2006; Letourneau et al., 2012). These include Markov Chains, empirical–statistical functions, optimization strategies, cellular automata, micro simulation and system dynamics models (Lambin et al., 2000; Parker et al., 2008), with a focus on biophysical or economic processes or both (Lotze-Campen et al., 2008; van Tongeren et al., 2001).

Integrated assessment models (IAMs) include interactions between society, the biosphere and the climate system (Heistermann et al., 2006; Hurtt et al., 2011). Land-use change and land management are one of the components of the broader IAM approach and has been the focus of considerable attention in recent years. For instance, in the IAM IMAGE (Integrated Model to Assess the Global Environment) the land-use module was refined by replacing a set of decision rules to allocate land cover change with an approach to represent land-use as land-use systems. Land-use systems are combinations of land cover, land-use (including livestock), population and accessibility, and represent the heterogeneity of landscape patterns and the interactions of humans with the environment (Letourneau et al., 2012). A similar approach is applied in other types of models, e.g. in the CLUMondo (Conversion of Land Use change Mondo) land-use allocation model. In CLUMondo changes in the land system are allocated to fulfil various demands at the same time and with multiple combinations of land systems (van Asselen and Verburg, 2013).

Other developments include the integration of the use of water for food production in global food security modelling, which is, for example, an integrated component of the Model of Agricultural Production and its Impact on the Environment (MAgPIE; Lotze-Campen et al., 2008). Another example of the integration of water for food production in global land use modelling is IMPACT (International Model for Policy Analysis of Agricultural Commodities and Trade), where cropland changes are modelled for food production units, accounting for interactions between land-use and the hydrology of the food producing watershed (Rosegrant et al., 2013). Furthermore, bioenergy is increasingly considered as a driver of global LUC, for instance in 7 out of 10 global agro-economic models that participated in a recent comparison of LUC trajectories up to 2050 (Schmitz et al., 2014). The GLOBIOM (Global Biosphere Management Model), for example, has been used to study the impact of first and second generation biofuel on direct and indirect land-use and its effects on greenhouse gas balances, but the model can also assess the impact of food production on land and water resources (Havlík et al., 2011; Kraxner et al., 2013; Schneider et al., 2011). While the above models are spatially explicit, the SIMPLE (Simplified International Model of agricultural Prices Land-use and the Environment) model is, by contrast, neither spatially explicit nor dynamic, but calculates cropland change at the global scale with a set of equations based on the supply and demand elasticity of agricultural products and land from one point in time (e.g., 1961) to another (e.g., 2006; Baldos and Hertel, 2013).

Hence, land-use models aspire to represent the effects and interactions of multiple global drivers, differing spatial extents and turnover times of key ecological and social processes, and connections between individual actions, institutional responses, and ecological changes across these multiple dimensions of scale. In practice, it is very challenging to develop global land-use models that simulate all of the key processes in the land system for both human activity and the biophysical environment (Schaldach and Priess, 2008; Schaldach et al., 2011). All models are imperfect representations, or simplifications, of a real world observation. The nature and extent of these simplifications usually reflects the judgement of individual model developers and the research question addressed by the model. While more complex models arguably have a better representation of system processes, they are less transparent than simpler models and there is a greater potential for error propagation between coupled sub-models. Transparency is important in communicating results, especially where the purpose of a model is to inform policy making (Giupponi et al., 2013; Rounsevell et al., 2012). Additionally, the associated computational overhead of complex models is such that applications are limited by the number of possible model runs within a reasonable amount of time. Simpler models are easier to formulate, reducing the probability of human-induced errors, and are computationally inexpensive which makes them amenable to sensitivity experiments using probabilistic estimates of uncertain parameters (Tao et al., 2009; Tebaldi and Lobell, 2008).

Sensitivity analysis (SA) increases the credibility and utility of models and assists in identifying areas of model improvement (Norton, 2015). SA should be applied more commonly as part of modelling exercises, but is rarely undertaken especially for global scale land-use modelling studies (Letourneau et al., 2012). SA has emerged as part of "post-normal science" when dealing with highly uncertain systems, such that: "uncertainty is not banished, but is managed, and values are not presupposed, but are made explicit" (Funtowicz and Ravetz, 1993, 1990, p. 740). However, model simplification risks neglecting important system processes or scales that contribute to real-world phenomena. A model that is unable to replicate these phenomena, at least in part, with appropriate levels of realism is of limited value. Moreover, simplification of a modelled system increases the number and arguably the complexity of assumptions (Wainwright and Mulligan, 2004). Conversely, identifying simple models that perform well, i.e., in comparison with observations and for which parameter sensitivities are known, can support understanding of the key processes that control system behaviour. In this paper we address the question: can the temporal trends and large-scale spatial patterns in global agricultural land-use be understood from simplified socio-economic processes at the country level? We do this by:

1. Developing a simple conceptual model of the socio-economic processes that determine global agricultural land-use change (represented here by cropland and grassland areas);

2. Representing this simple conceptual model within the formulation of a parsimonious land-use model (PLUM);-
3. Testing the concept by evaluating the model against observational data for the period 1991–2010; and
4. Exploring the uncertainty and sensitivity of global cropland to the variability in global input parameters using a variance-based global sensitivity analysis.

MATERIAL AND METHODS

The conceptual model

The conceptual model presented here aspires to follow the principle of Occam's razor, in which the parsimonious (as simple as possible, but not simpler) solution to a problem is preferred. The conceptual model is based on the understanding that changes in cropland occur based on changes in demand for agricultural products and changes in the productivity of cropland. It considers important socio-economic drivers such as population and income level development, consumption, trade, food production and technological change. Changes in consumption and yield changes (due to technological change) are assumed to result in agricultural land-use change. Here agricultural land is assumed to comprise grassland and cereal land, or cropland. Cereal land is chosen as a proxy for cropland, which globally consisted of 60% cereal land in 2010 (FAOSTAT, 2012), while the other 40% was divided between oil-crops, pulses, roots and tubers, and vegetables. The main unit in the conceptual model is cereals (see Fig. 1). Changes in cereal consumption are expected to be proportional to changes in population (Fig. 1, purple area), as historically (1961–1990) per capita cereal consumption was constant for the majority of countries, albeit with some yearly variability (FAOSTAT, 2012). Changes in the consumption of animal products by contrast are assumed to be influenced by population, income levels and life-styles (Fig. 1, purple area). The historical per capita consumption of animal products showed an increasing trend for most countries (1961–1990), which correlated with rising per capita income levels. Additionally, the changes in animal-product consumption are still very different for countries with different cultures and, therefore, life-style is assumed to also influence animal-product consumption. The total use of cereals within a country is assumed to consist of cereals for food and cereals used to feed animals that are used to produce meat and milk.

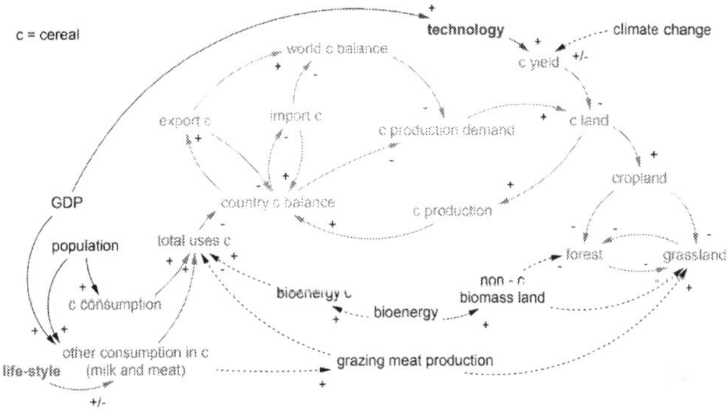

Figure 1: The consumption of food (purple) is driven by economic development (GDP per capita), population and culturally-determined lifestyle choices. The demand for production drives the conversion of forest or grassland to cereal land and cropland (green). GDP drives technology (blue), which in turn increases yields. Relationships indicated with dashed lines (e.g. bioenergy and climate change) are part of the conceptual model, but are not yet implemented in the present version of PLUM. Factors printed in bold are implemented as global variables in PLUM; all other variables are country specific.

The total use of cereals in a country depletes that country's cereal balance. It is assumed that the country cereal balance can be replenished by either importing cereals from the world cereal balance or by increasing domestic cereal production (Fig. 1, orange area). Countries that produce more cereals than are consumed domestically are assumed to export this production, and in so doing they contribute to the world cereal balance. Countries with a negative cereal balance import cereals from the world's cereal balance. A low world cereal balance and low country cereal balance is assumed to increase cereal production demand and vice versa.

An increased demand for cereal production is assumed to be satisfied by cereal yield increases in the first place (intensification). In the second place, a continued increase in cereal demand is assumed to lead to cereal land expansion (Fig. 1, green area). Cereal yield in turn is assumed to be affected by technological change and climate change. Technological change is considered to be indirectly related to income levels, as countries with higher income are more likely to invest in yield improving technologies, such as fertilizers, machinery and irrigation. Climate change is acknowledged to be an important future driver for changes in yield, but is not considered further in the work presented here (hence the dashed line in Fig. 1). Similarly, bioenergy production will become an important component of land use change to meet high energy requirements and strict mitigation targets (Lotze-Campen et al., 2014). However, for the model evaluation presented here bioenergy is not modelled explicitly. In future applications of the model, bioenergy will be represented as an influence on the agricultural system (Fig. 1, dashed lines). Grassland and forests are treated as residual land covers (following cropland changes) with grassland comprising managed and unmanaged components.

Translating the concept into a parsimonious land use model

Implementation of the conceptual model within the visual modelling environment Simile (Muetzfeldt and Massheder, 2003) was based on statistical relationships and a set of rules describing the functional relationships represented in Fig. 1 to create the Parsimonious Land Use Model (PLUM), outlined in the following subsections. All equations are executed yearly on a per country basis if not stated otherwise (model unit, 162 countries, see Appendix). The countries are connected through global variables, such as the world cereal balance.

Consumption module

For the calculation of cereal consumption, the per capita consumption from 1990 was multiplied by each country's population. Population is an exogenous input to PLUM, supplied on an annual basis. Variability within cereal consumption was accounted for by introducing the global variable cereal consumption variability *indCerealVar*(Table 1). However, in scenario experiments *indCerealVar* could also be used to account for scenario specific assumptions regarding food waste (decreased cereal consumption) or the amount of processed food (increased cereal consumption).

Table 1: Parameterizations used in the model evaluation runs (mean value) and sensitivity analysis (mean value and standard deviation). Classes 1–4 for meat and milk increase rates refer to 1: countries with traditionally high meat/milk consumption; 2: countries with traditionally low meat/milk consumption, despite high income levels; 3: emerging economy countries; 4: low income countries

Nr.	***Variable name* (unit)**		**Mean value**	**Standard deviation**
1	*Overproduction rate* (1/time)		0	0.035
2	*Cereal consumption variability* (1/time)		0	0.001
3	*Meat increase rate* (kg meat per capita/log(GDP per capita))	*class 1*	11.38	7.0
		class 2	6.56	4.3
		class 3	7.72	6.0
		class 4	1.14	4.4
4	*Milk increase rate* (kg milk per capita/log(GDP per capita))	class 1	15.95	23.1
		class 2	6.16	12.0
		class 3	4.37	16.4
		class 4	3.33	9.0
5	*Feed conversion ratio improvement rate* (1/time)		0.005	0.0025
6	*Yield improvement rate* (1/time)	*low income countries*	0.016	0.014
		middle income countries	0.022	0.011
		high income countries	0.020	0.006
7	*Abandonment rate developed countries* (1/time)		0.023	0.006–0.071
8	*Abandonment rate developing countries* (1/time)		0.015	0.004–0.054
9	*New cereal land rate developed countries* (1/time)		0.015	0.003–0.049
10	*New cereal land rate developing countries*(1/time)		0.029	0.002–0.046
11	*Grassland forest ratio* (unitless)		0.5	0.3
12	*Forest degradation rate* (1/time)		0.0001	0.00001
12	*Cereal land degradation rate* (1/time)		0.001	0.0001

Previous studies have reported a significant positive correlation between the consumption of animal products, milk (excluding butter) and meat (beef, pork, mutton, and poultry), and income levels (FAO and

LEAD, 2006; Keyzer et al., 2005; Smil, 2002; Zuidema et al., 1994). In PLUM a functional relationship is adopted that increases consumption rapidly with income levels when very low income levels are surpassed until a maximum is reached (see Appendix). Gross Domestic Product per capita (GDP per capita) is used as an indicator of income levels, and except for population, is the only continuous input variable used in PLUM. Although this relationship is valid for most countries, cultural factors and resource conditions influence the rate of change in consumption levels and the maximum level attained (York and Gossard, 2004). Thus countries are divided into four classes to take account of these cultural differences. Class 1 includes high-income countries with a high per capita consumption of animal products, but where the increase in consumption has been observed to slow down (e.g. Germany). Class 2 includes countries that in spite of high income levels consume little animal products (e.g., for meat consumption in Norway or Japan), due to cultural factors or the availability of other protein sources (e.g., fish). Class 3 summarises countries that have presently surpassed very low income levels and are in a rapid transition from low or moderate consumption of animal products towards moderate or high levels of consumption. Class 4 includes countries that do not yet have the economic means to increase the consumption of animal products rapidly (e.g., Nigeria). Conceptually, countries should be able to transform from Class 4 to Class 3, but this was not implemented in the present version of PLUM. The rate of consumption change is different for the four classes and is regulated by the global parameters meat increase rate class 1–4 (*meat1*, *meat2*, *meat3* and *meat4*) and milk increase rate class 1–4 (*milk1*, *milk2*, *milk3* and *milk4*), see Table 1.

Conversion and trade module

The simulated consumption of milk and meat drives the demand for cereal feed. The demand for feed is based on the composition of consumed animal products (milk, beef, poultry, pork and mutton), the cereal feed available for each country in 1990 (see Appendix), and globally applied feed conversion ratios for poultry, pigs, beef and mutton (*fcrPoul, fcrPig, fcrBeef, and fcrMut*, see Appendix for baseline values). Technological change is assumed to affect the feed conversion ratios through the food conversion ratio improvement rate (*fcrImp,*Table 1). It is also assumed that as demand for meat and milk changes the ratio of grass to fodder usage remains the same. Cereal consumption and cereal animal feed represent 70% of global cereal use, if compared with all uses listed in the FAOSTAT database (FAOSTAT, 2012). PLUM does not yet represent cereals used for bioenergy, seed production or waste, which globally account for about 15% of cereal use. Between 1990 and 2005 approximately 15% of global cereal production was produced without a specific purpose, other than as a buffer against harvest shortfalls (FAOSTAT, 2012). This known overproduction and other cereal uses, are accounted for in PLUM by introducing a global overproduction of 30%, which can vary over time with changes in *overProRate* (see Appendix).

The trade and production regulating mechanism in PLUM (Fig. 2) aims to decrease production in times of global overproduction and increase production in times of global underproduction. For each year and country, the balance of cereal production minus cereal use is calculated. Based on a free market assumption, countries with a negative balance import cereals, and countries with a surplus export this surplus. Each country's change in production demand is determined from the country's cereal balance and the world cereal balance (Fig. 2). The global variable world cereal balance connects all countries by summing up all country exports (positive) and imports (negative). In the case of global overproduction (i.e. positive world cereal balance, including the 30% overproduction) exporting countries are assumed to decrease their production (lower left box in Fig. 2). Exporting countries decrease production by their share of the global overproduction. Importing countries do not change production demand, except if they increase their cereal self-sufficiency by the amount of their domestic underproduction (lower middle left box in Fig. 2). In the model evaluation, only developing countries were given the possibility to increase cropland for self-sufficiency in times of global overproduction justified by the assumption that trade barriers and subsidies limit the full participation of developing countries in world trade. When the world is in a state of underproduction (world cereal balance is negative) both importing and exporting countries are assumed to increase their production (lower middle right and right box Fig. 2). Exporting countries are assumed to increase production by their share of the global underproduction, while importing countries increase production by either their share of global underproduction or their domestic underproduction, whichever value is the largest (for details see Appendix). This trade structure maintains parsimony by not including prices, nor being explicit about trade patterns between countries. Additionally, these simplified trade flows are based on a free market assumption, as previously implemented in a European land-use model (Rounsevell et al., 2006).

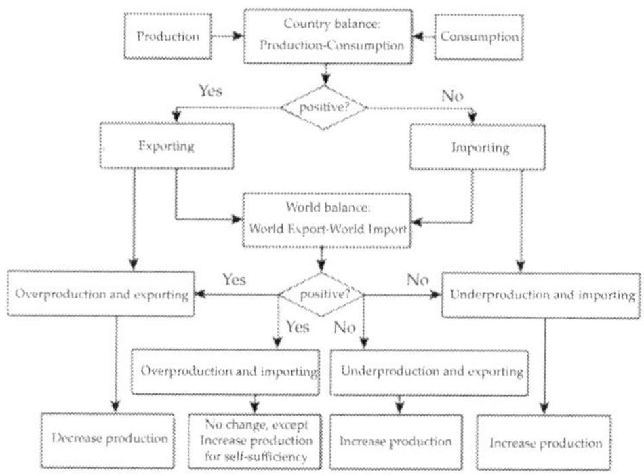

Figure 2: Flowchart of PLUM's trade and production regulating mechanism.

Land conversion module

For each country, the expected change in cereal production from the trade module is divided by the cereal yield to calculate the potential change in cereal land area. To avoid demand-driven expectations of changes in cereal production causing implausible year to year fluctuations, an annual maximum rate of land conversion is imposed. This rate is assumed to differ between developing and developed countries, on the basis that land-use change in developing countries is less regulated by policy. Additionally, the approach differentiates between abandonment of cereal land (and cropland) and the conversion of grassland or forest to new cereal land (and cropland; *abandonmentRich, abandonmentPoor, newCLRich, newCLPoor,* and *newCLSelfPoor*). However, in the evaluation runs these rates are maximum rates, which means that they are only applied to very large expected changes in production and do not have much influence on the model output during the evaluation period (more significant for scenarios).

Cereal yields increase at historic rates, assuming annual yield increase rates from 1.6% (low-income countries, *yieldR1*), through 2% (high-income countries, *yieldR3*) to 2.2% for medium-income countries (*yieldR2*). These rates were defined by analysing cereal yield data from FAOSTAT for the years 1961–1990 (FAOSTAT, 2012). Using the conceptual model the increase in yields is explained by rising income levels and technological change. In the relatively simple approach of PLUM, changes in the area of cereal land (either increases or decreases) are used to estimate cropland. It is assumed that the share of cereal land of cropland in 1990 remains constant over the simulation period. Cropland in turn is used to adjust forest and grassland areas. The proportional changes in forest and grassland areas are assumed to be 50:50 in the runs performed here. However, as this is a global parameter, it could be modified in scenario exercises e.g. to simulate stronger protection of forests. To avoid complete deforestation or conversion of grassland, a minimum area is maintained equal to 10% of a country's forested or grassland area in 1990.

Data and parameterization

Processes (e.g. consumption) and stocks (e.g. cereal land) were initialised with country specific data from the Food and Agricultural Organization (FAOSTAT, 2012; Appendix). Time series (1991–2010) for economic and population development at the country level were used as continuous input data and were derived from the World Bank database (WB, 2012). Coefficients of the statistical relationships implemented in PLUM were set as either global parameters (for example the production rate) or at the country class levels (for example for rates of consumption, see Table 1). The global (or class) values were retrieved from the statistical analysis of FAOSTAT data for the period 1961–1990 (FAOSTAT, 2012; Appendix), which resulted in the values shown in Table 1. This time period was chosen to minimise the effect of inter-annual variability on the long-term trends.

Model evaluation and sensitivity analysis

To evaluate model performance at the global aggregated level, ensemble runs using Monte Carlo (MC) sampling and probability distribution functions (PDFs) created from the values in Table 1 were performed (n = 5000). We assumed normal distributions for all parameters, except for the land conversion rates. The land conversion rates are maximum conversion ratios, so we assumed that all values have the same probability of occurring, which led to the use of a uniform distribution. The World Bank population and GDP per capita data for 1991–2010 (WB, 2012; Appendix) were used as model input. The model uncertainty, which is the variance of outputs given the PDFs of the input parameters, was confined by the standard deviation. The standard deviation and mean were calculated from the ensemble runs for global consumption, production and land use model output. In the description of the aggregated global results by "range" we refer to the range spanned by the standard deviation. We compared these results with FAO time series (1991–2010, or in some cases 2009; FAOSTAT, 2012).

At the country level, over- or underestimation of modelled consumption and land use (running the model with mean values from Table 1) was mapped in comparison with FAOSTAT in 2009. Over- or underestimation was calculated as the modelled output minus the observed data in % of the observed data. At the country level modelled and observed time series were compared for cereal land change in selected countries or regions.

To summarise model uncertainty we calculated the relative standard deviation (standard deviation normalised with the mean) for global consumption, production and land use variables in 2010. To explore the sources of model uncertainty we used sensitivity analysis. Traditional sensitivity analysis uses partial differentials or response plots to vary a single parameter while others are kept constant; the so-called "one at a time" (OAT) method. OAT methods have several drawbacks: they are only appropriate for linear models; they sample only a subset of parameter space, whose relative size decreases as the dimensionality increases; and they give special emphasis to the central point in parameter space (Saltelli et al., 2006). Derivative based or OAT methods are commonly used because of the computational costs in running some models (van Griensven et al., 2006), but other approaches such as variance-based methods have also been used for environmental modelling (e.g. Collins and Avissar, 1994).

Variance-based methods seek to estimate the output variance due to variability in each input parameter. The total sensitivity can be described by the importance S_{Ti} which includes first and higher order effects (interactions) of factor $X_{i.}$ ($X_{\sim i}$ being the matrix of all factors). The importance (S_{Ti}) of an input variable is defined as follows (Saltelli et al., 2008, 2010, p.21, p 260, (1)):

$$S_{Ti} = \frac{E_{X_{\sim i}}\left(V_{X_i}(Y|X_{\sim i})\right)}{V(Y)} \quad (1)$$

Here we applied a variance based Global Sensitivity Analysis (GSA), following the steps outlined in Lilburne and Tarantola (2009):

- The target functions of the study are global areas of the land-use classes in the model, which were treated as separate output variables for simplicity.
- The inputs of interest are given in Table 1. It should be noted that in all cases these are parameter values and do not relate to model structure. Hence the global sensitivity analysis only assesses uncertainties related to input parameters and not to the model structure per se.
- A PDF (Probability Density Function) was estimated for each input parameter using the values given in Table 1.
- The Soboljansen method (Saltelli et al., 2010; Sobol' et al., 2007) from the R "sensitivity" package (Pujol, 2008), was used to create a sampling design (n = 500, p = 26) requiring 14000 runs.

The Soboljansen method was applied since it is based on the Monte Carlo sampling technique and provides an efficient method for calculating S_{Ti}, the total importance of parameters.

- The runs were automated, using the R interface to Simile.
- For each output, the results were returned to the sensitivity package to compute the importance S_{Ti} of each input factor.

In addition to the GSA, the set of MC runs were used to generate response plots for each input to aid understanding of model behaviour. These MC response plots are different to an OAT response plot, as they include the variance of all parameters at once, and hence do not suffer from the limitations discussed previously.

RESULTS

Testing the concept by evaluating PLUM against observational data

Aggregated global outputs

At the global scale, the simulated model range covers the observed time series (1991–2010, FAOSTAT) for consumption and production variables as well as for agricultural land use (Fig. 3). The range of modelled cereal consumption contains observed cereal consumption, though during the first half of the evaluation period the observed consumption is slightly higher than the modelled range. In the second half of the evaluation period, by contrast, the modelled cereal consumption is overestimated compared to the observed values. The relative standard deviation of cereal consumption is 2.0% in 2010. For both milk and meat consumption, the modelled range is almost below the observed time series during 1990–2002, but picks up in the last years of the evaluation period. In 2010, the relative standard deviation of milk and meat consumption is 9.0% and 11.0% respectively. For milk consumption the observed value is at the upper end of the modelled range in 2010, while for meat consumption the observed value is in the middle of the modelled range. The relative standard deviation of cereal feed is 9.7% in 2010 and the modelled range of cereal feed corresponds well to cereal feed reported by the FAO. Only in the last years of the evaluation period does the simulated cereal feed show a stronger increase, while the observed cereal feed has a more fluctuating pattern. This fluctuating pattern might be due to high cereal prices, caused by weather-related harvest shortfalls (in several main cereal producing countries, such as the US, most European countries, Russia and Australia in 2006), policies to decrease cereal stocks, increased cereal demand for biofuel production, and investment in food commodities (Beddington, 2009; USDA, 2007).

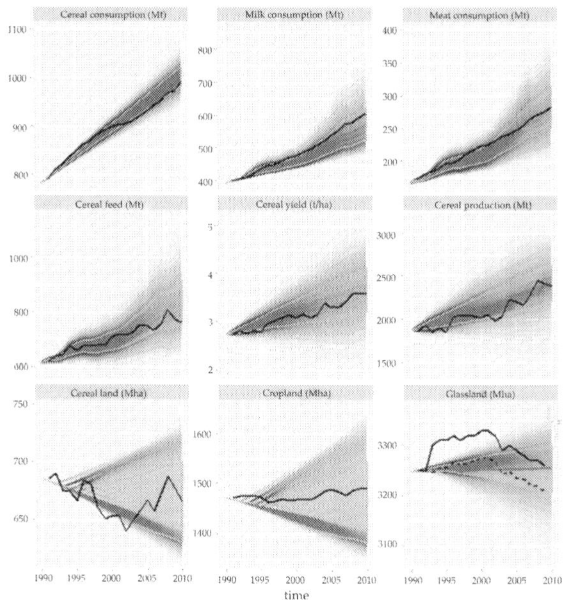

Figure 3: Model runs (faint grey lines) and thereof calculated standard deviation (green shaded area) for global consumption (cereals, milk and meat; Mt), global cereal feed (Mt), yield (t/ha) and production (Mt) and Cereal Land, Cropland and Grassland (Mha). Global observed time series (FAOSTAT) are indicated with black lines. For Grassland an adjusted time series (subtracted data reporting inconsistency for Saudi Arabia between 1992 and 1993) is displayed using a black dashed line.

The simulated range of cereal yield embraces the reported cereal yield during 1991 and 2010. In 2010 the relative standard deviation was 12.7% for cereal yield and 8.7% for cereal production. The observed cereal production is marginally lower than the modelled range in 1993–1995 and 2002, but in general the modelled range of cereal production contains the observed cereal production. Furthermore the modelled range in cereal land area is consistent with the observations of cereal land area, though it is mainly in the second half of the evaluation period that the modelled ranged contains the full extent of observed cereal land (Fig. 3). The relative standard deviation of cereal land is 5.2% in 2010. In PLUM, cropland is simply cereal land scaled by the ratio of cropland to cereal land in 1990. The modelled range of cropland includes the observed cropland. However, the observed cropland is closer to the upper end of the modelled range and it seems that the simple estimate of cropland in

PLUM has a tendency to underestimate cropland. In 2010, the relative standard deviation for cropland was 5.6% and 1.4% for grassland. For grassland, the modelled range covers the observations, if they are adjusted for a data reporting inconsistency in Saudi Arabia between 1992 and 1993 (Fig. 3, dashed line). Between 1992 and 1993 Saudi Arabia's reported grassland increased from 56% to 79% of its land area (FAOSTAT, 2012).

Model outputs at the country level

The temporal trends at the global scale do not show how the simulated outputs for single countries compare with the observational data. In Fig. 4 we compare simulated and observed cereal, milk and meat consumption, cereal yield, cereal land and cropland for the year 2009. For 131 out of 162 countries, the simulated cereal consumption is ±20% of the observations in 2009 (Fig. 4 panel a, countries displayed in light yellow and in the centre of the distributions in the histograms). For almost half of all the countries (77) the modelled milk consumption deviates by less than ±20% of the observed consumption (Fig. 4, panel b). Milk consumption was underestimated by 51.3% in China compared with observed milk consumption in 2009, as well as by 69.8% in the DRC. However, a clear geographical pattern of over- and underestimation does not emerge, as for other countries in Middle and Eastern Africa (i.e. Congo, Angola, Zambia and Mozambique) simulated milk consumption is overestimated compared to observed consumption. Similarly the comparison of modelled and observed meat consumption does not show a clear geographical pattern of countries that are over or underestimated (Fig. 4, panel c). Interestingly for some African countries where milk consumption was underestimated (DRC) or overestimated (Angola, Mozambique), the reversed pattern was observed for meat consumption. In general, consumption was simulated within the ±20% range for the five countries accommodating half of the world›s population (US, Indonesia, China, India and Brazil). Exemptions are the underestimated milk consumption for China and Brazil (51.3% and 27.8% respectively), the overestimation for meat consumption in India by 27.6% and the underestimated meat consumption in Brazil by 29.9% compared with observed values in 2009.

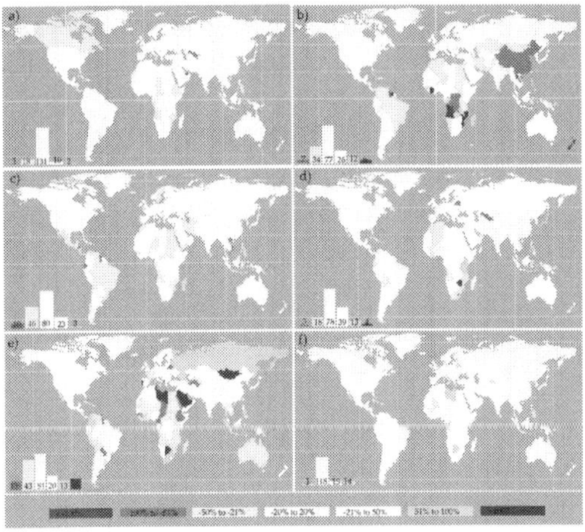

Figure 4: Underestimation (red) and overestimation (blue) in % for a) cereal consumption, b) milk consumption, c) meat consumption, d) cereal yield in 2009 and e) cereal land and f) cropland in 2010. The numbers in the histograms are the country-count of each bin. Countries that are not included in the model or the comparison (see Appendix) are displayed in light grey.

For most countries in America, Asia, and North, South and Western Europe the simulated cereal yield when compared to observed cereal yield was within the ±20% range (Fig. 4, panel d). Yield was overestimated for all Eastern European countries, except Belarus, but also Saudi Arabia and Australia. For African countries the pattern is more diverse, neighbouring countries are over- and underestimated, e.g. Zimbabwe and Mozambique (Fig. 4, panel d).

For cereal land the deviation between modelled and observed data in 2010 shows a large variation across all countries (Fig. 4, panel e). For two thirds of countries, cereal land deviates by more than 20% of observed cereal land. However, larger deviations occur mainly for island states and or countries with very little cereal land (e.g. Botswana, Libya and Saudi Arabia). Large cereal producers such as India, China, the US, Brazil and Nigeria, are all within the ±20% range of deviation. For two other large producers, the modelled cereal land area was overestimated by 65.3% in Russia and underestimated by 42.3% in Australia in 2010.

When the changes in cereal land are scaled to cropland (using the same ratio of cereal land to cropland as in 1990), for the majority of countries, including Russia and Australia, the simulated changes are within the ±20% range of over and underestimation (Fig. 4, panel f). In general, cropland changes are more frequently overestimated than underestimated, as the skewed distribution in Fig. 4 panel f shows.

Looking at the entire evaluation period, the model-data discrepancy for Russia and Australia starts prior to 1995 and continues to grow until 2010 (Fig. 5). While it is not clear why PLUM does not capture the increase in cereal land in Australia, the decrease of cereal land in Russia can be attributed to structural changes following the end of the Soviet Union, which are not included in PLUM.

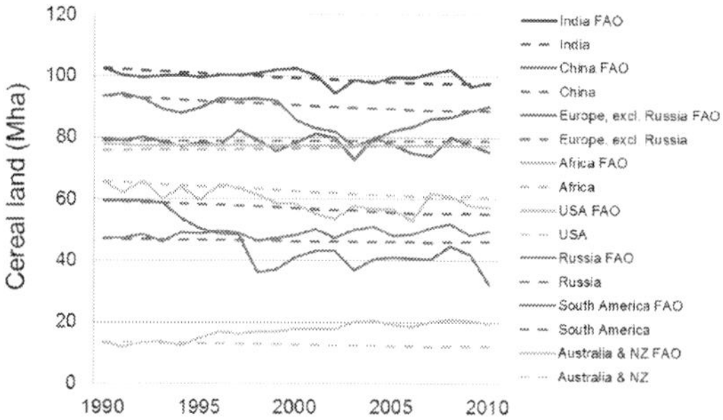

Figure 5: Simulated (solid lines) and reported (dashed lines; FAOSTAT, 2012) cereal land (Mha) during the evaluation period.

Responses to input variables and sensitivity analysis

The variability of the input parameters resulted in relative standard deviations calculated from the multiple model runs between 1.4% for grassland and 12.7% for cereal yield in 2010. The global sensitivity analysis reveals the parameters and their uncertainties that have large effects on model outputs as shown by the 'importance' values (S_{Ti}) for each input, see Fig. 6.

Of the consumption input parameters (*indCerealVar, meat, milk*), *meat1* has the highest importance for cereal demand. Countries that belong to meat class 1 have a meat-rich diet and variability in this diet (e.g. a decrease or further increase in meat consumption) proves to be very important for global cereal demand. The improvement of the feed conversion ratios (*fcrImp*) is also important for global cereal demand, as is the level of global cereal stocks (influenced by *overProRate*, Fig. 6).

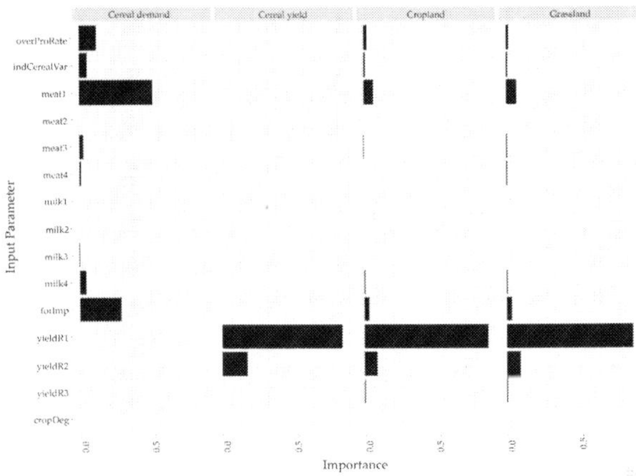

Figure 6: Importance S_{Ti} of input parameters (left) for selected outputs (top). For an importance plot and response plot including all input parameters and outputs see supporting material.

The global sensitivity analysis shows that the variability in the technological development and thus yield improvement rates of low income countries (*yieldR1*) is essential for the outcome of global cereal yield. The current yields in developing countries are often very low in comparison with yields achieved in high input agriculture practised in mostly developed countries (Licker et al., 2010). The potential to increase yields, but also the variability in yield increase is highest in low income countries. During the evaluation period most countries with large areas of cereal land belong to low income countries, such as India, China, Kazakhstan and Nigeria. Low income countries together account for two thirds of global cereal land in 2010, while the high income countries only account for 2.1% of global cereal land. The variability in yield for low income countries has a very strong effect on global yield, while variability of yields of high income countries has no effect (Fig. 6). The model output cropland responded to consumption (*indCerealVar, meat1, meat3* and *milk4*) and production input parameters (*overProRate, fcrImp, yieldR1* and *yieldR2*), with clearly the largest effect being the uncertainty of the input parameter *yieldR1*, as the importance of *yieldR1* is as large as the sum of importance of all of the other input parameters combined (Fig. 6).

DISCUSSION

General model performance

Successfully representing global agricultural land-use change depends firstly on whether the model includes the dynamics that drive global agricultural land-use change at the level of detail adequate for the research question, and secondly on the parameterisation applied in the model. In this study we found that the socio-economic processes of food consumption and production, trade and yield development are sufficient to derive global outputs of cereal land that are consistent with the main trends in the observational data. The comparison of simulated cereal land with observed cereal land showed that the land-use allocation mechanism in PLUM works for several large cereal producers, such as India, China and the US. For other countries over and underestimation of cereal land occurred, but for aggregated regions, e.g. Africa and Europe, simulated cereal land is consistent with observed data. The observed cereal, milk and meat consumption and cereal feed are all within the range of one modelled standard deviation at the global scale.

Overall, PLUM performs best at the global and regional scale and less well for single countries. Better model performance at a higher level of aggregation is a common phenomenon, as exemplified by the validation of the global partial equilibrium model SIMPLE (Baldos and Hertel, 2013). Despite model evaluation being crucial in identifying potential misrepresentations of concepts, missing processes or improper parameterisation, it is not commonly carried out. Exceptions are the evaluation of MAgPIE (Lotze-Campen et al., 2008) and the global land use allocation model by Meiyappan et al. (2014). MAgPIE simulations during the period 1970–1995 of cropland patterns (with the exception of the regions of Sub-Saharan Africa and Middel East/North Africa) and changes in cropland area and yield were shown to agree well with observed (from FAOSTAT (2005) in Lotze-Campen et al., 2008) and observed and fitted data (from Ramankutty and Foley (1999) in Lotze-Campen et al., 2008). Meiyappan et al. (2014) evaluated their global land use allocation model against historical reconstruction data (based on Ramankutty (2012) in Meiyappan et al., 2014). They found that their integrative framework of economic theory, observed land use history and socio-economic and bio-physical land use change related data, is able to replicate general land use patterns, but also the timing and magnitude of spatial shifts observed in historical land use for the time period 1901–2005 (Meiyappan et al., 2014). However, this land use allocation model requires more input data (i.e. regionally aggregated land use information) than PLUM, which explains why e.g. the land use allocation model was able to replicate spatial patterns of land use changes during the end of the Soviet Union, while PLUM was not. Model evaluation, including sensitivity assessments, increases confidence where models do well and helps to identify shortcomings and potential reasons for these shortcomings (Norton, 2015). Confronting model output with observations needs to become standard in the LUC modelling community as a necessary (but not always sufficient) step to strengthen the credibility of models used to simulate the future.

Appropriated level of model structure complexity

PLUM performs well in reproducing temporal trends at the global level, even though it has a simplified representation of trade processes compared with general equilibrium models (e.g. AIM and FARM (Schmitz et al., 2014)). It was shown previously that simple supply and demand functions can be applied successfully to modelling agricultural land-use change at the continental scale for Europe (Rounsevell et al., 2005, 2006). The study presented here demonstrates that this is also possible at the global scale. Additionally, depending on the purpose of the modelling exercise, simulating global agricultural land-use change does not necessarily require an explicit treatment of prices. However, the period of time over which a model can reasonably be projected into the future, is dependent on the validity and stability of the underlying assumptions through time. This is a challenge that applies to all modelling approaches. For empirical and rule-based models, such as PLUM, results must be interpreted with care when the model is used far into the future when conditions might not resemble those under which the model was developed. However, more process-oriented modelling approaches, such as those used in the climate science community (Collins et al., 2013; Friedlingstein et al., 2006) or agro-economic partial and general equilibrium models (Schmitz et al., 2014) also show large divergence in projections beyond the coming few decades. One advantage of the parsimonious approach in PLUM is the transparency of the model's underlying assumptions and algorithms, which are reported in the Appendix of this paper. Their validity under future conditions can therefore be assessed if data to do so become available.

The model evaluation showed that while for several large producing countries cereal land was reproduced by less than 20% deviation of the observed values, the spatial performance of PLUM could be improved. Information about the bio-physical components of land suitability and yield development are commonly part of more complex land use models. GLOBIOM, for instance, requires geographical information (slope, soil, and elevation) and land profitability of crop allocation (based on information about potential cropland management options based on policies and biophysical processes) for its land allocation (Havlík et al., 2011; Schmitz et al., 2014). CLUMondo includes 30 combinations of land cover and land intensity that can represent the heterogeneity of landscape patterns and takes into account the dynamic interactions between humans and the environment (van Asselen and Verburg, 2013). The addition of information about the suitability of land for agriculture and bio-physical potential of yield development could improve three parts of PLUM. First, the trade and regulating mechanism could be informed by land suitability and the land allocation procedure refined. Second, the yield development process could be complemented with information on country specific maximal potential yield in order to ensure realistic yield projections (this could be extended to include different climate scenarios). Third, information about land suitability could support the process that describes the intensive (cereal fed) and extensive (grazing) livestock production.

Overall, complementing socio-economic processes with bio-physical information on land suitability and yield development is an approach commonly adopted by global land use models (Schmitz et al., 2014) and would also be a valuable addition to PLUM. In PLUM, bio-physical information about land suitability would also allow potential arable land to be derived independently from the FAOSTAT data for permanent pastures and meadows. The FAOSTAT data for permanent pasture and meadows is limited, since this land use class includes everything from very productive and intensively managed pastures to nomadic grazing areas (Ramankutty et al., 2008). Additionally, there are reporting inconsistencies in these data (see Appendix for details), which could explain the discrepancy between the modelled and observed grassland during the evaluation period. In general, comparing model output to observation can be problematic where errors occur in the observational data.

Other conditions that explain discrepancy between modelled and observed properties include extreme climatic events and unforeseen events such as geo-political and structural change, e.g. the collapse of the Soviet Union. However, it would be problematic to represent the complex processes that influence extreme weather events or geo-political and structural changes in a model such as PLUM. Moreover these are often surprise events that are difficult to anticipate in practice. Overall, the presented parsimonious concept of socio-economic processes driving land-use change highlights the trade-offs that exist between selection and simplification of processes and model performance when confronted with observations. For instance, the aggregation of several cereal types is a simplification and masks potential higher variation and dynamics in the supply and demand of specific cereal types. There is a fine line in deciding what to include within a model and what not, and this is highly dependent on the aim of the modelling exercise. Using a model such as CLUMondo allows the study of research questions related to land-use intensity at detailed/disaggregated levels, while the PLUM model is better suited to the development of global rather than local land-use scenarios. PLUM can also be useful in providing global scale boundary conditions for regional and local scale studies (e.g. see Rounsevell et al., 2012) to study environmental issues, such as the impact of land-use change on isoprene emissions (Hardacre et al., 2012).

Parameterisation and sensitivity analysis

The global parameterization of PLUM is a methodological trade-off. On the one hand, the global parameterization compromises the variability across countries. On the other hand, the reduction to global (or class related) parameters for all countries makes the model parsimonious and allows global sensitivity experiments to be undertaken more easily. With the global sensitivity analysis uncertainties in model input parameters are made explicit and the uncertainty ranges for each model output are efficiently quantified. This is important, as for instance, small differences in income elasticities, which are used in global models such as GCAM,

GLOBIOM and IMPACT (von Lampe et al., 2014) have been shown to have large effects on projected food demand (Schmitz et al., 2014; von Lampe et al., 2014). Changes in income elasticity are conceptually similar to changes in the parameters of the logarithmic functions that determine food consumption in PLUM, with the advantage that these changes in global parameters (lifestyle, technological change) can be systematically tested with the sensitivity analysis. In this way, the variability across countries was covered by the multiple runs.

The quantification of the uncertainty in output related to the uncertainty in input is also very important for global parameters that describe processes that we do not understand very well or lack data to quantify in the first place, but which nonetheless are conceptually important. In PLUM this is especially the case for the impact of technological change on the feed conversion ratios, which were assumed to increase by 0.5% per year. One could argue that this improvement rate is over-parameterized, but we believe that this it is not the case since the parameter is linked to a conceptually important process. However, as the estimation of the feed conversion improvement rate is uncertain, we estimated the standard deviation to be rather high at 50% of the value.

From the global time-series of the main model outputs, which also displayed all single runs, it becomes apparent that the quantification of uncertainties related to input parameters is relevant because they can change model outputs by the same magnitude as the observed changes in the evaluation period. The quantification of uncertainties in land-use change is important in understanding the future demand for land-use and to develop strategies that allow the efficient and sufficient supply of food and other land-use based products to the global population. For example, if future land-use for food production, including the attached uncertainties, can be projected, the remaining land can be used for other purposes without compromising food security. The global sensitivity analysis performed here showed the importance of yield developments for agricultural land-use change. Parameters that have the largest influence on model output should be explored in more detail (Norton, 2015), which supports the need to complement the yield development process with bio-physical information, as discussed above.

Implications and contextualization of findings

The importance of using land productively, and thus the importance of changes in cereal yields were highlighted by the global sensitivity analysis. More than half of the total variation in cereal land is explained by the sensitivity of the yield improvement rate for developing countries. The role of increases in yields for developing countries in meeting (future) food demand has been demonstrated by others (Balmford et al., 2005; Fischer et al., 2009; Foley et al., 2011). Also, PLUM showed a greater sensitivity in the variability of meat consumption for countries with meat-rich diets, compared to developing countries. The opposite pattern was observed for milk consumption, which reflects a greater potential to increase consumption in developing countries. The greater potential in developing countries is due to low initial consumption levels, as well as the large part of the global population belonging to the developing country group.

In general, the potential for increases in consumption and yields in developing countries is positive as malnutrition and yield gaps are reduced. However, it should be noted that the simulated food production increase does not automatically imply an increase in food security, which requires more than making more food available through production. The distribution and trade of food, as well as access to food and its utilization are equally important in achieving food security (Ericksen, 2008). Also, the modelled increase in yield is based on the assumption that technological change improves yields, but specific management options that are stimulated by technological change and result in yield improvements are not distinguished. In principle, the increases in yield can be achieved through various approaches, some being more desirable than others. Historically, yield increases were often accompanied by negative consequences for the environment, such as ground water and air pollution through excessive fertilization, decreased biodiversity in large monocultures and loss of soil organic carbon (Foley et al., 2011). A more desirable path towards increasing yields in developing countries and sustaining high yields in developed countries is that of sustainable intensification (Campbell et al., 2014; Tilman et al., 2011). The path chosen will depend on a country's capacity and institutional support to implement sustainable intensification (Kuyper and Struik, 2014; Titonell, 2014). Thus, for future simulations, model results and their feasibility need to be discussed consistently within the context of the scenario narratives. It is also important to note that the improved sustainability of agricultural production needs to co-evolve with improvements in the whole food system, including shifts in diets for countries with meat-rich diets and a reduction in food loss and waste (Smith, 2013).

CONCLUSION AND OUTLOOK

We conclude that the generalised process representations of the PLUM model captures the essential dynamics of global agricultural land-use in response to our initial research question: "Can the temporal trends and spatial patterns in global agricultural land-use be understood from simplified socio-economic processes at the country level?" In spite of the limitations in modelling unforeseen, rapid and extreme events, simple supply and demand processes, driven by population, economic development, life style choices and technological change offer insights into understanding the temporal trends of global agricultural land-use change. However, we acknowledge that in pursuit of model parsimony some important processes may potentially not be represented. The inclusion of bio-physical information about land suitability and yield development would be a valuable addition to the model in order to improve performance at the country level. PLUM is currently less well able to capture the variability of agricultural land-use change across all countries, while it still captures the trend for many large cereal producers. However, it is important to note that the purpose of this study was not to represent each country precisely. Moreover, the parsimonious nature of the model allows the efficient exploration of the uncertainty range of model input parameters, which demonstrated that global cereal land is strongly affected by changes in cereal yields in low income countries. Because of its capacity to efficiently explore parameter uncertainty and transparency of underlying assumptions (see Section 4.2), the model could be used to develop probabilistic land-use futures. Probabilistic futures have already been used for population and emissions projections conditional on a broader set of scenarios (O'Neill, 2005; van Vuuren et al., 2008), and could be used to assess the importance of scenario parameters such as rates of consumption increase and technological change for land-use change. Due to its low computational overhead the model is also suitable for rapid scenario assessments including normative (target orientated) visions.

ACKNOWLEDGEMENTS

This study was initiated by a study supported by NERC (Natural Environment Research Council, CH was supported by grant number NE/F003919/1) and was carried out under the Formas Strong Research Environment grant to AA, Land use today and tomorrow (LUsTT; dnr: 211-2009-1682). AA and MR acknowledge support from the EU FP7 LUC4C (grant no. 603542). The author would like to thank Sara Broogard and Jonathan Seaquist for valuable discussions during the model building process.

APPENDIX AS UPPLEMENTARY DATA

The following are the supplementary data related to this article:

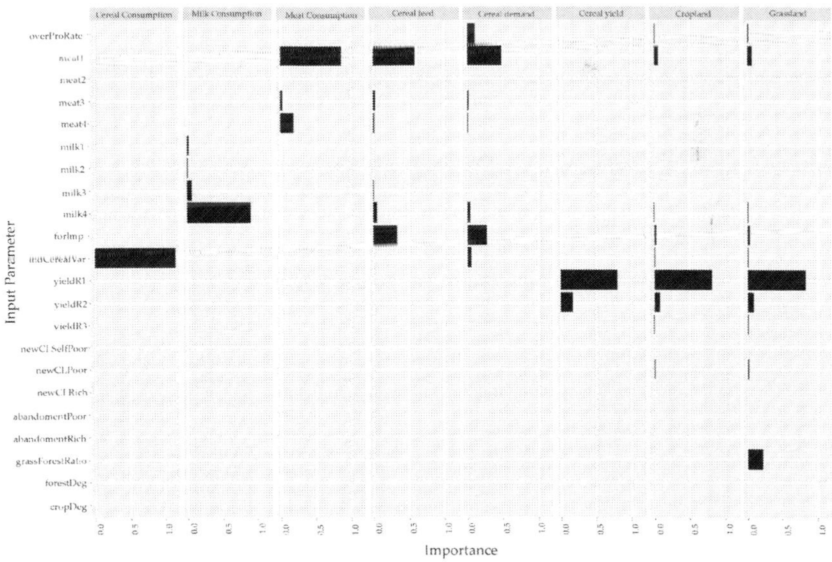

Supporting material 1: Importance S_{Ti} of input parameters (left) for all outputs (top).

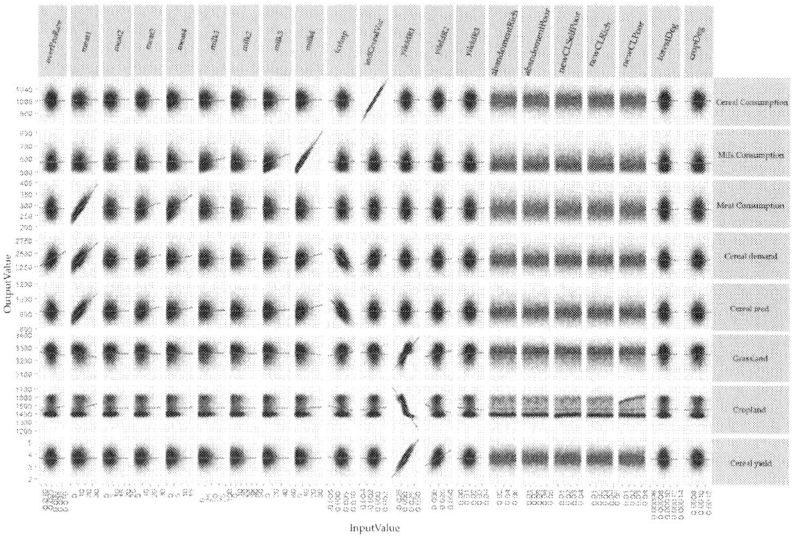

Supporting material 2: Response of model outputs (consumptions and production in ton, land in 1000 ha) to the variation of the input parameters (for units: see Table 1).

Appendix: Technical model description

All equations are applied on an annual basis, for the results described in this paper from 1990 to 2010. If not stated otherwise, equations are calculated for each of the 162 countries. The model is initialized with FAOSTAT data for the year 1990, but for variables with high inter-annual variation a 10 year average (1986–1995) was used. Additionally, for countries formed after the end of the Soviet Union, data were retrieved from the earliest year available in FAOSTAT. (FAOSTAT, 2012). Data were available for 162 countries. Afghanistan, Kiribati, Iraq, Oman, Singapore, Somalia, Qatar, Equatorial Guinea, Bahrain and the Democratic Republic of Congo were excluded because of a lack of data for consumption variables. The small island states were also excluded because GDP per capita and population data were not available.

For population and income World Bank data was used for the evaluation runs (indicator: Population, total; indicator code: SP. POP. TOTL; WB, 2012) but can be replaced with data based on the SRES (Special Report on Emission Scenarios; CIESIN, 2002) for scenario runs (results are not reported here).

Consumption module

The relative income level of a country compared to other countries (*gdpPcRatio*) was calculated to establish differences between countries in, for example, land conversion rates and yield increases (A.1) [unitless].

$$gdpPcRatio = \frac{gdpPc}{\sum_{i=0}^{n} max(gdpPc)} \quad (A.1)$$

gdpP = Annual GDP per capita [US$].

Data was retrieved from World Bank for evaluation runs (indicator: GDP (current US$); indicator code: NY. GDP.MKTP.CD; 1990–2010; WB, 2012) and country-level downscaled GDP projections in 5 year time-steps for the SRES (1990–2100; CIESIN, 2002).

Cereal consumption (*cerealCon*) was assumed to be proportional to population development, but a consumption variability (*indCerealVar*) was included (A.2), [ton].

$$cerealCon = \left(\frac{cerealConPc1990}{1000} + \frac{cerealConPc1990 * indCerealVar}{1000}\right) * population \quad (A.2)$$

cerealConPc1990 = cereal consumption per capita in 1990 [kg person-1 year-1] *population* = population [person]

indCerealVar = variability in cereal consumption [1/time]

Data was retrieved from FAOSTAT Food supply, Crops Primary equivalent, Cereals – Excluding Beer + (Total) for cereal consumption; and for population from World Bank for evaluation runs (1990–2010; WB, 2012) and country-level downscaled population projections in 5 year time-steps for SRES (1990–2100; CIESIN, 2002). For *indCerealVar* the mean value is set 0 (following the assumption that cereal consumption is proportional to population), but for the sensitivity analysis the standard deviation was estimated to be 0.01. This was based analysing the inter-annual variation of cereal consumption during the time period 1961–1990.

The consumption of milk (A.3) and meat (A.4) [ton] was modelled using a logarithmic function:

$$milkCon = \left(\frac{milk_j*(log(gdpPc) - log(gdpPc1990)) + milkConPc1990}{1000}\right)*population \quad \text{A.3)}$$

$$meatCon = \left(\frac{meat_k*(log(gdpPc) - log(gdpPc1990)) + meatConPc1990}{1000}\right)*population \quad \text{A.4)}$$

j = 1–4; number of milk classes

k = 1–4; number of meat classes

$milk_j$ = global parameter, rate of increase in milk consumption [kg milk per capita/log(GDPpc)]

$meat_k$ = global parameter, rate of increase in meat consumption [kg meat per capita/log(GDPpc)]

milkConPc1990 = milk consumption per capita 1990 [kg person-1 year-1]

meatConPc1990 = meat consumption per capita 1990 [kg person-1 year-1]

The four classes are differentiated based on cultural factors, separating countries that historically ate much meat (class 1) from countries that historically ate less meat (class 2), countries which cannot afford high meat consumption due to low income levels (class 4) and the rest (class 3), see Table A1. Conceptually countries in class 4 were assumed to progress to class 3 countries when their income levels surpass 2200 US$ per capita, but this was not implemented in the current version of PLUM. The rate of increase in milk and meat consumption is different for the four classes, see Table A1. To find the parameters linear regression ($y = a*\ln x + b$) was performed for all countries with a complete dataset available over the time period 1961–1990) for each class (see Table A1). However, the slope of the regression overestimates the value drastically if the division into classes is rough. Therefore the slope was calculated for each country and the median of all slopes for one class used as value. Similarly, the standard deviation was calculated from the regression values retrieved for each country within each class (1–4).

Table A1: Methods and values of mean and standard deviation of $meat_j$ and $milk_j$.

Variable	Class division	Median	Regression	SD
meat1	meatConPc1990 ≥ 80 kg/person, incl. China	11.38	13.54	7.08
meat2	meatSpending ≥ 300 US$/kg meat	6.56	9.02	4.31
meat3	meatConPc1990 < 80 kg/person, meatSpending <300 US$/kg meat, gdpPc1990 ≥ 2200 US$	7.72	11.10	6.01
meat4	gdpPc1990 < 2200 US$	1.14	6.02	4.45
milk1	milkConPc1990 ≥ 255 kg/person	15.95	9.26	23.13
milk2	meatSpending ≥ 200 US$/kg milk	6.16	27.01	11.99
milk3	meatConPc1990 < 255 kg/person, meatSpending <200 US$/kg milk, gdpPc1990 ≥ 2200 US$	4.37	35.89	14.09
milk4	gdpPc1990 < 2200 US$	3.33	15.71	9.04

For countries with very low or decreasing GDP per capita the function could result in negative values. This was avoided by including a GDP per capita minimum. If GDP per capita was smaller than GDP per capita minimum, the per capita consumption was kept constant. To account for individual consumption levels per country, the consumption per capita in 1990 year 1990 was used as a starting value. Data was retrieved from FAOSTAT Food supply, Livestock and Fish Primary Equivalent, Milk – Excluding Butter + (Total) and Meat + (Total) respectively (FAOSTAT, 2012).

Conversion and trade module

The potential demand for cereal feed due to milk and meat production was calculated using (A.5) and (A6) [ton].

$$milkConverted = milkCon^{*}\left(\frac{meatYield1990}{milkYield1990}\right) \tag{A.5}$$

meatYield1990 = livestock productivity in 1990 [hg animal-1 (carcass weight)]

milkYield1990 = livestock productivity in 1990 [hg animal-1 yr-1]

Data was retrieved from FAOSTAT, Production, Livestock Primary, Cattle meat, Yield and Cow milk, whole fresh, Yield respectively (FAOSTAT, 2012).

$$potentialCrealFeed = \left(\left(\sum_{i=Poul}^{Beef} r_i{}^{*}\frac{meatConPc1990^{*}population1990}{1000}\right)\right.$$

$$\left. + \; milkConverted^{*}fcr1990_{Beef}\right)^{*}fcr1990_i \tag{A.6}$$

i = Poul, Pig, Mut, Beef

$fcr1990_i$ = baseline values for feed conversion ratios for poultry, pigs, mutton and beef of 4.5, 9, 10 and 25 [kg cereals kg carcass weight-1] respectively.

The baseline values for poultry, pigs and beef were adopted from Smil (Smil, 2002) and the value for mutton was estimated.

The ratio of meat and milk that was produced using cereal feed 1990 with (A.7) [unitless].

$$feedRatio1990 = \frac{cerealFeed1990}{potentialCerealFeed} \tag{A.7}$$

cerealFeed1990 = Cereal feed in 1990 [ton].

Data downloaded from FAOSTAT, Food Balance Sheets, Food Balance Sheets, Cereals – Excluding Beer + (Total), Feed (FAOSTAT, 2012).

The feed conversion ratio improvement rate (*fcrImp*) was used to calculate cereal feed (A.8) [ton].

$$cerealFeed(t) = \left(\left(\sum_{i=Poul}^{Beef} r_i * meatCon * fcr1990_i - (fcr1990_i * fcrImp * t)\right) + milkConverted * fcr1990_{Beef} - \left(fcr1990_{Beef} * fcrImp * t\right)\right) * feedRatio1990 \tag{A.8}$$

i = Poul, Pig, Mut, Beef

$t = 0, 1, 2, \ldots, 20$ (time, corresponds to year 1990–2010) and

fcrImp = feed conversion rate Improvement rate [1/time]

r_i = share of bovine, pig, poultry and mutton meat on total meat consumption in 1990 [unitless]

The mean value for *fcrImp* was estimated to be 0.5% per year, based on the improvement in feed conversion observed during the evaluation period. However, as this estimation is a rather uncertain, we estimated the standard deviation to be rather high with 50% of the mean value, which equals 0.25% per year. r_i was calculated using data from FAOSTAT, Food supply, Livestock and Fish Primary Equivalent, Bovine Meat, Pigmeat, Poultry Meat and Mutton & Goat Meat and Meat + (Total) (FAOSTAT, 2012).

The country's cereal balance was calculated with (A.9) [ton].

$$cerealBalance = cerealProduction - cerealCon - cerealFeed \tag{A.9}$$

cerealProduction = see (A.30).

Depending on the cereal balance, countries were assumed to import cereals (negative cereal balance) or export cereals (positive cereal balance) and contribute to world import (A.10) and world export (A.11), both being global variables [ton].

$$worldImport = \sum_{i=1}^{260} min(cerealBalance_i, 0) \tag{A.10}$$

$$worldExport = \sum_{i=1}^{260} max(cerealBalance_i, 0) \tag{A.11}$$

i = FAO country code.

The global variable world cereal balance (A.12) in the model was described by the state of the cereal world market [ton].

$$worldCerealBalance = worldExport + worldImport \tag{A.12}$$

The cereal consumption for seeds, biofuels, and waste and overproduction on purpose was not directly included in the model. But this was accounted for by the overproduction demand (A.13) [ton].

$$overproDem = \left(\sum_{i=1}^{260} cerealCon + cerealFeed\right)$$

$$*(overProRate1990 + overProRate1990*overProRate*t) \tag{A.13}$$

i = FAO country code

overProRate1990 = overproduction rate in 1990 (see text below)

overProRate = rate that changes *overProRate1990* with time [1/time], 0 in evaluation runs.

The overproduction rate in 1990 was estimated, comparing global cereal production with cereal consumption and cereal used as feed for the period of 1961–1990. The average rate of surplus production is the value for *overProRate* 0.3, that is 30% per year. Standard deviation was calculated from the time series (1961–1990) of *overProRate* and resulted in 0.035 [1/time].

The variable *expectedProduction* was included to describe how much production should be potentially decreased (negative value) or increased (positive value). This depends on the cereal balance in the country and the cereal balance on the world market (see also Fig. 2).

Negative value for importing countries when world cereal balance is larger than overproduction demand (A.14) [ton]:

$$expectedProduction_{impHigh} = cerealBalance \tag{A.14}$$

Negative value for exporting countries when world cereal balance is larger than overproduction demand (A.15) [ton]:

$$expectedProduction_{expHigh} = max(-(cerealBalance + cerealBalance*(overProRate1990 + overProRate1990*overProRate*t)), -(newDemand*worldCerealBalance + newDemand*worldCerealBalance*(overProRate1990 + overProRate1990*overProRate*t))) \tag{A.15}$$

newDemand = country's share on worldExport or worldImport [unitless].

Positive value for importing countries when world cereal balance is smaller than overproduction demand (A.16) [ton]:2

$$expectedProduction_{impLow} = max(-(cerealBalance + cerealBalance*(overProRate1990 + overProRate1990*overProRate*t)), -(newDemand*worldCerealBalance + newDemand*worldCerealBalance*(overProRate1990 + overProRate1990*overProRate*t)) \tag{A.16}$$

Positive value for exporting countries when world cereal balance is smaller than overproduction demand (A.17) [ton]:

$$expectedProduction_{expLow} = newDemand*worldCerealBalance + newDemand*worldCerealBalance*(overProRate1990 + overProRate1990*overProRate*t) \tag{A.17}$$

Land conversion module

The cereal yield was calculated using the baseline average cereal yield for each country and accounting for yield improvements over time depending on income levels. A maximum of cereal yield was defined by the *cerealYieldCeiling* (A.18) [hg ha-1].

$$erealYield(t) = \min((cerealYield1990 * yieldR_{1,2,3} * t), (cerealYieldCeiling + cerealYieldCeiling * yieldR3)) \quad (A.18)$$

$cerealYield1990$ = cereal yield in 1990 [hg ha-1]
$yieldR_{1,2,3}$ = yield increase rate [1/time]
$cerealYieldCeiling$ = upper limit for average cereal yield [hg ha-1]
For *cerealYield1990* data was retrieved from FAOSTAT, Production, Crops, Cereal, Total (+Total), Yield (FAOSTAT, 2012) and the average (1986–1995) calculated and used as 1990 value. The yield increase rates were derived from the average yearly increase of cereal yield over the time period 1961–1990 for 3 classes (class 1: gdpRatio ≤ 0.1; class 2: gdpRatio >0.1 and ≤ 0.5; class 3: gdpRatio>0.5). Standard deviation was calculated from the yearly increase values for each country within each class (1–3). As upper limit for the average cereal yield we assumed 75000 hg ha-1, which is little above the yields reported for the country with highest yields (the Netherlands, with yields around 70 000 hg ha-1 in 1990) (FAOSTAT, 2012).
The variable *landConversion* was included to introduce the areal changes of each countries (A.19)-(A.24) [1000 ha]. To avoid implausible large land conversions, we assumed that the land conversion rates multiplied with the cereal land area would provide the maximum yearly land conversion. The land conversion rates were assumed to be different for developed and developing countries (Table A2).

$$landConversion_{abandExportingRich} = \max\left(\frac{expectedProduction}{cerealYield * 10}, -abandonmentRich * cropland\right) \quad (A.19)$$

$$landConversion_{abandExportingPoor} = \max\left(\frac{expectedProduction}{cerealYield * 10}, -abandonmentPoor * cropland\right) \quad (A.20)$$

$$landConversion_{newCLSelfImportingRich} = \min\left(\frac{-expectedProduction}{cerealYield * 10}, newCLSelfRich * cropland\right) \quad (A.21)$$

$$landConversion_{newCLSelfImportingPoor} = \min\left(\frac{-expectedProduction}{cerealYield * 10}, newCLSelfPoor * cropland\right) \quad (A.22)$$

$$landConversion_{newCLRich} = \min\left(\frac{expectedProduction}{cerealYield * 10}, newCLRich * cropland\right) \quad (A.23)$$

$$landConversion_{newCLPoor} = \min\left(\frac{expectedProduction}{cerealYield * 10}, newCLPoor * cropland\right) \quad (A.24)$$

abandonmentRich, abandonmentPoor, newCLSelfRich, newCLSelfPoor, newCLRich, newCLPoor = see Table A2, [1/time].

Table A2: Methods and values of mean and standard deviation of land conversion rates

Land conversion rate	Abbreviation	Method and data sources	Mean	Min & max value
Abandonment rate developed countries	*abandonmentRich*	Average observed yearly decrease in cereal land in developed/developing countries (1961–1990). The maximum is the observed significant maximal negative rate of change for cereal land in all developed/developing countries. With data from FAOSTAT (cereal production; area in ha)	0.023	0.006 −0.071
Abandonment rate developing countries	*abandonmentPoor*		0.015	0.004 −0.054
New cereal land rate developed countries	*newCLRich*	Average observed yearly increase in cereal land in developed/developing countries (1961–1990). The maximum is the observed significant maximal positive rate of change for cereal land in all developed/developing countries. With data from FAOSTAT (cereal production; area in ha)	0.015	0.003 −0.049
New cereal land rate developed countries for self sufficiency	*newCLRich Self*		0.015	0.003 −0.049
New cereal land rate developing countries	*newCLPoor*		0.029	0.002 −0.046
New cereal land rate developing countries for self sufficiency	*newCLPoor Self*		0.029	0.002 −0.046[a]

As the land conversion rate are upper boundary values, instead of a normal distribution around the standard deviation a uniform distribution was assumed when performing the sensitivity analysis. Additionally, values from the statistical analysis that were below 0.005 (0.5%) were replaced with 0.005 in the model runs. The division between developed and developing countries was made on the basis of the *gdpPcRatio*. If *gdpPcRatio* <0.2 countries are assumed to be developing countries, otherwise if *gdpPcRatio* ≥0.2 then countries are considered as developed countries in the model.

The land area which is converted to cereal land was assumed to be taken from grassland and forest. Vice versa, if cereal land is abandoned, the land area is allocated to grassland and forest. The proportion between forest and grassland was introduced by the grassland to forest ration *grassForestR*, which was 0.5 [unitless] for the evaluation runs (A.25) [1000 ha].

$$\frac{cerealland}{dt} = \frac{cerealland1990}{1000} + (grassForestR{*}landconversion_i) + ((1 - grassForestR){*}landconversion_i) - (cropDeg{*}cerealLand) \quad \text{(A.25)}$$

i = abandExportingRich, …, newCLPoor; (A19-A24);

cropDeg = cropland degradation [1/time]

cereal land1990 = cereal land in 1990 [ha]

Cropland degradation (soil erosion, salinization, desertification), was assumed to be on average 0.1% of cereal land. Degraded cropland is assumed to become grassland. For standard deviation 10% of the mean value was assumed. For cereal land in 1990 data was retrieved from FAOSTAT, Production, Crops, Cereal, Total (+Total), Area (FAOSTAT, 2012).

A minimum of 10% of the initial land cover type was assumed to be sustained at any time.

Cropland was estimated using the share of cereal land of cropland in 1990 and applying this share on cereal land (

$$shareCropland = \frac{((arableLand1990 + permanentCrops1990)/100)*((agriculturalLand1990/100)*landArea)}{cerealland1990} \quad (A.26)$$

arableLand1990 = FAOSTAT [% of agriculturalLand1990]

permanentCrops1990 = FAOSTAT [% of agriculturalLand1990]

agriculturalLand1990 = FAOSTAT Agricultural area [% of *LandArea*]

landArea = FAOSTAT [1000 ha]

$$cropland = shareCropland * cerealland \quad (A.27)$$

Grassland and forest were treated as residuals of cropland (A.28) and (A.29) [1000 ha]. Natural forest degradation was assumed to contribute to the formation of new grassland. Cereal land degradation was assumed to fall into the forested area.

$$\frac{grassland}{dt} = grassland1990 - ((1 - grassForestR)*landconversion) + (forestDeg*forest) \quad (A.28)$$

$$\frac{forest}{dt} = forest1990 - (grassForestR*landconversion) + (cropDeg*cerealLand) \quad (A.29)$$

forestDeg = forest degradation rate [1/time]

grassland1990 = grassland in 1990 [1000 ha]

forest1990 = forest in 1990 [1000 ha]

The forest degradation rate was assumed to be due to natural loss of forest land to grassland with a rate 0.01% of forest per year. For standard deviation 10% of the mean value were assumed. For *grassland1990* data was downloaded, from FAOSTAT, Resources, Resources, Land,

Permanent meadows and pastures, Area (FAOSTAT, 2012). The "permanent pastures and meadows" category in FAOSTAT is problematic as it includes intensive and extensive managed grasslands, and suffers from reporting inconsistencies. For the evaluation period in this paper the mismatch of global modelled and reported grassland starting in 1993 is partly due to reporting inconsistencies in the observed data, in particular for Saudi Arabia. In 1990, 56% of Saudi Arabia's land area was classified as grassland (FAOSTAT category "permanent pastures and meadows"), but from 1993 onwards 79% was reported as grassland (FAOSTAT, 2012), which accounts for most of the discrepancy (FAO adj. in Fig. 3, = FAO global – change in grassland in Saudi Arabia from 1992 to 1993). Data for forest was downloaded from FAOSTAT, Resources, Resources, Forest area, Area (FAOSTAT, 2012). Finally, cereal production was calculated as the product of cereal land and cereal yield (A.30) [ton].

$$cerealProduction = \frac{cerealLand * cerealYield}{10} \tag{A.30}$$

Aggregated Global variables

The global variables Cereal consumption, Milk consumption, Meat consumption, Cereal feed, Cereal production, Cereal land, Cropland and Grassland are the sums of the country level variables cerealCon, milkCon, meatCon, cerealFeed, cerealProduction, cereal landcereal land, cropland and grassland respectively and were converted from ton to Mt (megaton) and 1000 ha to Mha (megahectare). Cereal yield was estimated by dividing global Cereal production with Cereal land (ton/ha).

REFERENCES

1. Alcamo et al., 2011 J. Alcamo, R. Schaldach, J. Koch, C. Kölking, D. Lapola, J. PriessEvaluation of an integrated land use change model including a scenario analysis of land use change for continental Africa Environ. Model. Softw., 26 (2011), pp. 1017-1027
2. Baldos and Hertel, 2013 U.L.C. Baldos, T.W. HertelLooking back to move forward on model validation: insights from a global model of agricultural land use Environ. Res. Lett., 8 (2013), p. 034024
3. Balmford et al., 2005 A.Balmford, R.E. Green, J.P.W.

ScharlemannSparing land for nature: exploring the potential impact of changes in agricultural yield on the area needed for crop production Glob. Change Biol., 11 (2005), pp. 1594-1605

4. Barth et al., 2005 M. Barth, J.P. McFadden, J. Sun, C. Wiedinmyer, P. Chuang, D. Collins, R. Griffin, M.Hannigan, T. Karl, S.-W. Kim, S. Lasher-Trapp, S. Levis, M. Litvak, N. Mahowald, K. Moore, S.Nandi, E. Nemitz, A. Nenes, M. Potosnak, T.M. Raymond, J. Smith, C. Still, C. StroudCoupling between land ecosystems and the atmospheric hydrologic cycle through biogenic aerosol pathways Bull. Am. Meteorol. Soc., 86 (2005), pp. 1738-1742
5. Beddington, 2009 J. BeddingtonFood security: contributions from science to a new and greener revolution Philos. Trans. R. Soc. B Biol. Sci., 365 (2009), pp. 61-71
6. Brovkin et al., 2006 V. Brovkin, M. Claussen, E. Driesschaert, T. Fichefet, D. Kicklighter, M.F. Loutre, H.D. Matthews, N. Ramankutty, M. Schaeffer, A. SokolovBiogeophysical effects of historical land cover changes simulated by six earth system models of intermediate complexity Clim. Dyn., 26 (2006), pp. 587-600
7. Brown and Funk, 2008 M.E. Brown, C.C. FunkFood security under climate change Science, 319 (2008), pp. 580-581
8. Campbell et al., 2014 B.M. Campbell, P. Thornton, R. Zougmoré, P. van Asten, L. LipperSustainable intensification: what is its role in climate smart agriculture? Curr. Opin. Environ. Sustain., 8 (2014), pp. 39-43
9. Cochrane and Barber, 2009 M.A. Cochrane, C.P. BarberClimate change, human land use and future fires in the amazon Glob. Change Biol., 15 (2009), pp. 601-612
10. Collins and Avissar, 1994 D.C. Collins, R. AvissarAn evaluation with the Fourier amplitude sensitivity test (FAST) of which land-surface parameters are of greatest importance in atmospheric modeling J. Clim., 7 (1994), pp. 681-703
11. Collins et al., 2013 M. Collins, R. Knutti, J. Arblaster, J.-L. Dufresne, T. Fichefet, P. Friedlingstein, X.Gao, W.J. Gutowski, T. Johns, G. Krinner, M. Shongwe, C. Tebaldi, A.J. Weaver, M. WehnerLong-term climate change: projections, commitments and irreversibility T.F. Stocker, D. Qin, G.-K. Plattner, M. Tignor, S.K. Allen, J. Boschung,

A. Nauels, Y. Xia, V. Bex, P.M. Midgley (Eds.), In: Climate Change 2013: the Physical Science Basis. Contribution of Working Group I to the Fifth Assessment Report of the Intergovernmental Panel on Climate Change, Cambridge University Press, Cambridge, United Kingdom and New York, NY, USA(2013)

12. de Chazal and Rounsevell, 2009 J. de Chazal, M.D.A. RounsevellLand-use and climate change within assessments of biodiversity change: a review Glob. Environ. Change, 19 (2009), pp. 306-315
13. Deffuant et al., 2012 G. Deffuant, I. Alvarez, O. Barreteau, B. Vries, B. Edmonds, N. Gilbert, N. Gotts, F.Jabot, S. Janssen, M. Hilden, O. Kolditz, D. Murray-Rust, C. Rougé, P. SmitsData and models for exploring sustainability of human well-being in global environmental change Eur. Phys. J. Spec. Top., 214 (2012), pp. 519-545
14. Ericksen, 2008 P. EricksenConceptualizing food systems for global environmental change research Glob. Environ. Change, 18 (2008), pp. 234-245
15. FAO, 2006 FAO, LEADLivestock's Long Shadow: Environmental Issues and Options (2006)
16. FAOSTAT, 2012 FAOSTATFood and Agriculture Organization Statistics (2012)
17. Fischer et al., 2009 R.A. Fischer, D. Byerlee, G.O. EdmeadesCan Technology Deliver on the Yield Challenge to 2050? Expert Meeting on how to feed the world in 2050, Rome June 2009 (2009)
18. Foley et al., 2005 J.A. Foley, R.S. DeFries, G.P. Asner, C. Barford, G. Bonan, S.R. Carpenter, S.Chapin, M.T. Coe, G.C. Daily, H.K. Gibbs, J.H. Helkowski, T. Holloway, E.A. Howard, C.J.Kucharik, C. Monfreda, J.A. Patz, I.C. Prentice, N. Ramankutty, P.K. SnyderGlobal consequences of land use Science, 309 (2005), pp. 570-574
19. Foley et al., 2011 J.A. Foley, N. Ramankutty, K.A. Brauman, E.S. Cassidy, J.S. Gerber, M. Johnston, N.D. Mueller, C. O'Connell, D.K. Ray, P.C. West, C. Balzer, E.M. Bennett, S.R. Carpenter, J. Hill, C. Monfreda, S. Polasky, J. Rockström, J. Sheehan, S. Siebert, D. Tilman, D.P.M. ZaksSolutions for a cultivated planet Nature, 478

(2011), pp. 337-342

20. Fresco, 2009 L.O. FrescoChallenges for food system adaptation today and tomorrow Environ. Sci. Policy, 12 (2009), pp. 378-385
21. Friedlingstein et al., 2006 P. Friedlingstein, P. Cox, R. Betts, L. Bopp, W. von Bloh, V. Brovkin, P.Cadule, S. Doney, M. Eby, I. Fung, G. Bala, J. John, C. Jones, F. Joos, T. Kato, M. Kawamiya, W.Knorr, K. Lindsay, H.D. Matthews, T. Raddatz, P. Rayner, C. Reick, E. Roeckner, K.-G.Schnitzler, R. Schnur, K. Strassmann, A.J. Weaver, C. Yoshikawa, N. ZengClimate–carbon cycle feedback analysis: results from the C4MIP model intercomparison J. Clim., 19 (2006), pp. 3337-3353
22. Funtowicz and Ravetz, 1990 S.O. Funtowicz, J.R. RavetzUncertainty and Quality in Science for Policy Kluwer Acad. Publ., Dordrecht (1990)
23. Funtowicz and Ravetz, 1993 S.O. Funtowicz, J.R. RavetzScience for the post-normal age Futures, 25 (1993), pp. 739-755
24. Giupponi et al., 2013 C. Giupponi, M.E. Borsuk, B.J.M. de Vries, K. HasselmannInnovative approaches to integrated global change modelling Environ. Model. Softw., 44 (2013), pp. 1-9
25. Guo and Gifford, 2002 L.B. Guo, R.M. GiffordSoil carbon stocks and land use change: a meta analysis Glob. Change Biol., 8 (2002), pp. 345-360
26. Hardacre et al., 2012 C.J. Hardacre, P.I. Palmer, K. Baumanns, M. Rounsevell, D. Murray-RustProbabilistic estimation of future emissions of isoprene and surface oxidant chemistry associated with land use change in response to growing food needs Atmos. Chem. Phys. Discuss., 12 (2012), pp. 33359-33410
27. Havlík et al., 2011 P. Havlík, U.A. Schneider, E. Schmid, H. Böttcher, S. Fritz, R. Skalský, K. Aoki, S.D.Cara, G. Kindermann, F. Kraxner, S. Leduc, I. McCallum, A. Mosnier, T. Sauer, M. ObersteinerGlobal land-use implications of first and second generation biofuel targets Energy Policy, 39 (2011), pp. 5690-5702
28. Heistermann et al., 2006 M. Heistermann, C. Müller, K. RonnebergerLand in sight?Achievements, deficits and potentials of continental to global scale land-use modeling Agric. Ecosyst. Environ., 114 (2006), pp. 141-158

29. Hurtt et al., 2011 G.C. Hurtt, L.P. Chini, S. Frolking, R.A. Betts, J. Feddema, G. Fischer, J.P. Fisk, K.Hibbard, R.A. Houghton, A. Janetos, C.D. Jones, G. Kindermann, T. Kinoshita, K. Klein Goldewijk, K. Riahi, E. Shevliakova, S. Smith, E. Stehfest, A. Thomson, P. Thornton, D.P. Vuuren, Y.P. WangHarmonization of land-use scenarios for the period 1500–2100: 600 years of global gridded annual land-use transitions, wood harvest, and resulting secondary lands Clim. Change, 109 (2011), pp. 117-161
30. Keyzer et al., 2005 M.A. Keyzer, M.D. Merbis, I.F.P.W. Pavel, C.F.A. van WesenbeeckDiet shifts towards meat and the effects on cereal use: can we feed the animals in 2030? Ecol. Econ., 55 (2005), pp. 187-202
31. Kraxner et al., 2013 F. Kraxner, E.-M. Nordström, P. Havlík, M. Gusti, A. Mosnier, S. Frank, H. Valin, S.Fritz, S. Fuss, G. Kindermann, I. McCallum, N. Khabarov, H. Böttcher, L. See, K. Aoki, E. Schmid, L. Máthé, M. ObersteinerGlobal bioenergy scenarios – future forest development, land-use implications, and trade-offs Biomass Bioenergy, 57 (2013), pp. 86-96
32. Kuyper and Struik, 2014 T.W. Kuyper, P.C. StruikEpilogue: global food security, rhetoric, and the sustainable intensification debate Curr. Opin. Environ. Sustain., 8 (2014), pp. 71-79
33. Lal, 2008 R. LalSoils and sustainable agriculture: a review E. Lichtfouse, M. Navarrete, P. Debaeke, S. Véronique, C. Alberola (Eds.), Sustainable Agriculture, Springer, London (2008)
34. Lambin et al., 2000 E.F. Lambin, M.D.A. Rounsevell, H.J. GeistAre agricultural land-use models able to predict changes in land-use intensity? Agric. Ecosyst. Environ., 82 (2000), pp. 321-331
35. Letourneau et al., 2012 A.Letourneau, P.H. Verburg, E. StehfestA land-use systems approach to represent land-use dynamics at continental and global scalesEnviron. Model. Softw., 33 (2012), pp. 61-79
36. Licker et al., 2010 R. Licker, M. Johnston, J.A. Foley, C. Barford, C.J. Kucharik, C. Monfreda, N.RamankuttyMind the gap: how do climate and agricultural management explain the 'yield gap' of croplands around the world? Glob. Ecol. Biogeogr., 19 (2010), pp. 769-782

37. Lilburne and Tarantola, 2009 L. Lilburne, S. TarantolaSensitivity analysis of spatial models Int. J. Geogr. Inf. Sci., 23 (2009), pp. 151-168
38. Lobell et al., 2008 D.B. Lobell, M.B. Burke, C. Tebaldi, M.D. Mastrandrea, W.P. Falcon, R.L. NaylorPrioritizing climate change adaptation needs for food security in 2030 Science, 319 (2008), pp. 607-610
39. Lotze-Campen et al., 2008 H. Lotze-Campen, C. Müller, A. Bondeau, S. Rost, A. Popp, W. LuchtGlobal food demand, productivity growth, and the scarcity of land and water resources: a spatially explicit mathematical programming approach Agric. Econ., 39 (2008), pp. 325-338
40. Lotze-Campen et al., 2014 H. Lotze-Campen, M. von Lampe, P. Kyle, S. Fujimori, P. Havlik, H. van Meijl, T. Hasegawa, A. Popp, C. Schmitz, A. Tabeau, H. Valin, D. Willenbockel, M. WiseImpacts of increased bioenergy demand on global food markets: an AgMIP economic model intercomparison Agric. Econ., 45 (2014), pp. 103-116
41. Meiyappan et al., 2014 P. Meiyappan, M. Dalton, B.C. O'Neill, A.K. JainSpatial modeling of agricultural land use change at global scale Ecol. Model., 291 (2014), pp. 152-174
42. Muetzfeldt and Massheder, 2003 R. Muetzfeldt, J. MasshederThe simile visual modelling environment Eur. J. Agron., 18 (2003), pp. 345-358
43. Norton, 2015 J. NortonAn introduction to sensitivity assessment of simulation models Environ. Model. Softw., 69 (2015), pp. 166-174
44. O'Neill, 2005 B.C. O'NeillPopulation scenarios based on probabilistic projections: an application for the millennium ecosystem assessment Popul. Environ., 26 (2005), pp. 229-254
45. Parker et al., 2008 D.C. Parker, A. Hessl, S.C. DavisComplexity, land-use modeling, and the human dimension: fundamental challenges for mapping unknown outcome spaces Geoforum, 39 (2008), pp. 789-804
46. Pujol, 2008 G. Pujol Sensitivity Package, 1 (2008) 4–0
47. Ramankutty and Foley, 1999 N. Ramankutty, J.A. FoleyEstimating historical changes in global land cover: Croplands from 1700 to

1992 Glob. Biogeochem. Cycles, 13 (1999), pp. 997-1027

48. Ramankutty et al., 2007 N. Ramankutty, H.K. Gibbs, F. Achard, R. Defries, J.A. Foley, R.A. HoughtonChallenges to estimating carbon emissions from tropical deforestation Glob. Change Biol., 13 (2007), pp. 51-66
49. Ramankutty et al., 2008 N. Ramankutty, A.T. Evan, C. Monfreda, J.A. FoleyFarming the planet: 1. Geographic distribution of global agricultural lands in the year 2000 Glob. Biogeochem. Cycles, 22 (2008)
50. Ramankutty, 2012 N. RamankuttyGlobal Cropland and Pasture Data from 1700–2007
51. Retrieved from. http://www.geog.mcgill.ca/~nramankutty/Datasets/Datasets.html. from the LUGE (Land Use and the Global Environment) Laboratory Department of Geography, McGill University, Montreal, Quebec, Canada (2012) [25 September 2012]
52. Rockström et al., 2009 J. Rockström, M. Falkenmark, L. Karlberg, H. Hoff, S. Rost, D. GertenFuture water availability for global food production: the potential of green water for increasing resilience to global changeWater Resour. Res., 45 (2009)
53. Rosegrant et al., 2013 M.W. Rosegrant, C. Ringler, T. Zhu, S. Tokgoz, P. BhandaryWater and food in the bioeconomy: challenges and opportunities for development Agric. Econ., 44 (2013), pp. 139-150
54. Rounsevell et al., 2005 M. Rounsevell, F. Ewert, I. Reginster, R. Leemans, T. CarterFuture scenarios of European agricultural land use II. Projecting changes in cropland and grassland Agric. Ecosyst. Environ., 107 (2005), pp. 117-135
55. Rounsevell et al., 2006 M. Rounsevell, I. Reginster, M. Araujo, T. Carter, N. Dendoncker, F. Ewert, J.House, S. Kankaanpaa, R. Leemans, M. MetzgerA coherent set of future land use change scenarios for Europe Agric. Ecosyst. Environ., 114 (2006), pp. 57-68
56. Rounsevell et al., 2012 M.D.A. Rounsevell, B. Pedroli, K.-H. Erb, M. Gramberger, A.G. Busck, H. Haberl, S. Kristensen, T. Kuemmerle, S. Lavorel, M. Lindner, H. Lotze-Campen, M.J. Metzger, D. Murray-Rust, A. Popp, M. Pérez-Soba, A. Reenberg, A.

Vadineanu, P.H. Verburg, B. WolfslehnerChallenges for land system science Land Use Policy, 29 (2012), pp. 899-910

57. Saltelli et al., 2006 A.Saltelli, M. Ratto, S. Tarantola, F. CampolongoSensitivity analysis practices: strategies for model-based inference Reliab. Eng. Syst. Saf., 91 (2006), pp. 1109-1125
58. Saltelli et al., 2008 A.Saltelli, M. Ratto, T. Andres, F. Campolongo, J. Cariboni, D. Gatelli, M. Saisana, S.TarantolaGlobal Sensitivity Analysis: the Primer Wiley Online Library, West Sussex (2008)
59. Saltelli et al., 2010 A.Saltelli, P. Annoni, I. Azzini, F. Campolongo, M. Ratto, S. TarantolaVariance based sensitivity analysis of model output. Design and estimator for the total sensitivity index Comput. Phys. Commun., 181 (2010), pp. 259-270
60. Scanlon et al., 2005 B.R. Scanlon, R.C. Reedy, D.A. Stonestrom, D.E. Prudic, K.F. DennehyImpact of land use and land cover change on groundwater recharge and quality in the southwestern US Glob. Change Biol., 11 (2005), pp. 1577-1593
61. Schaldach and Priess, 2008 R. Schaldach, J.A. PriessIntegrated models of the land system: a review of modelling approaches on the regional to global Living Rev. Landsc. Res., 2 (2008)
62. Scharlemann and Laurance, 2008 J.P.W. Scharlemann, W.F. LauranceHow Green are biofuels? Science, 319 (2008), pp. 43-44
63. Schaldach et al., 2011 R. Schaldach, J. Alcamo, J. Koch, C. Kölking, D.M. Lapola, J. Schüngel, J.A.PriessAn integrated approach to modelling land-use change on continental and global scales Environ. Model. Softw., 26 (2011), pp. 1041-1051
64. Smith, 2013 P. SmithDelivering food security without increasing pressure on land Glob. Food Secur., 2 (2013), pp. 18-23
65. Schmitz et al., 2014 C. Schmitz, H. van Meijl, P. Kyle, G.C. Nelson, S. Fujimori, A. Gurgel, P. Havlik, E.Heyhoe, D.M. d'Croz, A. Popp, R. Sands, A. Tabeau, D. van der Mensbrugghe, M. von Lampe, M.Wise, E. Blanc, T. Hasegawa, A. Kavallari, H. ValinLand-use change trajectories up to 2050: insights from a global agro-economic model comparison Agric. Econ., 45 (2014), pp. 69-84
66. Schneider et al., 2011 U.A. Schneider, P. Havlík, E. Schmid, H. Valin, A. Mosnier, M. Obersteiner, H.Böttcher, R. Skalský, J. Balkovič, T. Sauer, S. FritzImpacts of population growth, economic

development, and technical change on global food production and consumption Agric. Syst., 104 (2011), pp. 204-215

67. Smil, 2002 V. SmilWorldwide transformation of diets, burdens of meat production and opportunities for novel food proteins Enzyme Microb. Technol., 30 (2002), pp. 205-311
68. Sobol' et al., 2007 I.M. Sobol, S. Tarantola, D. Gatelli, S.S. Kucherenko, W. MauntzEstimating the approximation error when fixing unessential factors in global sensitivity analysis Reliab. Eng. Syst. Saf., 92 (2007), pp. 957-960
69. Tao et al., 2009 F. Tao, Z. Zhang, J. Liu, M. YokozawaModelling the impacts of weather and climate variability on crop productivity over a large area: a new super-ensemble-based probabilistic projection Agric. For. Meteorol., 149 (2009), pp. 1266-1278
70. Tebaldi and Lobell, 2008 C. Tebaldi, D.B. LobellTowards probabilistic projections of climate change impacts on global crop yields Geophys. Res. Lett., 35 (2008)
71. Tilman et al., 2011 D. Tilman, C. Balzer, J. Hill, B.L. BefortGlobal food demand and the sustainable intensification of agriculture PNAS, 108 (2011), pp. 20260-20264
72. Tittonell, 2014 P. TittonellEcological intensification of agriculture—sustainable by nature Curr. Opin. Environ. Sustain., 8 (2014), pp. 53-61
73. Turner et al., 2007 B.L. Turner, E.F. Lambin, A. ReenbergLand change science special Feature: the emergence of land change science for global environmental change and sustainability Proc. Natl. Acad. Sci., 104 (2007), pp. 20666-20671
74. USDA (United States Departement of AgricultureJoint Agricultural Weather Facility), 2007 USDA (United States Departement of Agriculture, Joint Agricultural Weather Facility)Global crop production review, 2006 WMO Bull., 56 (2007), pp. 220-225
75. van Asselen and Verburg, 2013 S. van Asselen, P.H. VerburgLand cover change or land-use intensification: simulating land system change with a global-scale land change model Glob. Change Biol., 19 (2013), pp. 3648-3667
76. van Griensven et al., 2006 A.van Griensven, T. Meixner, S. Grunwald, T. Bishop, M. Diluzio, R.SrinivasanA global sensitivity

analysis tool for the parameters of multi-variable catchment models J. Hydrol., 324 (2006), pp. 10-23

77. van Tongeren et al., 2001 F. van Tongeren, H. van Meijl, Y. SurryGlobal models applied to agricultural and trade policies: a review and assessment Agric. Econ., 26 (2001), pp. 149-172
78. van Vuuren et al., 2008 D.P. van Vuuren, B. de Vries, A. Beusen, P.S.C. HeubergerConditional probabilistic estimates of 21st century greenhouse gas emissions based on the storylines of the IPCC-SRES scenarios Glob. Environ. Change, 18 (2008), pp. 635-654
79. von Lampe et al., 2014 M. von Lampe, D. Willenbockel, H. Ahammad, E. Blanc, Y. Cai, K. Calvin, S.Fujimori, T. Hasegawa, P. Havlik, E. Heyhoe, P. Kyle, H. Lotze-Campen, D. Mason d'Croz, G.C.Nelson, R.D. Sands, C. Schmitz, A. Tabeau, H. Valin, D. van der Mensbrugghe, H. van MeijlWhy do global long-term scenarios for agriculture differ? an overview of the AgMIP global Economic model Intercomparison Agric. Econ., 45 (2014), pp. 3-20
80. Verburg et al., 2009 P.H. Verburg, J. van de Steeg, A. Veldkamp, L. WillemenFrom land cover change to land function dynamics: a major challenge to improve land characterization J. Environ. Manag., 90 (2009), pp. 1327-1335
81. Wainwright and Mulligan, 2004 J. Wainwright, M. MulliganEnvironmental Modelling: Finding Simplicity in Complexity John Wiley & Sons Ltd, Chichester (2004)
82. WB, 2012 WBThe World Bank DATA (2012) Retrieved from http://data.worldbank.org/indicator [22 August 2012]
83. Xu et al., 2008 J. Xu, R. Sharma, J. Fang, Y. XuCritical linkages between land-use transition and human health in the himalayan region Environ. Int., 34 (2008), pp. 239-247
84. York and Gossard, 2004 R. York, M.H. GossardCross-national meat and fish consumption: exploring the effects of modernization and ecological context Ecol. Econ., 48 (2004), pp. 293-302
85. Zuidema et al., 1994 G. Zuidema, G.J. van den Born, J. Alcamo, G.J.J. KreilemanSimulating changes in global land cover as affected by economic and climatic factors Water, Air, Soil Pollut., 76 (1994), pp. 163-198

CITATIONS

CHAPTER 2

Alexander Popp, Katherine Calvin, Shinichiro Fujimori, Petr Havlik, Florian Humpenöder, Elke Stehfest, Benjamin Leon Bodirsky, Jan Philipp Dietrich, Jonathan C. Doelmann, Mykola Gusti, Tomoko Hasegawa, Page Kyle, Michael Obersteiner, Andrzej Tabeau, Kiyoshi Takahashi, Hugo Valin, Stephanie Waldhoff, Isabelle Weindl, Marshall Wise, Elmar Kriegler, Hermann Lotze-Campen, Oliver Fricko, Keywan Riahi, Detlef P. van Vuuren, Land-use futures in the shared socio-economic pathways, In Global Environmental Change, Volume 42, 2017, Pages 331-345, ISSN 0959-3780, https://doi.org/10.1016/j.gloenvcha.2016.10.002.

CHAPTER 3

Ilpo Tammi, Kaisa Mustajärvi, Jussi Rasinmäki, Integrating spatial valuation of ecosystem services into regional planning and development, In Ecosystem Services, Volume 26, Part B, 2017, Pages 329-344, ISSN 2212-0416, https://doi.org/10.1016/j.ecoser.2016.11.008.

CHAPTER 4

Gianbattista Bussi, Simon J. Dadson, Christel Prudhomme, Paul G. Whitehead, Modelling the future impacts of climate and land-use change on suspended sediment transport in the River Thames (UK), In Journal of Hydrology, Volume 542, 2016, Pages 357-372, ISSN 0022-1694, https://doi.org/10.1016/j.jhydrol.2016.09.010.

CHAPTER 5

I.P. Holman, C Brown, V Janes, D Sandars, Can we be certain about future land use change in Europe? A multi-scenario, integrated-assessment analysis, In Agricultural Systems, Volume 151, 2017, Pages 126-135, ISSN 0308-521X, https://doi.org/10.1016/j.agsy.2016.12.001.

CHAPTER 6

Lourdes López-Merino, Oscar Serrano, María Fernanda Adame, Miguel Ángel Mateo, Antonio Martínez Cortizas, Glomalin accumulated in seagrass sediments reveals past alterations in soil quality due to land-use change, In Global and Planetary Change, Volume 133, 2015, Pages 87-95, ISSN 0921-8181, https://doi.org/10.1016/j.gloplacha.2015.08.004.

CHAPTER 7

C. Gutzler, K. Helming, D. Balla, R. Dannowski, D. Deumlich, M. Glemnitz, A. Knierim, W. Mirschel, C. Nendel, C. Paul, S. Sieber, U. Stachow, A. Starick, R. Wieland, A. Wurbs, P. Zander, Agricultural land use changes – a scenario-based sustainability impact assessment for Brandenburg, Germany, In Ecological Indicators, Volume 48, 2015, Pages 505-517, ISSN 1470-160X, https://doi.org/10.1016/j.ecolind.2014.09.004.

CHAPTER 8

Alessandro De Pinto, Man Li, Akiko Haruna, Glenn Graham Hyman, Mario Andrés Londoño Martinez, Bernardo Creamer, Ho-Young Kwon, Jhon Brayan Valencia Garcia, Jeimar Tapasco, Jesus David Martinez, Low Emission Development Strategies in Agriculture. An Agriculture, Forestry, and Other Land Uses (AFOLU) Perspective, In World Development, Volume 87, 2016, Pages 180-203, ISSN 0305-750X, https://doi.org/10.1016/j.worlddev.2016.06.013.

CHAPTER 9

Kerstin Engström, Mark D.A. Rounsevell, Dave Murray-Rust, Catherine Hardacre, Peter Alexander, Xufeng Cui, Paul I. Palmer, Almut Arneth, Applying Occam's razor to global agricultural land use change, In Environmental Modelling & Software, Volume 75, 2016, Pages 212-229, ISSN 1364-8152, https://doi.org/10.1016/j.envsoft.2015.10.015.

INDEX

D

E

F

G

H

I

L

M